Theory of Solid-Propellant Nonsteady Combustion

Wiley-ASME Press Series

Computer Vision for Structural Dynamics and Health Monitoring
Dongming Feng, Maria Q. Feng

Theory of Solid-Propellant Nonsteady Combustion
Boris V. Novozhilov, Vasily B. Novozhilov

Introduction to Plastics Engineering
Vijay K. Stokes

Fundamentals of Heat Engines: Reciprocating and Gas Turbine Internal Combustion Engines
Jamil Ghojel

Offshore Compliant Platforms: Analysis, Design, and Experimental Studies
Srinivasan Chandrasekaran, R. Nagavinothini

Computer Aided Design and Manufacturing
Zhuming Bi, Xiaoqin Wang

Pumps and Compressors
Marc Borremans

Corrosion and Materials in Hydrocarbon Production: A Compendium of Operational and Engineering Aspects
Bijan Kermani and Don Harrop

Design and Analysis of Centrifugal Compressors
Rene Van den Braembussche

Case Studies in Fluid Mechanics with Sensitivities to Governing Variables
M. Kemal Atesmen

The Monte Carlo Ray-Trace Method in Radiation Heat Transfer and Applied Optics
J. Robert Mahan

Dynamics of Particles and Rigid Bodies: A Self-Learning Approach
Mohammed F. Daqaq

Primer on Engineering Standards, Expanded Textbook Edition
Maan H. Jawad and Owen R. Greulich

Engineering Optimization: Applications, Methods and Analysis
R. Russell Rhinehart

Compact Heat Exchangers: Analysis, Design and Optimization using FEM and CFD Approach
C. Ranganayakulu and Kankanhalli N. Seetharamu

Robust Adaptive Control for Fractional-Order Systems with Disturbance and Saturation
Mou Chen, Shuyi Shao, and Peng Shi

Robot Manipulator Redundancy Resolution

Yunong Zhang and Long Jin

Stress in ASME Pressure Vessels, Boilers, and Nuclear Components

Maan H. Jawad

Combined Cooling, Heating, and Power Systems: Modeling, Optimization, and Operation

Yang Shi, Mingxi Liu, and Fang Fang

Applications of Mathematical Heat Transfer and Fluid Flow Models in Engineering and Medicine

Abram S. Dorfman

Bioprocessing Piping and Equipment Design: A Companion Guide for the ASME BPE Standard

William M. (Bill) Huitt

Nonlinear Regression Modeling for Engineering Applications: Modeling, Model Validation,

and Enabling Design of Experiments

R. Russell Rhinehart

Geothermal Heat Pump and Heat Engine Systems: Theory and Practice

Andrew D. Chiasson

Fundamentals of Mechanical Vibrations

Liang-Wu Cai

Introduction to Dynamics and Control in Mechanical Engineering Systems

Cho W.S. To

Theory of Solid-Propellant Nonsteady Combustion

Boris V. Novozhilov
The Semenov Institute of Chemical Physics
Russian Academy of Sciences
117977 Moscow, Russia

Vasily B. Novozhilov
Professor of Mathematics
Victoria University
Melbourne Victoria 8001
Australia

WILEY

ASME PRESS

This edition first published 2021

Registered Offices
John Wiley & Sons, Inc., 111 River Street, Hoboken, NJ 07030, USA
John Wiley & Sons Ltd, The Atrium, Southern Gate, Chichester, West Sussex, PO19 8SQ, UK

Editorial Office
The Atrium, Southern Gate, Chichester, West Sussex, PO19 8SQ, UK

For details of our global editorial offices, customer services, and more information about Wiley products visit us at www.wiley.com.

Wiley also publishes its books in a variety of electronic formats and by print-on-demand. Some content that appears in standard print versions of this book may not be available in other formats.

Library of Congress Cataloging-in-Publication Data

Names: Novozhilov, Boris V., author. | Novozhilov, Vasily B., author.
Title: Theory of solid-propellant nonsteady combustion / Boris V. Novozhilov, Vasily B. Novozhilov.
Description: Hoboken, NJ : Wiley-ASME, [2020] | Series: Wiley-ASME press series | Includes bibliographical references and index.
Identifiers: LCCN 2020017338 (print) | LCCN 2020017339 (ebook) | ISBN 9781119525707 (cloth) | ISBN 9781119525646 (adobe pdf) | ISBN 9781119525585 (epub)
Subjects: LCSH: Solid propellants–Combustion.
Classification: LCC TL785 .N68 2020 (print) | LCC TL785 (ebook) | DDC 621.43/56–dc23
LC record available at https://lccn.loc.gov/2020017338
LC ebook record available at https://lccn.loc.gov/2020017339

Cover Design: Wiley
Cover Image: Designed by Inga Novozhilov

Set in 9.5/12.5pt STIXTwoText by SPi Global, Chennai, India

Printed and bound by CPI Group (UK) Ltd, Croydon, CR0 4YY

10 9 8 7 6 5 4 3 2 1

To
Ludmila Novozhilova
Natalia Golubnichaya
Inga Novozhilov
Natalia Novozhilova

Contents

About the Authors

Professor Boris V. Novozhilov

Professor Boris V. Novozhilov (1930–2017) was born in Alma-Ata (Kazakhstan, which at that time was part of the Soviet Union). He graduated with honors in Applied Physics from the Leningrad (currently Peter the Great St. Petersburg) Polytechnic Institute in 1953. He received his PhD (1959) and DrSc (1968) degrees in Physical and Mathematical Sciences.
From 1954 to 2017, Professor B.V. Novozhilov worked at the Institute of Chemical Physics (currently the Semenov Institute of Chemical Physics) of the USSR (later Russian) Academy of Sciences in various roles, including the Head of Laboratory of Mathematical Methods in Chemical Physics (1976–1992) and Chief Researcher.
Professor B.V. Novozhilov is best known for his outstanding fundamental contribution to the theory of propellant combustion and, together with Ya. B. Zeldovich, is a founder of the Zeldovich–Novozhilov theory of nonsteady solid propellant combustion.
Professor B.V. Novozhilov's other research interests include nuclear physics (propagation of gamma quanta in matter), the theory of spin combustion, and the theory of "cold" flame propagation.
Professor B.V. Novozhilov is a recipient of the Ya. B. Zeldovich Gold Medal from The Combustion Institute "for outstanding contributions to the theory of combustion" (1996). He has also received a number of Russian Federation Government Awards in Science and Technology.
Professor B.V. Novozhilov is the author of over 150 journal papers and 12 books.

Professor Vasily B. Novozhilov

Professor Vasily B. Novozhilov was born in 1963 in Moscow. He graduated with an MSc in Applied Mathematics from the Russian State University of Oil and Gas in 1986. He later received a PhD in Physical and Mathematical Sciences (Mechanics of Fluid, Gas and Plasma) from the Moscow Aviation Institute in 1993.
Professor V.B. Novozhilov held research positions at the Russian Academy of Sciences (Institute for Problems in Mechanics) and the University of Sydney. Furthermore, he held academic appointments at Nanyang Technological University (Singapore), as a Professor in Fire Dynamics at The University of Ulster (UK), and as a Professor of Mathematics

at Victoria University (Australia). From 2014 to 2017 he was a Director of the Centre for Environmental Safety and Risk Engineering at Victoria University, Australia.

Major research interests of Professor V.B. Novozhilov include combustion (solid propellants, combustion theory and fire research) and the theory of heat transfer. He is a leading expert in theoretical and computational methods in the areas of combustion and fire research, in particular, computational fluid dynamics modelling of compartment fires. He has also made important contributions to the application of dynamical system methods in fire dynamics, and has also been greatly involved with analytical methods of the heat transfer theory in application to ultra-fast heat transfer processes.

Professor V.B. Novozhilov is the author of over 60 journal papers and four book chapters. He was a Keynote Speaker at the 68th International Astronautical Congress (2017) delivering an overview of the fundamentals of the Zeldovich–Novozhilov propellant combustion theory, as well as of other contributions by Professor B.V. Novozhilov to the physics of combustion.

Preface

Nonsteady operating regimes, where fuel burning rates vary in time, are common for solid rocket motors. Under such conditions, combustion chamber pressure and, consequently, specific impulse are also functions of time. Some examples of such processes are combustion under variable pressure, a transition from one operating regime to another, oscillating combustion, erosion combustion in a dynamical regime, and propellant charge ignition and extinction under rapid depressurization.

In contrast to a steady-state regime, propellant burning rate in such situations depends not only on instantaneous parameters (initial pressure, temperature, and the velocity of a tangential gas stream), but is also determined by the full history of the process. This is due to the inertia of the combustion wave, which includes a heated layer of the condensed phase, a chemical reaction zone, and a certain region in space that is occupied by combustion products.

The natural way of describing nonsteady propellant combustion would be to use the theory of steady-state burning regimes. The transition to unsteady theory would require a simple addition of time derivatives into the relevant set of differential equations. At the current level of available computational resources, this additional mathematical complexity does not present a problem. However, the described hypothetical approach is not possible for a very simple reason: the consistent and universal theory of steady-state propellant combustion describing experimental observations does not exist.

Each physical and chemical process which occurs during the combustion of propellants is immensely complex. For the overwhelming majority of substances, the burning rate is determined by chemical kinetics. Therefore, the kinetic parameters of reactions are a substantial part of practically any combustion theory. However, with some exception, this knowledge of the kinetics of combustion reactions is as yet incomplete. In particular, information on chemical transformations which occur during the combustion of condensed substances is very scarce so it is necessary to involve model kinetic schemes, which only remotely resemble real chemical processes (typically, Arrhenius dependence on temperature and power dependence on reactant concentration are adopted).

There is a large body of steady-state homogeneous and composite propellant combustion models presented within this work. Naturally, such models contain a large number of parameters (reaction rate constants, activation energies, heat of combustion, transfer coefficients, thermophysical properties of gas, condensed phases, etc.) which are unknown in most cases. Evidently, an adjustment of numerous parameters allows the experimental data to be approximated, which of course does not imply a proper description of real fuel. Such studies are therefore of qualitative nature only, and are hardly suitable for comparison with experiments. Moreover, it would probably be impossible to develop a quantitative steady-state combustion theory applicable to a wide range of substances due to the large variation in their properties.

A drastically different approach to the development of nonsteady theory (avoiding the necessity to create a detailed description of the steady-state regime) was proposed by Ya.B. Zeldovich in 1942. It invokes an elegant and powerful idea of using the experimentally determined steady-state dependence of a propellant burning rate on pressure and an initial temperature for studying nonsteady combustion regimes. It was demonstrated that this is only possible taking into account the thermal inertia of the condensed phase. The idea was formulated in the original paper by Zeldovich (1942) in the following way: 'Since the relaxation time of combustion in gas is very small, we have the right to consider gaseous combustion as determined by the thermal condition of the thin condensed phase layer adjacent to the interface; the temperature distribution within deeper layers does not have a direct effect on the processes near the surface. The conditions of gas must be fully determined by instantaneous values of the surface temperature, and a temperature gradient in the condensed phase at the surface. Consider the surface temperature as being constant.'

Thus, the nonsteady propellant combustion theory was reduced to a consideration of a relatively slow variation of temperature distribution in the condensed phase. This is achieved by the solution of the heat transfer equation, combined with the known dependency of the burning rate on instantaneous values of pressure and the temperature gradient in a condensed phase at the surface. The latter may be obtained from the (theoretical or experimental) steady-state dependency of the burning rate on pressure and initial temperature.

This theory explained some nonsteady combustion phenomena qualitatively, but its quantitative comparison with experimentation leads to contradiction. Most remarkably, this contradiction manifests itself in the conclusion that, according to this theory, a steady-state burning regime of real systems is actually unstable. The reason for this discrepancy is an oversimplification of the theory, which considered propellant surface temperature as constant.

For all practically used compositions, the temperature at the interface between the condensed and gas phases depends on external conditions: the pressure, initial temperature, and velocity of the tangential gas stream. The theory proposed by Zeldovich (1942) was generalized and transformed to its contemporary state by B.V. Novozhilov in 1965 (Novozhilov 1965a,b). An additional function characterizing the steady-state regime was introduced: the dependence of the propellant surface temperature on external parameters. Such a generalization of the Zeldovich theory allowed the majority of phenomena related to nonsteady combustion regimes to be explained.

It is important to note that this generalization of the theory by Zeldovich (1942) preserved its main idea, that is, the possibility of investigating nonsteady phenomena using steady-state dependencies. The theory contains experimental dependencies related to a steady-state combustion regime. These dependencies carry all the information on kinetics of chemical reactions as well as on various physical processes (thermal conduction and diffusion in the gas phase, devolatilization of the fuel, etc.).

Even in the absence of such dependencies, the theory still turns out to be helpful when considering a comparison of its conclusions with experimental data on nonsteady combustion. Thus, the comparison of theoretical and experimental data on acoustic combustion instability enables prediction of the behaviour of the same fuel under different nonsteady conditions, such as during combustion in a semi-enclosed volume.

Moreover, the theory allows some fuel parameters, related to the kinetics of reactions at its surface, to be extracted from experimental data. For example, it is possible to obtain the effective activation energy of the chemical reaction at the fuel surface from the data on acoustic admittance.

The theory developed on the basis of results by Zeldovich (1942) and Novozhilov (1965a,b) is usually called the Zeldovich–Novozhilov theory (the ZN theory). The other titles that are used include the phenomenological theory of nonsteady combustion or the t_c approximation. The latter emphasizes that the time of thermal relaxation of the condensed phase is the only fuel characteristic time. In the following text terms such as energetic material, solid rocket fuel, volatile condensed combustion system, and propellant are used as synonyms.

Within the framework of the theory presented in this monograph, the nonsteady process of propellant combustion is investigated by means of solving the heat transfer equation in the condensed phase with relevant initial and boundary conditions. Other necessary elements of the theory are the steady-state dependencies of the burning rate and surface temperature on the pressure and initial temperature. These may be obtained experimentally or by considering a specific theoretical propellant combustion model. It is clear that all conclusions of the theory are applicable to real systems as the aforementioned dependencies are obtained from experiments with the exactly same systems.

Let us discuss briefly the assumptions which form the foundation of the theory. It should be noted that, with a few exceptions, these assumptions are adopted in all studies of the nonsteady combustion of solid rocket fuels. First of all, fuel is assumed to be homogeneous and isotropic. The scale of nonhomogeneity must be much smaller than the characteristic scale following from steady-state theory, that is, the Michelson length. This requirement is undoubtedly fulfilled for ballistites. In the case of composite fuels, this assumption is valid for sizes of fuel and oxidizer particles much smaller than the thickness of the heated layer of the condensed phase. In the following discussion, one-dimensional problem formulation assuming flat flame front and interface between the phases is considered nearly everywhere. Second, the basic assumption of the discussed theory is that thermal decomposition of the condensed phase and combustion in the gaseous phase occur much faster than the heating up of the condensed phase. This proposition may be justified by simple estimations, which are presented in the main body of the monograph.

The first review of the proposed theory was presented by Novozhilov (1968). It considered the major results obtained by that time: combustion stability at constant pressure, linear oscillating combustion regimes, acoustic admittance of the surface of burning propellant, combustion stability in a semi-enclosed volume, nonlinear oscillations of burning rate, transitional combustion regimes, and propellant extinction. Later, the monographs (in Russian) by Novozhilov (1973a) and Zeldovich et al. (1975), as well as the more recent review by Novozhilov (1992a) appeared.

It seemed initially (to many, including the first author) that the phenomenological theory would be replaced soon by a more sophisticated approach to nonsteady propellant combustion. Such an approach could be built on the consideration (by numerical methods) of a set of differential equations, complete and consistent from the point of view of macroscopic chemical physics, describing the specific problem. This may eventually happen. However, progress beyond the framework of the ZN theory has been very slow owing to the significant complexity (even for homogeneous systems) of propellant combustion. This complexity, in the first place, is due to phase transition from condensed to gaseous, complicated even further by chemical reactions.

All the considerations below are applicable, strictly speaking, to homogeneous propellants only. The theory of composite systems is at a rudimentary stage since the processes in the combustion wave of such substances are much more complicated compared to homogeneous propellants. Apart from a nearly complete absence of data on the kinetics of chemical reactions, there are additional obstacles to a quantitative consideration of the combustion process of composite systems. Although during steady-state conditions the mean burning rate is constant in time, the processes occurring in the vicinity of the surface are nonsteady. The geometry of the surface continuously changes in time as burnt particles are replaced by virgin ones at other locations at the surface. Temperature distribution in the vicinity of the interface between the phases, and on the interface itself, is a random function of time. It should be noted that attempts were made (Romanov 1976) to expand the theory to heterogeneous systems. It was proposed, in addition to steady-state dependencies of the burning rate and surface temperature on pressure and initial temperature, to use dependencies of average values of these quantities on fuel (or oxidizer) mass fraction at the interface. Unfortunately, significant difficulties in obtaining such dependencies experimentally did not allow this approach to proceed.

Nevertheless, one may hope that some of the results obtained for homogeneous propellants would also be qualitatively applicable to composite systems. For example, the resonance response of the burning rate to periodically varying pressure may be expressed in the same terms as for homogenous systems. Naturally, the parameters characterizing such composite systems would have to be considered as adjustable values.

Let us discuss briefly the comparison of conclusions (which are discussed within the book) of the theory, with experimentation. As with any other theory, ZN theory requires, first of all, some experimental input data. These are steady-state burning laws, that is, steady-state dependencies of the burning rate and surface temperature (and, in some cases, of other properties of the combustion wave, e.g. combustion temperature) on external parameters, that is, on pressure and initial temperature. The theory demands a rather high accuracy of input experimental data as its conclusions follow from peculiarities of steady-state burning laws. For example, the study of linear nonsteady phenomena is only possible if first derivatives of burning laws, with respect to external parameters, are known. Calculation of these derivatives obviously involves large errors as such a mathematical operation is ill-posed.

The same applies to experimental data which is used for comparison with the theory outcomes. Observation of various unsteady combustion phenomena are associated with significant difficulties. At best, relative errors are of the order of tens of percent. The theory, however, predicts a number of quite distinctive qualitative effects, for example the existence of the natural frequency of propellant combustion and, as a consequence, a resonance response of the nonsteady burning rate to harmonically oscillating pressure. Such effects are actually being observed, and it is usually possible to reconcile the theory and experimental results quantitatively for the values of parameters within the experimental uncertainty region.

Before briefly outlining the monograph content, let us notice that the overwhelming majority of presented results are obtained using a universal approach. The following is its mathematical formulation.

A one-dimensional unsteady heat transfer equation

$$\frac{\partial \theta}{\partial \tau} = \frac{\partial}{\partial \xi}\left(\frac{\partial \theta}{\partial \xi} - v\theta\right), \quad -\infty < \xi \leq 0, \quad \tau \geq 0$$

is considered, along with the relevant boundary and initial conditions

$$\xi \to -\infty, \quad \theta = 0$$

$$\xi = 0, \quad \theta = \vartheta(\tau)$$

$$\theta(\xi, 0) = \theta_i(\xi)$$

Nonsteady relations between the burning rate $v(\tau)$ and the surface temperature $\vartheta(\tau)$, on the one hand, and some external parameter $\eta(\tau)$ (most often pressure) and the temperature gradient $\varphi(\tau) = (\partial\theta/\partial\xi)_{\xi=0}$, on the other

$$v = \Phi_u(\varphi, \eta), \quad \vartheta = \Phi_s(\varphi, \eta)$$

are prescribed.

There must also be prescribed the function

$$\eta = \Pi(\tau)$$

or some auxiliary equation which determines this function.

A specification of the functions $\Phi_u(\varphi, \eta)$ and $\Phi_s(\varphi, \eta)$, and the external parameter dependence on time $\eta(\tau)$ lead to a class of problems that are related to rapid development in recent decades in the multidisciplinary area of synergetics (Mikhailov 2011).

The majority of results in this area are obtained by a numerical analysis of various model sets of differential equations. Most often, the systems with a finite number of degrees of freedom are considered.

The formulation discussed above is probably the simplest for distributed dynamical systems. Despite this simple form, however, the set of system behaviour scenarios is quite rich. For example (and this is demonstrated in the relevant section of the book), studying the system behaviour under constant external conditions is directly related to the problem of turbulence. Variation of one of the control parameters leads to successive bifurcations of combustion regimes, ending up with chaotic behaviour.

In contrast to the majority of studied examples of dynamical systems, the model described above expresses real physical and chemical processes. Within the framework of the presented formulation, such practically important phenomena as propellant burning interaction with the acoustics of the combustion chamber, combustion extinction under depressurization, burning stability under constant pressure, etc. may be investigated.

Since experimental data on steady-state combustion are an essential input into the theory, this book begins with an establishing chapter describing the steady-state burning regime and presenting various fuel property dependencies on external conditions. The theoretical estimations and experimental results provided in the first chapter are necessary for a justification of nonsteady combustion theory. Analytical dependencies of the burning rate and surface temperature on external parameters (i.e. steady-state burning laws) are useful for quantitative analysis. Their physically sensible form may only be obtained from the current understanding of steady-state combustion regimes of the simplest systems. These issues are also discussed in the first chapter.

The second chapter, which is fundamental for the monograph, formulates major assumptions of the theory of nonsteady combustion of solid rocket fuels. A simplified case of constant propellant surface temperature (Zeldovich theory) is considered in detail. The two formulations of the ZN theory, differential and integral, are then presented.

In the first of these formulations, temperature distribution in the condensed phase is a necessary element of consideration. This profile, however, is rarely used in practice. It turns out that an alternative, integral formulation, involving only the most relevant quantities – pressure and burning rate, as an example – may be developed. Readers interested in a formal mathematical justification of the theory should pay attention to the last section of this chapter.

The third chapter considers propellant combustion at a constant pressure. In one-dimensional problem formulation, the stability criterion for the steady-state combustion regime, relating burning rate and surface temperature derivatives with respect to initial temperature is obtained. The possibility of two-dimensional instability is discussed using the example of the simplest combustion system. Numerical modelling is used to investigate nonsteady modes of propellant combustion under constant pressure beyond the stability boundary of the steady-state regime. The simplest propellant combustion model containing just two control parameters is considered. With a fixed value of one of these, the other plays the role of a bifurcation parameter. It is demonstrated that on variation of the bifurcation parameter, the system may transit from the steady-state to the chaotic combustion regime following the Feigenbaum scenario. A sequence of period doubling bifurcations of the burning rate oscillations, leading eventually to a chaotic combustion regime, is studied. A comparison with experimental data is discussed in the last section of the chapter.

The following three chapters are devoted to the influence of external parameters, that is, pressure, erosive gas stream, and thermal radiation, on propellant combustion under periodically varying pressure. Practical interest in these processes is due to the need to understand the causes of the development of various nonsteady effects, for example soft or hard excitation of burning rate and pressure oscillations, superimposed on the designed steady-state solid rocket motor regime.

First, the problem of combustion and acoustics interaction (Chapter 4) is considered in the linear approximation with the burning rate amplitude being proportional to pressure amplitude. An analytical expression for the response function of the burning rate to oscillating pressure is obtained. Furthermore, its properties and relation with the most important property of burning propellant surface (acoustic admittance) are discussed. Then, the notion of response functions of higher orders, with respect to pressure amplitude, is discussed. These would find an application in investigations of sustained and transitional regimes with finite burning rate and pressure amplitudes. A conducted nonlinear analysis reveals a fundamentally new phenomenon: period doubling bifurcations of burning rate oscillations that occur on an increase of amplitude or change of frequency of pressure oscillations. A sequence of bifurcations leading eventually to the chaotic combustion regime is investigated for a propellant combustion model containing a minimal number of parameters in nonsteady burning laws.

The fifth chapter considers nonsteady propellant burning in the tangential stream of combustion products. Within the framework of the phenomenological theory, the propellant burning rate response to periodically varying pressure and tangential mass flux of combustion products is investigated. An elementary acoustic perturbation in the form of a monochromatic travelling sound wave is considered. Analytical and numerical results are obtained for the simplest propellant model described by a minimal number of parameters. The role of steady-state and nonsteady erosion contributions at small and large values of the erosion ratio is also revealed.

The following chapter deals with nonsteady combustion under external radiation. In this case, the heat transfer equation includes an additional source term, while both steady-state and nonsteady burning laws also change. The stability of the steady-state combustion regime is investigated in the linear approximation. The response functions of the burning rate to harmonically oscillating pressure in the presence of constant radiative flux, as well as to harmonically oscillating radiative flux, are obtained. In the linear approximation of the ZN theory, an analytical relationship between the response function to oscillating pressure, obtained at a certain initial temperature, and response function to oscillating radiative flux, obtained at the same pressure but different (lower) initial temperature, is established. The difference of initial temperatures satisfies the requirement that steady-state burning rates with and without radiative flux are equal, and is directly proportional to radiative flux. It is likely that this relationship will be useful for obtaining experimental data on the response function to oscillating pressure.

Chapter 7 presents theoretical considerations and experimental data related to combustion regimes where pressure varies according to the law, which is different from harmonic oscillations. Such processes include, for example, propellant combustion during transition from one operational regime to another (at higher or lower pressure), extinction on rapid and deep depressurization, and others.

The eighth chapter describes nonsteady propellant combustion regimes in the combustion chamber of a rocket engine. There are three time scales that are relevant for this problem: the thermal relaxation time of the heated layer of the condensed phase t_c, the acoustic time t_a, and the time of combustion products efflux from the chamber t_{ch}.

If the relaxation time of the condensed phase is close to the efflux time, $t_c \sim t_{ch}$ (which occurs in small engines at low pressures), then such regimes may be called nonacoustic. Time scales in such problems are much larger than the acoustic time. This area of research may also be referred to as propellant combustion in semi-enclosed volume.

On the other hand, over the last few decades a specific and dedicated area of research which may be termed 'acoustics and combustion' has taken shape. It deals with the case where acoustic time is close to the condensed phase thermal relaxation time $t_a \sim t_c$, which leads to a possibility of sonic (in the general case nonlinear) oscillations development in the engine. The latter relation between the time scales applies to engines of large size with high pressure values in combustion chambers. The theory of such processes is still at a rudimentary stage. As an example, possible combustion regimes in a solid rocket engine with end burner grain geometry are investigated. A set of equations which allows the interaction between combustion and acoustic processes in a combustion chamber to be modelled is presented. The specific feature of the problem is the existence of the two distinctive time scales, namely the acoustic time and the time of pressure oscillation amplitude variation. These time scales differ by approximately three orders of magnitude, which demands high computational accuracy. A simpler solution method is developed in the quadratic approximation with respect to the amplitude of oscillations. This method accounts only for the effects related to the time scale of oscillation amplitude variation. Numerical results are obtained for the simplest propellant combustion model in the absence of entropic waves in combustion products. Stable and unstable combustion regimes are identified. In the latter regime, nonlinear effects may trigger shock waves in the combustion chamber.

The possibility of expanding the theory beyond the phenomenological framework is discussed in the final chapter. This development requires a more detailed combustion model that would adequately describe processes occurring in low-inertia zones of a combustion wave. The influence of low-inertia zones (the reacting layer of the condensed phase, preheat and reaction zones in the gas phase, the half-space occupied by gaseous combustion products) on various nonsteady phenomena are investigated both analytically and numerically. The consideration is presented within the framework of the Belyaev model. It is demonstrated that under a weak dependence of surface temperature on initial temperature accounting for the above low-inertia zones (even if their thermal inertia is small compared to the inertia of the preheat layer of the condensed phase) leads to significant corrections to the t_c approximation.

Finally, it is our pleasure to acknowledge the significant contribution of the people who helped us in the preparation of this book.

We are very grateful to Professor Vladimir Marshakov, who discussed various topics throughout the book with us at great length.

Special thanks are given to Inga Novozhilov. It is certain that without her very careful and dedicated work the manuscript could not have been adequately prepared.

We are also incredibly thankful to Professor Vladimir Posvyanskii, Ludmila Novozhilova, and Natalia Golubnichaya for their help in preparing the manuscript.

The second author would like to thank his wife Natalia Golubnichaya again for her love and continuous support throughout the project.

Moscow – Belfast – Melbourne
2011–2019

Boris V. Novozhilov
Vasily B. Novozhilov

Important Notation and Abbreviations

Abbreviations

ADN	Ammonium dinitramide
BVP	Boundary value problem
c.c.	Complex conjugate
ZN	Zeldovich–Novozhilov
FM	Flame model
HMX	Cyclotetramethylene tetranitramine
ODE	Ordinary differential equation
PDE	Partial differential equation
PETN	Pentaerythritol tetranitrate
QSHOD	Quasi-steady, homogeneous, one-dimensional
RDX	Cyclotrimethylene trinitramine
SHS	Self-propagating high–temperature synthesis
SRM	Solid rocket motor

Mathematical Functions

L_n	Laguerre polynomials
lg	$\log_{10}$
erfc	Complimentary error function
He_n	Hermite polynomials
W	Whittaker function

Notation

Over-bar complex conjugate; Laplace–Carson transform

prime — time derivative, case-specific dimension; perturbed value, case-specific dimension

Basic Physical Dimensions	M (mass), L (length), T (time), θ (temperature), N (amount of substance, e.g. mole)
a	speed of sound, LT^{-1}; amplitude, case-specific dimension
a_f	amplitude of forced oscillations, case-specific dimension
A	nozzle discharge coefficient, $L^{-1}T$
b	combustion temperature, nondimensional; correction (Chapter 9)
c	specific heat at constant volume, $L^2T^{-2}\theta^{-1}$
c_p	specific heat at constant pressure, $L^2T^{-2}\theta^{-1}$
D, D_g	gas diffusion coefficient, L^2T^{-1}; amplitude of perturbation (Chapter 3), L; integration constant, nondimensional (Chapter 9)
E	activation energy, $ML^2T^{-2}N^{-1}$
f	temperature gradient at the surface, condensed phase side, θL^{-1}
$\bar{f}(p)$	Laplace–Carson transform of $f(t)$, case-specific dimension
J	Jacobian, $L^2M^{-1}T$
g	mass velocity of gas flow, $ML^{-2}T^{-1}$
G	response function of gas velocity to oscillating pressure, non-dimensional; integration constant, nondimensional (Chapter 9)
h	amplitude of perturbation of relative pressure, nondimensional
I	radiative heat flux, MT^{-3}
k	sensitivity coefficient, nondimensional
K	distortion factor, nondimensional; burning rate amplification coefficient, nondimensional; wave vector, L^{-1}
l	thickness of thermal layer of the condensed phase, L; mean free path for radiation absorption in the condensed phase, nondimensional
l_b	distance away from surface at which heat generation becomes negligible
L	length of cylindrical combustion chamber, L; latent heat of evaporation, L^2T^{-2}
m	mass burning rate, $ML^{-2}T^{-1}$
m_g	mass velocity of the gas stream, $ML^{-2}T^{-1}$
M	Mach number, nondimensional; mass of gas in the chamber, M
p	pressure, $ML^{-1}T^{-2}$
q	heat flux into the condensed phase, MT^{-3}
$Q = Q_s + Q_g$	total heat of combustion/reaction, L^2T^{-2}
Q_g	heat of combustion/reaction in the gas phase, L^2T^{-2}
Q_s	heat of reaction in the condensed phase, L^2T^{-2}
r	sensitivity parameter, nondimensional
r_b	sensitivity parameter, nondimensional

R	Universal gas constant, $ML^2T^{-2}\theta^{-1}N^{-1}$
s	relative change of the nozzle cross-sectional area, nondimensional; correction (Chapter 9), nondimensional; surface temperature in the steady-state regime at initial pressure (Chapter 9), nondimensional
S	cross-sectional area of cylindrical combustion chamber, L^2
t	time, T
T	temperature, θ; period of oscillations, nondimensional
T_k	reference temperature (Chapter 7), θ
u	propellant linear burning rate, LT^{-1}
u_g	dimensional gas velocity normal to the surface, LT^{-1}
u_p	gas velocity normal to the surface in the combustion products zone (Belyaev model), LT^{-1}
U	response function of burning rate to oscillating pressure, non-dimensional; mass burning rate dependence on temperature and pressure (Chapter 9), $ML^{-2}T^{-1}$
v	propellant linear burning rate, nondimensional
v_g	gas velocity normal to the surface, nondimensional
v_p	gas velocity normal to the surface in the combustion products zone (Belyaev model), nondimensional
V	volume of combustion chamber, L^3
w	tangential gas velocity, LT^{-1}
w_t	tangential gas velocity, nondimensional
W	chemical reaction rate, $ML^{-3}T^{-1}$
x	cartesian coordinate, L
y	cartesian coordinate, L, or nondimensional; correction (Chapter 9), nondimensional
Y	reactant mass fraction, nondimensional

Greek

α	linear absorption coefficient, L^{-1}; numerical parameter (Chapter 7), nondimensional
β	burning rate temperature sensitivity coefficient, θ^{-1}
γ	specific heat ratio, nondimensional; frequency (Chapter 9), nondimensional
$\widehat{\gamma}$	frequency, T^{-1}
δ	Jacobian, nondimensional; Feigenbaum constant, nondimensional
$\delta_{m,n}$	Kronecker delta
ε	erosion ratio (coefficient), nondimensional
ζ	channel coefficient of resistance, nondimensional; acoustic admittance, nondimensional
$\widehat{\zeta}, \widehat{\zeta}_1$	nozzle gain coefficient, nondimensional

η	pressure, nondimensional; relative concentration of the product, nondimensional; progress variable of chemical transformation, nondimensional
θ	temperature, nondimensional
Θ	response function of temperature of combustion products to oscillating pressure, nondimensional
ϑ	surface temperature, nondimensional
ι	sensitivity coefficient, nondimensional
κ	propellant thermal diffusivity, L^2T^{-1}
λ	thermal conductivity, $MLT^{-3}\theta^{-1}$; oscillation damping decrement, nondimensional
μ	sensitivity coefficient, nondimensional
μ_b	sensitivity coefficient, nondimensional
$\widetilde{\mu}, \mu_g$	molecular weight, MN^{-1}
ξ	cartesian coordinate, nondimensional
ρ	density, ML^{-3}
σ	ratio of instantaneous nozzle cross-sectional area to its initial value, nondimensional; ratio of relaxation times of the gas and condensed phases, nondimensional
τ	time, nondimensional
$\tau_{U,V,r}$	lag times, nondimensional
φ	temperature gradient at the surface, at the condensed phase side, nondimensional
χ	apparatus constant, nondimensional
ψ	phase shift, nondimensional; velocity potential, L^2T^{-1}; auxiliary function of time (Chapter 7)
$\omega, \widetilde{\omega}$	complex frequency, nondimensional
Ω	complex frequency, T^{-1}, or no-dimensional (Chapter 9)

Gothic

$\aleph, \aleph_{g,p}$	corrections (Chapter 9); nondimensional
$\mathfrak{p}_{g,p}$	complex amplitude (Chapter 9); nondimensional

Superscripts

a	analytical
c	t_c approximation
cl	classical
f	final
i	initial; interval

I	radiation conditions
qs	quasi-steady-state
r	t_r approximation
0	steady-state regime

Subscripts

a	ambient; acoustic
b	burning; combustion temperature
bl	boiling
c	condensed phase
ch	combustion products efflux from the chamber
cr	chemical reaction
cs	condensed phase side, at the surface
d	delay
e	correction to steady-state value (Chapter 4); entropy; extremum
f	final; position of flame front; flame
g	gas; mass velocity
gs	gas phase side, at the surface
i	initial; infinite
I	radiation conditions
na	nonacoustic
p	pressure; products; time scale of unsteady process (e.g. a period of pressure variation); response to variable pressure
r	radiation; reference; relaxation
s	surface; solid; position of liquid–gas interface
ε	erosion; erosive combustion regime
φ	nondimensional temperature gradient at the surface (condensed phase side)

About the Companion Website

This book is accompanied by a companion website:

www.wiley.com/go/Novozhilov/solidpropellantnonsteadycombustion

The Website includes:

- Solution manual
- Chapter Abstracts and keywords

Scan this QR code to visit the companion website

1

Steady-state Combustion

1.1 General Characteristics of Solid Propellants

There are two types of propellants used in rocket motors: homogeneous and heterogeneous. They differ in both chemical composition and physical structure.

Homogeneous propellants have the fuel and oxidizer bound chemically at the molecular level. The basic component of this type of propellant is nitrocellulose, an ester obtained by nitration of cellulose in nitric acid. Nitrocellulose can gelatinize in various solvents, of which nitroglycerin is the most frequently used. Since such propellants invariably contain two basic components (nitrocellulose and nitroglycerin), they are often called double-base propellants. Sometimes they are also referred to as smokeless propellants (as they replaced black, or smoke, powder at the end of the nineteenth century) or ballistites.

Before World War II, priority was given to ballistites. Composite (heterogeneous) propellants found wide application in the post-war period. It should be noted, however, that the first solid propellant was of the composite type; it was the above-mentioned black powder, which is a mechanical mixture of charcoal, potassium nitrate, and sulphur.

The composite propellant is a mechanical mixture of two or more components. The fuel and oxidizer are not mixed at the molecular level and are separated from each other, therefore such compositions are called heterogeneous propellants. The components of a composite propellant are microscopic particles measuring from a few micrometres to tenths of a millimetre. Sometimes, however, particles of one component are interspersed in the body of the other.

One example of a homogeneous propellant is the American ballistite JPN. By weight, it contains 51.5% nitrocellulose, 43.0% nitroglycerin, 1% centralite, and 4.5% other additives. The centralite is added to improve the chemical stability of the propellant. Such materials are called stabilizers. Centralite reacts with the products of the spontaneous decomposition of a propellant in storage and thus prevents its autocatalytic decomposition. The density of the propellant JPN is 1.62 g/cm^3; other double-base propellants have about the same density.

Among Russian-made propellants, special mention should be made of ballistite N. This propellant has been widely investigated under laboratory conditions and contains 57% nitrocellulose, 28% nitroglycerin, 11% dinitrotoluene, and 4% other additives.

The burning of 1 g of a ballistite yields 3.3–5.0 kJ of energy and the temperature in a combustion chamber would reach 2000–2500 °C.

Theory of Solid-Propellant Nonsteady Combustion, First Edition. Boris V. Novozhilov and Vasily B. Novozhilov.

Companion website: www.wiley.com/go/Novozhilov/solidpropellantnonsteadycombustion

Since the composition of ballistites includes atoms of carbon, nitrogen, hydrogen, and oxygen, complete combustion would be expected to yield carbon dioxide, water, and molecular nitrogen. However, due to the lack of oxygen, the products of incomplete combustion – carbon monoxide and molecular hydrogen – are formed as well. At low pressures, nitrogen oxide is also released.

Composite propellants include high-caloric fuels rich in hydrogen. This ensures a low molecular weight of the combustion products. One of the basic requirements imposed on a fuel is that it should possess high binding properties. The commonly used fuels (or binders) are organic high-molecular-weight compounds such as polyurethane, polybutadiene, and other artificial polymers.

The weight content of the fuel in a composite propellant is about 10–25%. The bulk of the propellant consists of an oxidizer, a substance which is rich in oxygen and readily releases it. These requirements are met by strong perchlorates of ammonium, sodium, lithium, nitrates of ammonium and alkali metals, ammonium dinitramide, etc.

Composite propellants have a number of advantages over ballistites, including the possibility of wider variation in the propellant components, stable combustion at low pressures, and a high density. The specific impulse developed by composite propellants is also higher than that of ballistites.

It should be noted that in the last decade, particles of light metals – magnesium, sodium, boron, aluminium, beryllium, etc. – have been introduced into the composition of certain solid propellants (both ballistites and composites). Metal additions increase the combustion temperature and consequently the specific impulse.

Some double-base propellants also include ingredients used for composite propellants, for example ammonium perchlorate. The addition of ammonium perchlorate brings the fuel/oxidizer ratio closer to a stoichiometric one and therefore increases the burning temperature and the specific impulse. The flame temperature may also be increased by the addition of nitramine particles such as cyclotrimethylene trinitramine (RDX) or cyclotetramethylene tetranitramine (HMX). These compositions are called nitramine propellants.

Numerous experimental investigations have indicated that during the combustion of ballistites, the interface between the condensed and gas phases remains planar (provided the specimen diameter is large enough). Chemical reactions complicated by transfer processes and gas movement proceed in the near-interface regions of the gas and condensed substances.

Consider the general pattern of the processes taking place during the burning of ballistites and composite systems.

The one-dimensional burning process of a homogeneous propellant is illustrated in Figure 1.1, where the plane $x = 0$ is a condensed phase/gas phase interface. The entire transitional region (from initial solid propellant to combustion products) is divided into several zones. In the preheat zone of the condensed phase, where there are no chemical reactions, the substance is heated by heat conduction from the initial temperature T_a to a certain temperature $T^0(x_c)$ at which chemical reactions in the condensed phase commence.

In the gas phase one can also distinguish two zones. First, there is a zone of gas heating and chemical transformations. As a result of various chemical reactions, most of the heat is liberated here and is mainly used for heating the gas from $T^0(0)$ to the burning temperature T_b^0. Part of the heat is fed back to the condensed phase through heat conduction (and

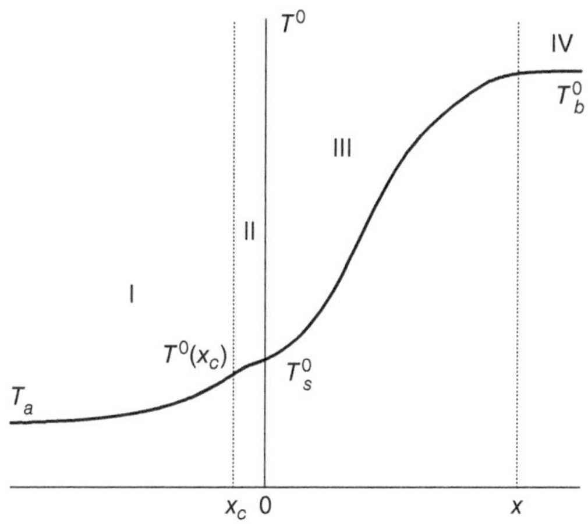

Figure 1.1 Combustion wave structure of a homogeneous propellant. I, preheat zone; II, chemical reaction zone in the condensed phase; III, zone of gas heating and chemical transformations; IV, zone of combustion products.

also through radiation, but to a much lesser extent). Moreover, there exists a zone of combustion products; the gas temperature in this zone is constant and is equal to the burning temperature.

The decisive contribution to unsteadiness is made by the condensed phase, in which thermal inertia is taken into account by the heat conduction equation. Keeping this in mind, consider now the temperature distribution in the preheated zone of the condensed phase.

In subsequent discussion a coordinate system rigidly connected with the propellant surface will always be used. The region $x \leq 0$ corresponds to the condensed phase and $x \geq 0$ to the gas phase. The origin of coordinates is fixed at the propellant surface, where the temperature is equal to T_s^0.

The chemical reaction rate usually increases abruptly with temperature, therefore the temperature change in the reaction layer of the condensed phase (zone II in Figure 1.1) is always small compared to $T_s^0 - T_a$. For instance, in the case of the Arrhenius relationship $u \propto \exp(-E/RT)$, where E represents the activation energy and R represents the universal gas constant, the reaction rate drops by a factor of e within the temperature range of RT^2/E. In other words, if the ratio RT/E is small (which is true in most cases), the chemical reaction will proceed only within a narrow temperature range near the surface temperature. As a good approximation, the spatial width of this zone may be assumed to be equal to zero.

In the adopted coordinate system, the condensed material moves from left to right with a speed equal to the linear burning rate of the propellant, $u(t)$. In the steady-state regime the burning rate is constant and the temperature distribution is time-independent. In the following equations the zero superscript corresponds to the steady-state value. The heat conduction equation is of the form:

$$\frac{\mathrm{d}}{\mathrm{d}x}\lambda\frac{\mathrm{d}T^0}{\mathrm{d}x} - \rho u^0 c\frac{\mathrm{d}T^0}{\mathrm{d}x} = 0 \tag{1.1}$$

Here, the first term corresponds to the conduction flow of heat and the second to the convection flow. In order to simplify the analysis, it is assumed that the density of the condensed phase ρ, its specific heat c, and the thermal conductivity λ are temperature-independent.

Therefore

$$\kappa \frac{d^2T^0}{dx^2} - u^0 \frac{dT^0}{dx} = 0 \quad (1.2)$$

where $\kappa = \lambda/(\rho c)$ is thermal diffusivity of the solid phase.

Elementary integration of this equation taking into account the boundary conditions

$$x \to -\infty, \quad T = T_a; \quad x = 0, \quad T = T_s^0 \quad (1.3)$$

leads to the temperature distribution

$$T^0(x) = T_a + (T_s^0 - T_a)\exp\left(\frac{u^0 x}{\kappa}\right) \quad (1.4)$$

which is known as the Michelson distribution. A discernible change in temperature in the condensed phase occurs at a distance of the order of the so-called heating thickness, or the thickness of the thermal layer of the condensed phase,

$$l = \frac{\kappa}{u} \quad (1.5)$$

The thermal diffusivity of the solid phase is of the order of $\kappa \propto 10^{-3}$ cm^2/s, and the burning rate varies from fractions of a millimetre per second to about 1 cm per second as the pressure increases from 1 to 100 atm. Therefore the thickness of the heated layer varies from 10^{-1} cm at low pressures to 10^{-3} cm at high pressures.

Below is a frequently used expression for the temperature gradient at the propellant surface (from the side of the condensed phase)

$$f^0 = \left.\frac{dT^0}{dx}\right|_{x=0}, \quad f^0 = \frac{u^0}{\kappa}(T_s^0 - T_a) \quad (1.6)$$

Sometimes it is important to know the excess (as compared to a cold specimen, $T^0 = T_a$) heat stored in the heated layer. This value is equal to

$$\rho c \int_{-\infty}^{0} (T - T_a)dx = \rho c l (T_s^0 - T_a) \quad (1.7)$$

Figure 1.2 illustrates the shape of the Michelson profile at different burning rates. Curve 1 refers to a low rate, while curve 2 refers to high pressure and burning rates. As the burning rate increases, the effective width of the Michelson distribution diminishes, the profile becomes steeper, and the excess amount of heat drops.

Let us proceed now to the description of the combustion pattern of heterogeneous systems, which is much more complicated compared to the case of homogeneous powders. Practically a complete lack of information on the kinetics of chemical reactions is compounded by other circumstances hindering the quantitative consideration of burning of composite systems. The principal difficulties are as follows:

a) Multidimensionality of burning. The heterogeneous nature of the system results in an uneven interface. Particles of fuel or oxidizer protrude from its surface at different places and up to different heights. The theory of burning of such systems cannot be one-dimensional as in the case of homogenous propellants.
b) Necessity of mixing. Prior to reacting, fuel and oxidizer must be mixed at the molecular level. If both components have equally high volatility, the mixing and burning occur in

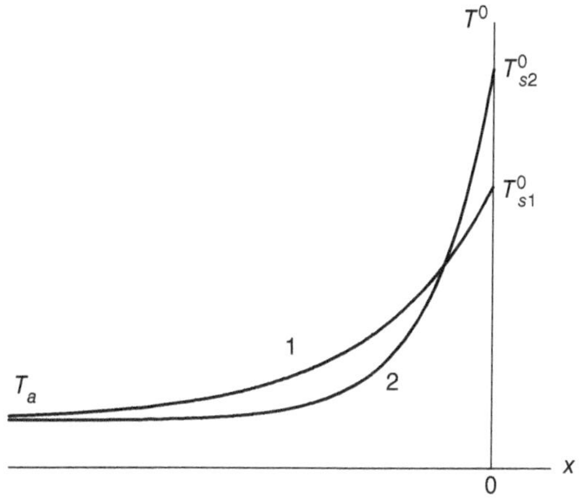

Figure 1.2 Michelson temperature distribution.

the gas phase. Otherwise the reaction proceeds on the surface of the particles of either fuel or oxidizer.

c) Unsteadiness of burning. Although the mean burning rate is constant in time during steady-state conditions, the processes occurring near the surface are nonsteady. Surface shape varies with time: the burnt-out particles are replaced by new ones at other sites on the surface. The temperature distribution at and near the interface is a random function of time. The combustion theory of such systems must evidently be statistical in a certain sense.

The theory presented in this book is one-dimensional, therefore it can only be applied to homogeneous propellants and composite systems burning with a plane flame front. It should be remembered that temperature distribution in the condensed phase can only be considered one-dimensional if its Michelson thickness exceeds the particle size.

For a sufficiently comprehensive survey of works dealing with experimental investigations of steady-state propellant combustion, the reviews by Kubota (1984), Price (1984), and Klager and Zimmerman (1992) are recommended.

The data presented in Tables 1.1 and 1.2 can be used to make numerical estimations. These tables contain the properties of the condensed phase and the combustion products of homogeneous propellants. More detailed data can be found in the review by Zanotti et al. (1992).

Table 1.1 lists the properties of the condensed phase; these are nearly independent of pressure and vary weakly with temperature.

Working pressure in solid propellant motors rarely exceeds 100 atm, therefore combustion product density can be found from the equation of state for an ideal gas

$$\rho_g = \frac{\tilde{\mu} p}{R T_g} \tag{1.8}$$

where $\tilde{\mu}$ is molecular weight. Thermal diffusivity is usually calculated as

$$\kappa_g(T_g, p) = \kappa_g(T_g^*, p^*) \frac{p^*}{p} \left(\frac{T_g}{T_g^*} \right)^n \tag{1.9}$$

where $\kappa_g(T_g^*, p^*)$ is given in Table 1.2.

Table 1.1 Properties of the condensed phase.

Parameter	Notation	Unit	Value
Density	ρ	g/cm^3	1.6–1.9
Specific heat	c	J/(g K)	1.3–2.1
Thermal conductivity	λ	J/(cm s K)	$(1.3–2.1) \cdot 10^{-3}$
Thermal diffusivity	κ	cm^2/s	$(0.5–1.5) \cdot 10^{-3}$

Table 1.2 Properties of combustion products.

Parameter	Notation	Unit	Value
Molecular weight	$\tilde{\mu}$	g/mol	25
Specific heat	c_p	J/(g K)	1.3–1.7
Specific heat ratio	γ	—	1.2–1.3
Thermal conductivity	λ	J/(cm s K)	$(8.4–16.8) \cdot 10^{-4}$
Reference thermal diffusivity $T_g^*=300°C$ $p^*=1$ atm	$\kappa_g(T_g^*, p^*)$	cm^2/s	10^{-1}

Table 1.2 lists the properties of the combustion products.

1.2 Burning Rate and Surface Temperature

For a given propellant composition, the burning rate may depend on the charge diameter, pressure, initial temperature, and tangential gas velocity near the surface. The effect of the charge diameter is of no interest here, since this factor is important only for combustion of small charges (less than about 1 cm in diameter). The effect of the tangential gas flow is considered in Section 1.4. Thus, only two external parameters remain: pressure and initial temperature.

Classical experimental studies of condensed system combustion regimes, as a function of assigned external conditions, have produced a significant volume of practically valuable information. By changing external conditions and charge parameters, important data concerning the effects of pressure, initial temperature, charge density, and other parameters on the burning rate have been obtained.

Ballistites have been investigated extensively. The burning rate of ballistites, as well as that of most of condensed substances (with few exceptions), increases with pressure and initial temperature.

Experimental data on the linear burning rate, u^0, for ballistite N at different values of pressure p and a constant initial temperature are given in Table 1.3. Tables 1.4 and 1.5, based on the data provided by Zenin (1980) in his review, list the values of u^0 for different initial temperatures.

Table 1.3 Burning rate and surface temperature of the propellant N at $T_a = 20\,°C$.

Parameter	Unit	Values						
p	atm	5	10	20	30	50	75	100
u^0	cm/s	0.15	0.19	0.34	0.48	0.67	0.85	1.06
T_s^0	°C	260	300	340	370	400	425	445

Table 1.4 Burning rate and surface temperature of the propellant N (Zenin 1980) at $p = 1$ atm.

Parameter	Unit	Values				
T_a	°C	−196	−100	0	50	100
u^0	cm/s	0.022	0.028	0.060	0.102	0.195
T_s^0	°C	190	200	230	250	290

Table 1.5 Burning rate and surface temperature of the propellant N (Zenin 1980) at $p = 20$ atm.

Parameter	Unit	Values						
T_a	°C	−150	−100	−50	0	50	100	120
u^0	cm/s	0.18	0.19	0.22	0.27	0.36	0.49	0.60
T_s^0	°C	310	315	320	325	345	360	370

Since there is no satisfactory theory of steady-state burning rate for condensed systems, its pressure dependence is usually represented by various empirical formulas, of which the most widely used is

$$u^0 = A + Bp^n \tag{1.10}$$

where A, B, and n are constants (sometimes formulas with $A = 0$ or $n = 1$ are used).

In considering small pressure variations, it is useful to introduce a value characterizing the relative change of the burning rate

$$\iota = \left(\frac{\partial \ln u^0}{\partial \ln p}\right)_{T_a} \tag{1.11}$$

This quantity depends both on pressure and initial temperature. For ballistite N at the initial room temperature $\iota \propto 0.4 - 0.7$.

The sensitivity of the burning rate to variations in initial temperature is characterized by the temperature coefficient of the burning rate, β, which is defined as the relative change of the burning rate with a temperature change of 1°:

$$\beta = \left(\frac{\partial \ln u^0}{\partial T_a}\right)_p \tag{1.12}$$

For ballistites this value is of the order of 10^{-2}–10^{-3} K^{-1}. The temperature coefficient of the burning rate depends rather strongly on pressure and initial temperature. For example, Korotkov and Leipunskii (1953) found that at $p = 1$ atm the temperature coefficient, which is equal to 2×10^{-3} K^{-1} at $T_a = -200\,°C$, increases by a factor of seven as the initial temperature rises to $T_a = 100\,°C$. The same is true for higher pressures. The experimental steady-state temperature sensitivity data are discussed by Kubota (1992).

Burning rate data, however is still insufficient for a clear insight into the physicochemical processes involved in propellant combustion. Efficient use of propellants in rocket motors requires a more refined study. Therefore, in the last few decades attempts have been made to study burning processes more rigorously, and this has provided an interesting information both on physical processes in the condensed and gas phases, and the nature of chemical transformations.

Most of the information concerning processes in various zones of burning propellant has been obtained by measuring temperature profiles in condensed and gas phases.

The thin-thermocouple method makes it possible to obtain the temperature profile in both the solid and gas phases. To do this, a thermocouple is embedded into a propellant specimen, and as the latter burns it goes through the entire temperature range from the initial to the combustion temperature. This method has been developed and substantiated by Zenin (1980), who demonstrated that the necessary condition for correct thermocouple measurements is the use of a specific version of Π-shaped thermocouples. These thermocouples considerably reduce measurement errors compared with the previously used V-shaped thermocouples. The method has been successfully applied in investigating the temperature profile in the condensed and gas phases during steady-state burning of ballistite N. Using high melting point tungsten-rhenium thermocouples, 3.5–7 μm thick, the complete temperature profile and the temperature values at various characteristic points, including those at the surface, were obtained for the first time.

It should be noted that experimental difficulties encountered in temperature measurements at high temperature gradients, inherent in propellant combustion, led to grave experimental errors (variations in temperature are often of the order of the error), which explains the considerable discrepancies between the results obtained by different scientists.

Of great interest are measurements of propellant surface temperature.

Tables 1.3–1.5 give a number of values for the surface temperatures of burning ballistite N at various pressures and initial temperatures. At a fixed initial temperature ($T_a = 20\,°C$), a rise in pressure from atmospheric to 100 atm increases the surface temperature by about 200 °C. The greatest change occurs at low pressures (up to 15–20 atm). With a further rise in pressure by 100 atm the surface temperature increases by about 100 °C.

If one describes the rise in surface temperature relative to pressure variation by the derivative

$$\mu_p = \left(\frac{\partial T_s^0}{\partial p} \right)_{T_a} \tag{1.13}$$

then the values of this quantity are from 5 K/atm at low pressures to 1 K/atm at high pressures.

In subsequent discussion the surface temperature–initial temperature relationship is described by the dimensionless derivative

$$r = \left(\frac{\partial T_s^0}{\partial T_a}\right)_p \tag{1.14}$$

Its values vary from 0.1 to about 0.5. It is convenient to introduce the following dimensionless parameters

$$k = \beta(T_s^0 - T_a) \quad r = \left(\frac{\partial T_s^0}{\partial T_a}\right)_p$$
$$\iota = \left(\frac{\partial \ln u^0}{\partial \ln p}\right)_{T_a} \quad \mu = \frac{1}{T_s^0 - T_a}\left(\frac{\partial T_s^0}{\partial \ln p}\right)_{T_a} \tag{1.15}$$

The parameter ι describes the dependence of the burning rate on pressure and has been used in most papers dealing with propellant combustion. The parameter k was introduced into nonsteady combustion theory by Zeldovich (1942). The derivatives r and μ, which describe surface temperature variations with respect to the initial temperature and pressure, were introduced into the theory by Novozhilov (1965a,b).

So far the dependences of the burning rate and surface temperature on pressure and initial temperature (i.e. the functions $u^0(T_a, p)$ and $T_s^0(T_a, p)$) have been considered. It would be instructive to reveal the relationship between the burning rate and the surface temperature, and check whether this relationship is single-valued (i.e. whether the burning rate is determined by the surface temperature uniquely). In other words, one would like to know whether the burning rate can be represented as a function $u^0(T_s^0)$ or if the latter function must include pressure as the second argument. The latter would mean that the processes in the reaction layer of the condensed phase are pressure-sensitive. This question was raised first by Zenin and Novozhilov (1973).

Figure 1.3 shows burning rate versus surface temperature, based on the data provided in Tables 1.3–1.5. It is evident that within the experimental error the unique relationship $u^0(T_s^0)$ exists. Mathematically this fact can be explained as follows. For a single-valued relationship $u^0(T_s^0)$ to exist, it is necessary that the functions $u^0(T_a, p)$ and $T_s^0(T_a, p)$ have a Jacobian

$$J = \frac{\partial(u^0, T_s^0)}{\partial(p, T_a)} = \left(\frac{\partial u^0}{\partial p}\right)_{T_a}\left(\frac{\partial T_s^0}{\partial T_a}\right)_p - \left(\frac{\partial u^0}{\partial T_a}\right)_p\left(\frac{\partial T_s^0}{\partial p}\right)_{T_a} \tag{1.16}$$

equal to zero.

It is obvious that experimental data cannot be used to show that the Jacobian is equal to zero, since there are always experimental errors present. On the other hand, if it is nonzero this could be, in principle, proved provided the experimental procedure is accurate enough. Unfortunately, large errors associated with the surface temperature measurements prohibit sufficiently accurate determination of its derivatives. At present, one can only state that the Jacobian may be equal to zero. In the future, with improved accuracy of the experiment, it may be possible to show that it is not equal to zero.

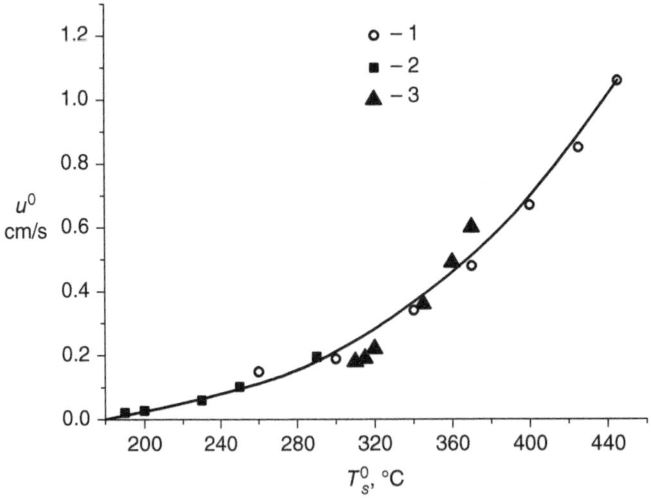

Figure 1.3 Dependence of burning rate on surface temperature for ballistite N. 1, $T_a = 20\,°\mathrm{C}$; different pressure values; 2, $p = 1$ atm, different temperature values; 3, $p = 20$ atm, different temperature values.

Consider, as an example, the value of the Jacobian at $p = 20$ atm and $T_a = 20°\mathrm{C}$. From the data in Tables 1.3 and 1.5, we find

$$\left(\frac{\partial u^0}{\partial p}\right)_{T_a} = 1.5 \cdot 10^{-2}\mathrm{cm/s\,atm} \quad \left(\frac{\partial T_s^0}{\partial p}\right)_{T_a} = 3.5\ \mathrm{K/atm}$$
$$\left(\frac{\partial u^0}{\partial T_a}\right)_p = 1.8 \cdot 10^{-3}\ \mathrm{cm/s\,K} \quad \left(\frac{\partial T_s^0}{\partial T_a}\right)_p = 0.4 \tag{1.17}$$

The Jacobian is represented by the difference of two numbers of the same order:

$$J = (6.0 - 6.3) \cdot 10^{-3}\ \mathrm{cm/s\,atm} \tag{1.18}$$

If one takes into consideration that the low accuracy of the surface temperature measurements lead to errors in its derivatives reaching tens of per cent, then it is clear that the data given in Tables 1.3 and 1.5 do not enable one to establish that the Jacobian is nonzero.

Let us introduce the dimensionless value

$$\delta = \iota r - \mu k \tag{1.19}$$

where the parameters k, r, ι, and μ are given by (1.15). It is easy to show that

$$\delta = \frac{p}{u^0} J \tag{1.20}$$

When the Jacobian is zero ($\delta = 0$), one of the four parameters k, r, ι, or μ can be found from the values of the remaining three. In many papers on nonsteady combustion a single-valued (Arrhenius) relation

$$u^0 \sim \exp\left(-\frac{E}{RT_s^0}\right) \tag{1.21}$$

between the surface temperature and the gasification rate of the propellant is postulated. In such models the propellant, in a linear approximation, is characterized by three parameters only.

1.3 Combustion Wave Structure. Burning Temperature

Let us consider now temperature distribution in the combustion zones under steady-state conditions. During propellant combustion, heat is released successively in three locations: in the condensed phase, near the propellant surface (in the so-called fizz or smoke-gas zone), and in the gas phase. The number of heat release zones and their contribution to the total heat balance depends on combustion conditions, that is, pressure and initial temperature.

In a vacuum, ballistite burns only at initial temperatures above 80–100 °C. At pressures exceeding $(2.6–6.5) \times 10^{-3}$ atm reaction starts in the fizz zone. A typical plot of the temperature distribution is given by curve 1 in Figure 1.4. The maximum temperature and the amount of heat liberated in this zone increases with pressure. As the pressure rises, the heat release region compresses and approaches the surface. This burning regime is called cold-flame, or single-flame. The flame in the fizz zone has a pale-blue hue and is visible only in the dark.

Further increase in pressure ($p > 15 - 20$ atm) gives rise to a third heat release zone characterized by a bright flame (curve 2, Figure 1.4). The distance of this luminous zone from the propellant surface rapidly decreases as the pressure rises, and at about 60–75 atm the second flame merges with the first. The temperature profile for burning at high pressure is given by curve 3 in Figure 1.4. Both the heat release and the maximum temperature in the gas zone (the burning temperature) increase with pressure. At about 60 atm the maximum possible heat release (complete burnout) and the maximum burning temperature, which remain constant on further pressure rise, are attained.

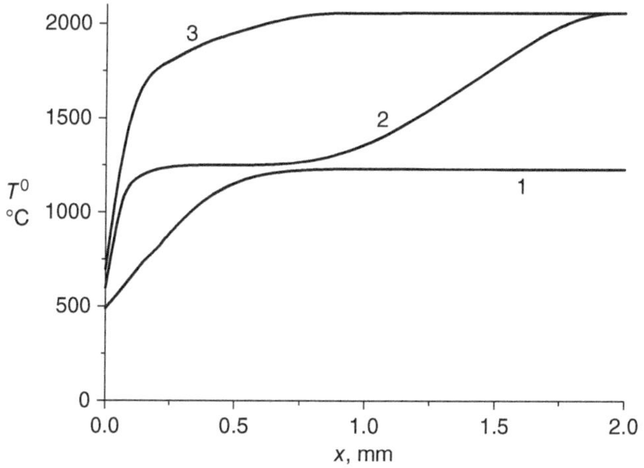

Figure 1.4 Sketch of the gas temperature profile during combustion of ballistite N. 1, $p = 10$ atm; 2, $p = 50$ atm; 3, $p = 100$ atm.

Thus, within certain pressure and initial temperature ranges, the reaction zone in the gas phase can be divided into the two regions. The first zone (closer to the surface) affects the burning rate; the temperature within this region, T^0_{b1}, is much lower than the combustion temperature T^0_b. In the second zone, the gas flame zone, which may be located several millimetres away from the surface, the temperature is equal to the combustion temperature.

The gas flame zone exists within the pressure range between 15 and 70 atm, and is characterized by the so-called induction burning regime (Zaidel and Zeldovich 1962; Merzhanov and Filonenko 1963) in which gases are heated up to the burning temperature not by heat conduction, but by internal self-heating with almost no heat flux from the gas flame to the smoke-gas zone. Therefore, the second flame practically does not affect the processes in the preceding zones or the burning rate.

Experimental values of the burning temperatures T^0_{b1} and T^0_b for ballistite N at different values of pressure and constant initial temperature, as obtained by Zenin (1980, 1992), are given in Table 1.6. Table 1.7 lists the values of T^0_{b1} and T^0_b for different initial temperatures.

To measure the change in T^0_b with initial temperature and pressure, it is useful to introduce the derivatives

$$r_b = \left(\frac{\partial T^0_b}{\partial T_a}\right)_p, \quad \mu_b = \left(\frac{\partial T^0_b}{\partial p}\right)_{T_a} \tag{1.22}$$

One can see from Tables 1.6 and 1.7 that r_b is approximately constant, $r_b \approx 1$. As for the value of μ_b, this ranges from 10 K/atm at low pressure to 1 K/atm at high pressure.

The temperature profiles can be used to determine various spatial and time parameters of the propellant combustion process, for example the extent of different zones or the duration of chemical reaction in them. Some of these characteristics are necessary for substantiation of the nonsteady theory.

Table 1.6 Burning temperatures of the propellant N at $T_a = 20\,°C$.

Parameter	Unit	Values						
p	atm	5	10	20	30	50	75	100
T^0_{b1}	°C	1000	1100	1180	1200	1250	–	–
T^0_b	°C	–	–	1650	1850	2010	2060	2060

Table 1.7 Burning temperatures of the propellant N at $p = 20$ atm.

Parameter	Unit	Values						
T_a	°C	−150	−100	−50	0	50	100	120
T^0_{b1}	°C	1040	1080	1130	1180	1230	1270	1300
T^0_b	°C	1500	1550	1600	1650	1690	1740	1750

Consider, for example, the gas phase characteristic time t_g

$$t_g = \int_0^{l_b} \frac{dx}{u_g^0(x)} \tag{1.23}$$

where $u_g^0(x)$ is the gas velocity distribution.

If the temperature distribution is known, the gas phase characteristic time t_g should be calculated from the formula

$$t_g = \frac{p\tilde{\mu}}{\rho u^0 R} \int_0^{l_b} \frac{dx}{T_g^0(x)} \tag{1.24}$$

At pressures higher than about 60 atm the two flame zones merge, therefore l_b should be understood as the distance at which the heat release nearly terminates. On the other hand, if combustion proceeds in a two-flame regime, it is natural, when calculating t_g, to take into account only the first (closer to the surface) flame, which actually affects the burning rate.

1.4 Combustion in Tangential Gas Stream

So far, propellant burning in the absence of gas flow along its surface has been considered. In rocket motors, however, such conditions are not always the case. Rather, as a rule, in a solid propellant rocket motor the propellant burns in the presence of the cross flow of combustion gases in the port of a rocket motor grain. A phenomenon of propellant burning rate increase on experiencing the flow of combustion products along its surface is called erosive burning.

The erosion effect is usually defined as the ratio of the steady-state burning rate u_ε^0 in the presence of a gas flow to that of u^0 without a flow. The ratio

$$\varepsilon^0 = \frac{u_\varepsilon^0}{u^0} \tag{1.25}$$

is termed the erosion coefficient (or the erosion ratio).

There is no satisfactory theory of erosive burning. This is a result of the absence of a steady-state theory of propellant combustion, and also of the complexity of the nonisothermal turbulent boundary layer flow with gas injection. All the semiempirical theories proposed so far explain erosive burning by an increase in the effective heat conduction of the gas due to the appearance of the turbulent boundary layer.

The first attempts to develop the theory of steady-state erosion were made by Vilyunov (1961) and Zeldovich (1971). Both explain burning rate increase by gas flow turbulization near the phase interface. The exact mechanism is believed to be an increase of effective gas conductivity under turbulent conditions, which in turn magnifies the heat flux from the combustion products to the propellant surface. It should be noted that both studies assume the simplest combustion wave structure comprising the two reaction zones, one at the phase interface and another in the gas phase.

The following relation is assumed (Vilyunov 1961) for tangential gas stream velocity

$$w_t = \frac{1}{U} \tanh(Uy) \tag{1.26}$$

where w_t and y are nondimensional gas velocity and distance from the propellant surface, respectively, and U is a constant parameter equal to 7×10^{-2}.

It was demonstrated (Vilyunov 1961) that the erosion ratio should depend on the dimensionless parameter

$$I = \frac{g^0}{m^0}\sqrt{\zeta} \tag{1.27}$$

where g^0 is the mass velocity of gas flow parallel to the burning surface, $m^0 = \rho u^0$ is the mass burning rate, and ζ is the channel coefficient of resistance.

The studies by Vilyunov (1961) and Zeldovich (1971) make rather crude assumptions, such as too simplified a description of the combustion front structure, isothermal approximation for the turbulent flow, neglect of injection, etc. With this in mind, Vilyunov (1961) has managed to obtain a fairly simple relation between the erosion coefficient and the parameter I above

$$\varepsilon = \sqrt{K(I) + LI} \tag{1.28}$$

Here $K(I)$ is nearly independent of I: $K(I) \approx 1$ and $L \approx$ constant. Unfortunately, the paper by Vilyunov (1961) contains a misprint: the second term under the root must be proportional to I^2. With this fact in mind and assuming the coefficient of resistance to be constant (its dependence on the Reynolds number is very weak), one obtains

$$\varepsilon^0 = \sqrt{1 + b\left(\frac{g^0}{m^0}\right)^2} \tag{1.29}$$

where $b \approx$ constant.

The following estimation can be derived from Vilyunov (1961)

$$b = \left(U\frac{\sqrt{\zeta}}{2\sqrt{2}}\ln\frac{c(T_b^0 - T_a) - Q_s}{c(T_s^0 - T_a) - Q_s}\right)^2 \tag{1.30}$$

where Q_s is heat of reaction in the condensed phase. Available values of parameters related to reaction zone in the condensed phase are not particularly accurate. Fortunately, they enter the formula above logarithmically, which allows rather reliable estimation for the parameter b to be made. It is approximately equal to 2–3.

The resistance coefficient can be estimated by the well-known formula due to Nikuradze (Loitsyanskii 1966)

$$\zeta = 0.0032 + \frac{0.221}{\mathrm{Re}^{0.237}} \tag{1.31}$$

Typical values of Reynolds number for the problem under consideration are of the order of 10^6, which yields $\zeta \approx 1.2 \times 10^{-2}$, and further

$$b \approx 5 \times 10^{-5} \tag{1.32}$$

The relation (1.29) with the coefficient (1.32) should really be considered as merely a one-parameter interpolation of existing data on erosion combustion. Nevertheless, some theoretical justification for this relation can be proposed.

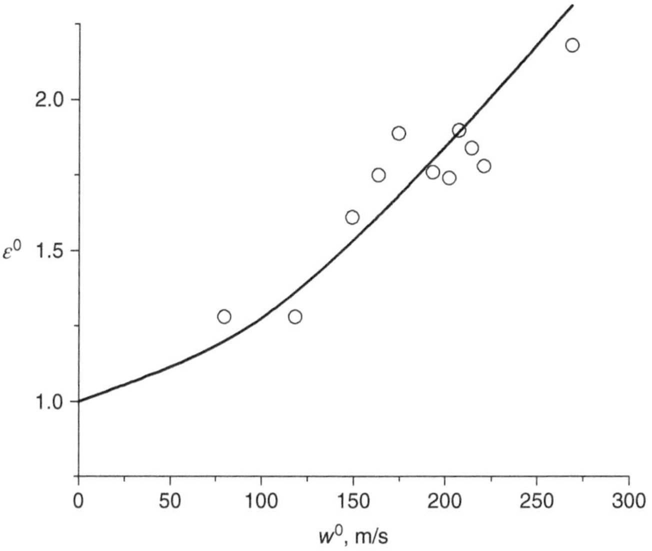

Figure 1.5 Erosion coefficient as a function of tangential gas velocity.

It is conventional to plot erosion coefficient versus tangential gas stream velocity w^0. In this way, (1.29) can be written as

$$\varepsilon^0 = \sqrt{1 + b_v(w^0)^2} \tag{1.33}$$

with

$$b_v = b\left(\frac{\rho_g}{\rho u^0}\right)^2 \tag{1.34}$$

If the units of w are taken as m/s, then the units of b_v are s^2/m^2. As a test for the validity of (1.33) consider, as an example, the experimental data by Leipunskii et al. (1982). These have been obtained by direct measurements of erosion coefficient for the ballistite N in a wide range of tangential gas flow velocities. Figure 1.5 shows experimental data, as well as the curve (1.33) at the pressure of 60 atm. The best agreement is achieved at $b_v \approx 6 \times 10^{-5}$ s^2/m^2, close to the value $b_v \approx 5 \times 10^{-5}$ s^2/m^2 obtained from (1.34).

On combustion of the internal surface (of radius R) of the hollow cylindrical sample, closed at one end ($x = 0$), combustion product velocity grows away from the closed end and reaches a maximum at the nozzle. This results in the following spatial dependence of the erosion coefficient

$$\varepsilon = \text{ch}2\sqrt{b}\frac{x}{R} \tag{1.35}$$

Uneven burnout of the propellant along its length allows such dependence to be verified directly in experiments (Leipunskii and Frolov 1982). The method employed in this study was a quick combustion termination after burning propellant for a certain time. Fuel left-out thickness at different cross-sections evidences the burning rate dependence on tangential gas stream velocity. Figure 1.6 shows data by Leipunskii and Frolov (1982) and

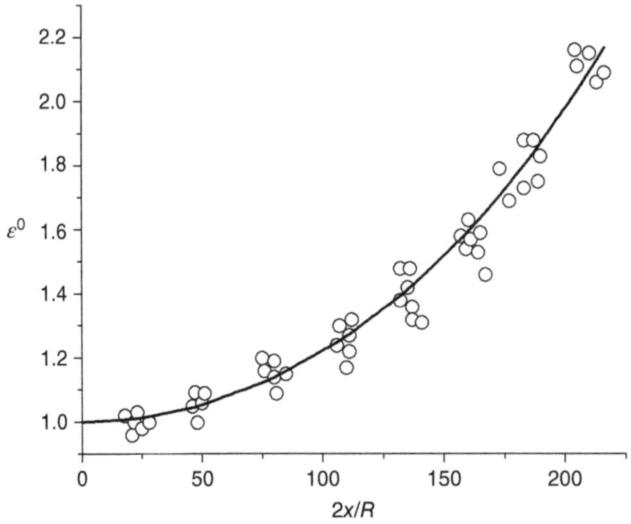

Figure 1.6 Spatial distribution of the erosion coefficient.

demonstrates good comparison with (1.35) (the value of the parameter b is $b \approx 4.25 \times 10^{-5}$, close to the estimation in (1.32)).

The effect of erosion is most significant on ignition of the charge at the start of a solid rocket motor. The growing burning rate leads to a so-called erosive pressure peak in the chamber. Prediction of this phenomenon in a real engine is extremely complicated. An alternative approach by Kohno et al. (1998) is an attempt to scale up the data obtained under laboratory conditions.

1.5 Gaseous Flame

The above data show that transformation of the condensed phase into gaseous combustion products occurs across several zones. At least one of these zones is within the gas phase. For this reason it is instructive to recall major facts related to constant-speed combustion wave propagation in a gas.

Such a theory, combining basic principles of heat and mass transfer with chemical kinetics, was formulated by Lewis and Von Elbe (1934). In the limit of large activation energy Zeldovich and Frank-Kamenetskii (1938) were able to find a closed form solution for the burning rate of a gas mixture, controlled by simple kinetics.

More specifically, let us consider a fresh mixture at the initial temperature $T_g^0 = T_a$ at $x \to -\infty$, and combustion products at the burning temperature T_b^0 at $x \to \infty$. Flame propagates from left to right.

Transformation to the coordinate system attached at the flame front leads to one-dimensional steady-state conservation equations for the energy and the fuel mass fraction

$(-\infty < x < \infty)$

$$\frac{\mathrm{d}}{\mathrm{d}x}\lambda_g\frac{\mathrm{d}T_g^0}{\mathrm{d}x} - m^0 c_p\frac{\mathrm{d}T_g^0}{\mathrm{d}x} + Q_g W(Y^0, T_g^0) = 0$$
$$\frac{\mathrm{d}}{\mathrm{d}x}D\rho_g\frac{\mathrm{d}Y^0}{\mathrm{d}x} - m^0\frac{\mathrm{d}Y^0}{\mathrm{d}x} - W(Y^0, T_g^0) = 0 \tag{1.36}$$

Here ρ_g, T_g^0, and Y^0 are gas density, gas temperature, and fuel mass fraction, respectively, λ_g and D are gas thermal conductivity and diffusion coefficient, respectively, Q_g is the heat of combustion, and W is the reaction rate.

In the adopted description, media moves from left to right with unknown velocity corresponding to the mass burning rate m^0.

For the global nth order reaction

$$W_n = \tilde{k}_n(\rho_g Y^0)^n \exp(-E_g/RT_g^0) \tag{1.37}$$

where $\tilde{k}_n$ and E_g are kinetic constants.

The required boundary conditions for (1.36) are

$$x \to -\infty, \quad T_g^0 = T_a$$
$$x \to \infty, \quad \frac{\mathrm{d}T_g^0}{\mathrm{d}x} = 0, \quad \frac{\mathrm{d}Y^0}{\mathrm{d}x} = 0 \tag{1.38}$$

Consider the case of unity Lewis number ($D\rho_g c_p = \lambda_g$). The reaction rate can be excluded from (1.36) to give the linear ordinary differential equation

$$\frac{\mathrm{d}}{\mathrm{d}x}\frac{\lambda_g}{c_p}\frac{\mathrm{d}H^0}{\mathrm{d}x} - m^0\frac{\mathrm{d}H^0}{\mathrm{d}x} = 0, \quad H^0 = Q_g Y^0 + c_p T_g^0 \tag{1.39}$$

The solution of this equation, satisfying the boundary conditions (1.38), is a constant function. Therefore, there exists a similarity between the temperature and the mass fraction (or concentration) fields

$$Y^0 = \frac{c_p(T_b^0 - T_g^0)}{Q_g} \tag{1.40}$$

with the burning temperature given by a simple and natural expression

$$T_b^0 = T_a + \frac{Q_g}{c_p} \tag{1.41}$$

Relation (1.40) between the two variables allows the first (heat transfer) of Eq. (1.36) to be reduced to the following equation containing an unknown temperature field only

$$\frac{\mathrm{d}}{\mathrm{d}x}\lambda_g\frac{\mathrm{d}T_g^0}{\mathrm{d}x} - m^0 c_p\frac{\mathrm{d}T_g^0}{\mathrm{d}x} + \Phi_n(T_g^0) = 0 \tag{1.42}$$

where

$$\Phi_n(T_g^0) = \frac{\tilde{k}_n}{Q_g^{n-1}}[\rho_g c_p(T_b^0 - T_g^0)]^n \exp\left(-\frac{E_g}{RT_g^0}\right) \tag{1.43}$$

Before proceeding to the burning rate calculation, recall that reaction rate grows rapidly with temperature. This justifies separate consideration of the two zones: the inert preheat zone and the actual reaction zone where chemical transformation occurs. The latter spreads, in temperature units, across the interval of the order of $R(T_b^0)^2/E_g$. In other words, chemical transformation occurs in a narrow temperature range, close to the combustion temperature.

At the notional boundary $x = x^*$ between the two zones, the temperature is close to that of combustion and the reaction rate is virtually zero. Therefore

$$\lambda_g \left.\frac{dT_g^0}{dx}\right|_{x=x*} = m^0 Q_g \tag{1.44}$$

Only the temperature gradient and the reaction rate change significantly across the reaction zone; the convective term in (1.42) can be neglected in the view of small temperature variation.

The heat transfer equation takes the form

$$\frac{d}{dx}\lambda_g \frac{dT_g^0}{dx} + \Phi_n(T_g^0) = 0 \tag{1.45}$$

where the transformation $\zeta(T_g^0) = \lambda_g dT_g^0/dx$ further reduces the order

$$\zeta \frac{d\zeta}{dT_g^0} + \lambda_g \Phi_n(T_g^0) = 0 \tag{1.46}$$

As the reaction is completed, the temperature gradient must turn into zero, that is,

$$T_g^0 = T_b^0, \quad \zeta = 0 \tag{1.47}$$

Taking this into account, (1.46) results in

$$\lambda_g \frac{dT_g^0}{dx} = \sqrt{2\int^{T_b^0} \lambda_g \Phi_n(T_g^0) dT_g^0} \tag{1.48}$$

A major contribution to the integral in (1.48) comes from the region of high temperatures (reaction rate strongly grows with temperature). The influence of the lower limit is negligible, and therefore it is omitted in (1.48).

Comparing (1.44) and (1.48), one obtains the mass burning rate

$$m^0 = \frac{1}{Q_g}\sqrt{2\int^{T_b^0} \lambda_g \Phi(T_g^0) dT_g^0} \tag{1.49}$$

Estimation of integrals containing Arrhenius exponent follows the methodology of Frank-Kamenetskii (1939), who proposed expansion

$$\frac{E}{RT} \approx \frac{E}{RT_b} + \frac{E(T_b - T)}{RT_b^2} \tag{1.50}$$

holding for $T_b - T << T_b$. This transformation moves the integration variable from the denominator to the numerator, and the integral in (1.49) is expressed analytically

$$\int^{T_b^0} \lambda_g \Phi_n(T_g^0) dT_g^0 =$$

$$\lambda_g(T_b^0)\tilde{k}\left[\frac{c_p \rho_g(T_b^0)}{Q_g}\right]^n \exp\left(-\frac{E_g}{RT_b^0}\right)\int^{T_b^0} (T_b^0 - T_g^0)^n \exp\left[-\frac{E_g(T_b^0 - T_g^0)}{R(T_b^0)^2}\right] dT_g^0 \tag{1.51}$$

Taking note of

$$\int_0^\infty z^n \exp(-\alpha z)\mathrm{d}z = \frac{n!}{\alpha^n} \tag{1.52}$$

the mass burning rate is finally obtained as

$$(m_n^0)^2 = 2n!\frac{\lambda_g(T_b^0)\tilde{k}_n[\rho_g(T_b^0)]^n}{c_p}\left(\frac{c_p R(T_b^0)^2}{Q_g E_g}\right)^{n+1}\exp\left(-\frac{E_g}{RT_b^0}\right) \tag{1.53}$$

Since the influence of temperature close to the combustion temperature is of major importance, gas density and gas thermal conductivity in (1.53) must be estimated at the combustion temperature. The combustion temperature itself is calculated from (1.41).

It should be noted that (1.53) holds only for low values of the reaction order. Global reactions with $n > 2$ are unlikely to occur. The simplest case is given by the first-order reaction where

$$(m_1^0)^2 = 2\frac{\lambda_g(T_b^0)\tilde{k}_1\rho_g(T_b^0)}{c_p}\left(\frac{cR(T_b^0)^2}{Q_g E_g}\right)^2\exp\left(-\frac{E_g}{RT_b^0}\right) \tag{1.54}$$

The major conclusion of the presented theory is that the burning rate is a power function of pressure and the Arrhenius function of the initial temperature. Indeed, (1.53) gives

$$m_n^0 \propto p^{n/2}\exp\left[-\frac{E_g}{2R\left(T_a + \frac{Q_g}{c_p}\right)}\right] \tag{1.55}$$

Since usually $T_a << Q_g/c_p$, then a simpler expression can be deduced

$$m_n^0 \propto p^{n/2}\exp(\beta T_a) \tag{1.56}$$

where β = constant.

1.6 Combustion Waves in the Condensed Phase

The experimental data presented in this chapter prove unequivocally the existence of a chemical reaction in the condensed phase. Therefore the problem of predicting the rate of an exothermal reaction front propagation in solid or liquid is of considerable interest.

Consider the case similar to the combustion of mixed gases, with no heat flux from combustion products to the reaction zone. Such a problem, formulated and considered by Novozhilov (1961), may be of use either in investigating burning of compositions whose combustion products are condensed, or in studying the conditions of flameless burning. In both cases combustion is sustained exclusively due to the reaction in the condensed phase: there is no external heat supply. The thermal theory of flame propagation in gases is greatly simplified by the similarity between the processes of diffusion and heat transfer (gas diffusion coefficient and thermal diffusivity can be assumed to be equal).

This fact makes it possible to relate concentration and temperature, and therefore to reduce the set of two equations (diffusion and heat conduction) to a single equation containing temperature distribution only. Unfortunately, this is obviously not the case for condensed fuels where the diffusion coefficient and thermal diffusivity may differ by several orders of magnitude. In subsequent discussion the diffusion coefficient is assumed to be zero.

Consider the one-dimensional case, assuming at $x \to -\infty$ a substance with density ρ = constant and temperature T_a, which undergoes chemical transformation with heat of reaction Q_s. The problem is considered in a coordinate frame attached at the flame front. In such a frame, fresh substance moves with velocity u^0 which has to be determined. Let us introduce the following additional notation: ρ_1 and ρ_2 for the densities of initial substance and product, respectively, $\rho = \rho_1 + \rho_2$ and $\eta^0 = \rho_2/\rho$ for the relative concentration of the product, and W for the chemical reaction rate.

The equations of energy and mass conservation have the form

$$\kappa\frac{d^2T^0}{dx^2} - u^0\frac{dT^0}{dx} + \frac{Q_s}{\rho c}W(T^0,\eta^0) = 0 \tag{1.57}$$

$$-u^0\frac{d\eta^0}{dx} + \frac{1}{\rho}W(T^0,\eta^0) = 0 \tag{1.58}$$

with boundary conditions

$$\begin{aligned} &x \to -\infty \quad T^0 = T_a \quad \eta^0 = 0 \\ &x \to \infty \quad \frac{dT^0}{dx} = 0 \quad \frac{d\eta^0}{dx} = 0 \end{aligned} \tag{1.59}$$

Eliminating the reaction rate from the above equations and taking into account the boundary conditions for $x \to -\infty$, one finds

$$\kappa\frac{dT^0}{dx} - u^0(T^0 - T_a) + \frac{u^0\eta^0}{c}Q_s = 0 \tag{1.60}$$

The relation (1.60), taken at the limit $x \to \infty$, provides the following expression for the combustion temperature

$$T_b^0 = T_a + \frac{Q_s}{c} \tag{1.61}$$

Due to the strong temperature dependence of the reaction rate the reaction occurs at temperatures close to T_b^0. Therefore one can assume $T^0 = T_b^0$ within the reaction zone, and thus (1.60) yields inside this zone

$$\kappa\frac{dT^0}{dx} = \frac{u^0Q_s}{c}(1-\eta^0) \tag{1.62}$$

This expression replaces in the present case the temperature–concentration relationship (1.40), which holds when the diffusion coefficient and thermal diffusivity are equal. The absence of diffusion leads to a relationship between temperature gradient and concentration. Following (1.62), at one boundary of the reaction zone ($x \to \infty$) $dT^0/dx = 0$, and at the other (start of chemical transformation at $\eta^0 \ll 1$)

$$\frac{dT^0}{dx} = \frac{u^0Q_s}{\kappa c} \tag{1.63}$$

Using the relationship (1.62) one can express the concentration, affecting the reaction rate W, through the temperature gradient and reduce the problem to a single heat conduction equation. As in the theory of gas flame propagation, the second term of this equation can be neglected since the temperature is nearly constant across the reaction zone. Introducing $\zeta(T^0) = dT^0/dx$, (1.57) is reduced to the first-order equation

$$-\zeta\frac{d\zeta}{dT^0} = \frac{Q_s}{\kappa\rho c}W(T^0,\zeta) \tag{1.64}$$

Consider two particular cases.

(1) A zero-order reaction

$$W_0 = \rho\tilde{k}_0\exp(-E/RT^0) \tag{1.65}$$

Eq. (1.64) with the condition (1.63) yields the burning rate

$$u_0^0 = \sqrt{\frac{2\tilde{k}_0\kappa cR(T_b^0)^2}{Q_sE}\exp\left(-\frac{E}{RT_b^0}\right)} \tag{1.66}$$

which coincides with the flame propagation rate in a gas (1.53) at $n = 0$. This is natural since the presence or absence of diffusion in no way affects the rate of a zero-order reaction.

(2) A first-order reaction

$$W_1 = \rho\tilde{k}_1(1-\eta^0)\exp(-E/RT^0) \tag{1.67}$$

Eq. (1.64) is readily integrated and, with account taken of (1.63), yields the front propagation rate

$$u_1^0 = \sqrt{\frac{\tilde{k}_1\kappa cR(T_b^0)^2}{Q_sE}\exp\left(-\frac{E}{RT_b^0}\right)} \tag{1.68}$$

which differs from u_0^0 only by a factor of $1/\sqrt{2}$.

The presented theory of combustion front propagation in infinite condensed media was considered (Novozhilov 1961) in relation with the possibility of polymerization front propagation. Indeed, the envisaged polymerization process was observed in the experiment of Chechilo et al. (1972) a few years later.

Purely gasless combustion was studied first by Belyaev and Komkova (1950) using thermites. Both fuel and combustion products in this case are hardly volatile and chemically stable. The reaction, therefore, occurs predominantly in the condensed phase and its rate must be pressure-independent. The latter fact was confirmed by the experiments of Belyaev and Komkova (1950).

Over time, widespread gasless combustion has become evident, especially after the development of self-propagating high-temperature synthesis (SHS) (Merzhanov 1994). A vast class of combustion systems undergoing gasless chemical transformation has been discovered which, in turn, has stimulated further theoretical developments.

A different problem arises when considering condensed systems undergoing gasification. Semi-infinite space $-\infty < x \leq 0$ must be considered, with $x = 0$ demarking the interface between the condensed and gas phases. Another important difference is that, in general, the heat flux from the gas phase to the condensed one is present. The temperature gradient at the interface is therefore different from zero.

Let us consider first the case with the heat flux to condensed phase absent.

Heat transfer and mass conservation (Eqs. (1.57) and (1.58)) still hold while the boundary conditions take the form

$$x \to -\infty, \quad T^0 = T_a, \quad \eta^0 = 0 \tag{1.69}$$

$$x = 0, \quad \frac{dT^0}{dx} = 0, \quad \eta^0 = \eta_s^0 \tag{1.70}$$

where η_s^0 is a progress variable (fraction of virgin fuel converted to product).

Similar to (1.60)

$$\kappa\frac{dT^0}{dx} - u^0(T^0 - T_a) + \frac{u^0\eta^0}{c}Q_s = 0 \tag{1.71}$$

At $x = 0$ the relation (1.71) gives the temperature at the interface

$$T_s^0 = T_a + \frac{Q_s}{c}\eta_s^0 \tag{1.72}$$

Furthermore, (1.70)–(1.72) allow the temperature gradient and the concentration within the reaction zone to be related

$$\kappa\frac{dT^0}{dx} = \frac{u^0 Q_s}{c}(\eta_s^0 - \eta^0) \tag{1.73}$$

Substitution of (1.73) into (1.58) yields

$$(u^0)^2\frac{Q_s(\eta_s^0 - \eta^0)}{c\kappa}\frac{d\eta^0}{dT^0} = \frac{1}{\rho}W(T^0, \eta^0) \tag{1.74}$$

For the zero-order reaction (1.65) integration of (1.74) gives

$$u_0^0 = \frac{1}{\eta_s^0}\sqrt{\frac{2\tilde{k}_0\kappa cR(T_s^0)^2}{Q_s E}\exp\left(-\frac{E}{RT_s^0}\right)} \tag{1.75}$$

while for the first-order reaction (1.67)

$$u_1^0 = \sqrt{\frac{\tilde{k}_0\kappa cR(T_s^0)^2}{[\eta_s^0 + (1 - \eta_s^0)\ln(1 - \eta_s^0)]Q_s E}\exp\left(-\frac{E}{RT_s^0}\right)} \tag{1.76}$$

For small values of the progress variable, the burning rates for the zero-order and first-order reactions are nearly equal, indeed at $\eta_s^0 << 1$

$$\frac{u_0^0}{u_1^0} \approx 1 + \frac{\eta_s^0}{6} \tag{1.77}$$

Propagation of the front of an exothermal reaction in condensed medium is considered in the papers by Merzhanov and Dubovitsky (1959), Manelis et al. (1996), and Novikov and Ryazantsev (1965a,b), which cover the case involving external heat supply to the reaction zone, that is, nonzero temperature gradient at the point of maximum temperature. Let q be the heat flux into the condensed phase

$$q = \lambda\left.\frac{dT^0}{dx}\right|_{x=0} \tag{1.78}$$

With no heat supply, the burning temperature is determined by heat release in the reaction zone: $Q_s = c(T_s^0 - T_a)$. With the heat flux of density q from the gas phase, the burning temperature T_{sq}^0 increases and can be found from the expression

$$q + \rho u_q^0 Q_s = \rho u_q^0 c(T_{sq}^0 - T_a) \tag{1.79}$$

where u_q^0 is the burning rate with heat supply.

For a zero-order reaction, this is (Chechilo et al. 1972; Belyaev and Komkova 1950)

$$u_{q0}^0 = \frac{\sqrt{[u_0^0(T_{sq}^0)\rho Q_s]^2 + q^2}}{\rho c(T_{sq}^0 - T_a)} \tag{1.80}$$

Here, $u_0^0(T_{sq}^0)$ denotes expression (1.66) with T_{sq}^0 replacing T_b^0.

1.7 The Two Approaches to the Theory of Nonsteady Propellant Combustion

The present monograph deals with the theory of nonsteady burning of solid propellants. Such a theory, obviously, must involve the properties of the steady-state combustion process. There are two vastly different approaches to implementing such properties into the theory.

The first is called the flame model (FM) method. The second approach is the Zeldovich–Novozhilov (ZN) theory.

Within the framework of the FM method one starts with developing a model for steady-state combustion. Unfortunately, this goal is practically unachievable for realistic propellant compositions (Miller and Anderson 2004).

The reasons for this are as follows: combustion always involves a large number of chemical reactions, and in most cases burning rate depends on chemical kinetics. Therefore, practically every combustion theory essentially relies on the kinetics of various reactions. However, our present understanding of combustion kinetics is far from satisfactory. Little is known, for example, about the kinetics of reactions involved in the combustion of condensed substances. As a result, theoretical calculations use certain reaction models which, more often than not, only slightly resemble real chemical processes. It is obvious that such investigations are only qualitative and hardly suitable for comparison with experiments. For instance, the steady-state burning theory for condensed substances has been developed exclusively for the simplest types of chemical reactions. While it can supply a qualitative explanation of burning rate dependencies on, say, pressure or initial temperature of the propellant, its experimental validation is practically impossible due to the highly idealized reaction model. Real physicochemical processes are much more complicated than the theoretical models that are currently being proposed. Moreover, it is evidently impossible to develop a quantitative steady-state burning theory which would cover a broad class of substances with widely different properties. De Luca (1992) offers a review devoted to the FM approach. The principal disadvantage of investigations made in this direction is that they do not provide a description of the phase transition mechanism. In particular, one has to deal with the transition from the condensed to the gaseous phase coupled with chemical reactions occurring at the interface. This fundamental problem is far from a satisfactory solution.

Often, use is made of the model proposed by Denison and Baum (1961). It assumes that the mass burning rates in the condensed and gas phases are both described by the Arrhenius law

$$m_s^0 = C_s \exp\left(-{}^{E_s}/_{RT_s^0}\right) \tag{1.81}$$

$$m_g^0 = C_g p^{n/2} (T_b^0)^{\left(n/2+1\right)} \exp\left(-{}^{E_g}/_{2RT_b^0}\right) \tag{1.82}$$

In these relationships the parameters C_s and C_g are constants. As follows from the results in Section 1.6, if E_s denotes the activation energy of the chemical reaction, a multiplier 2 must be introduced into the denominator of the exponent. For the burning rate in the gas phase, the expression (1.82) conforms with (1.53). It should be noted that the gas thermal diffusivity is assumed to be independent of the pressure and temperature (Denison and Baum 1961).

Furthermore, Denison and Baum (1961) assume specific heats of both the phases to be equal. Therefore, the combustion temperature is simply

$$T_b^0 = T_a + \frac{Q}{c} \tag{1.83}$$

Under steady-state conditions, mass burning rates are equal as well

$$m_g^0 = m_s^0 \tag{1.84}$$

The FM method has been also employed by some other investigators, for example Belyaev (1938,1940), Istratov and Librovich (1964), and Novikov and Ryazantsev (1966). The most convincing is the model by Belyaev (1938, 1940). It treats transformation of the condensed phase to the gas as an evaporation with equilibrium conditions at the interface. This model is considered in detail in the next section.

Most of the problems treated by the theory of nonsteady combustion of homogeneous propellants are solved under the assumption that only the preheat layer in the condensed phase possess thermal inertia. The theory includes the only quantity with units of time, that is, the relaxation time of the heated layer, $t_c \propto \kappa/(u^0)^2$. Finite relaxation times of chemical reaction zones in the gas and condensed phases, as well as that of heating regions in the gas are ignored. This approach is named QSHOD (quasi-steady, homogeneous, one-dimensional) analysis or t_c approximation.

Such an approach to the nonsteady theory was proposed by Zeldovich (1942). He demonstrated that, up to a pressure of the order of 100 atm, the relaxation time of the preheat layer is much bigger than the relaxation time of the gas phase. This was also confirmed later by Akiba and Tanno (1959) and Hart and McClure (1959).

Zeldovich (1942) showed that the t_c approximation allows a phenomenological nonsteady theory to be deduced in such a way that the kinetics of chemical reactions and all the complex physical processes are included through only one dependence, that is, the steady-state burning rate as a function of pressure and initial temperature. The latter can be determined experimentally and, importantly, it is sufficient to measure this dependence under steady-state conditions only. Taking account of the thermal inertia of the condensed phase, which is described by heat conduction equation, does not represent any substantial difficulty.

A great advantage of this theory is that it permits consideration of nonsteady combustion without involving a steady-state theory. The Zeldovich theory needs only dependence of the burning rate on initial temperature and pressure $u^0(T_a, p)$. This dependence can be introduced into the theory from experiments. Atwood et al. (1999a,b) discuss relevant experimental techniques.

The Zeldovich theory explains qualitatively certain phenomena inherent in nonsteady burning, but its quantitative comparison with experiment leads to a contradiction. The latter is manifested most vividly in the fact that according to this theory steady-state burning in real systems turns out to be unstable. This discrepancy is caused by the excessive simplification of the combustion model, where the surface temperature of the condensed phase is considered to be constant.

Zeldovich (1942) neglected chemical reactions in the condensed phase and the associated variation in the propellant surface temperature T_s^0. Experimental data show that the surface temperature is a function of initial temperature and pressure. Therefore, such neglect leads to a discrepancy between theoretical and experimental results.

Nonsteady theory with a variable surface temperature was developed by Novozhilov (1965a,b), who demonstrated that besides the function $u^0(T_a, p)$ the dependencies of surface and burning temperatures $T_s^0(T_a,p)$ and $T_b^0(T_a,p)$ on the same parameters are needed. These steady-state dependences can also be determined experimentally.

It is important to note that the mathematical implementation of the ZN theory is much simpler than that required in the FM method. For example, in the linear approximation (where the nonsteady process deviates only slightly from the steady state) the fuel properties are always reflected simply through the linear sensitivity coefficients of the burning rate, surface temperature, and burning temperature to the initial temperature and pressure.

1.8 Steady-state Belyaev Model

As discussed above, the major obstacle to development of the steady-state combustion theory of energetic materials undergoing gasification is the immense complexity of combustion mechanisms, even for homogeneous systems. In particular, this concerns phase transition from condensed to gas matter, coupled with chemical transformations.

The first realistic combustion model for condensed propellants was proposed by Belyaev (1938, 1940). He also studied the problem experimentally and compared theoretical predictions with measurements. The original model was developed for liquid substances, but due to similarity between vaporization and sublimation processes, it is equally applicable to solid propellants. In this model chemical reaction occurs in the vapours of the liquid that is heated to a certain temperature and evaporated (with a latent heat of evaporation L) by the heat flux from the combustion zone. The simplest reaction scheme $A \to P + Q_g$ is assumed, where A and P are the initial propellant and reaction products, respectively, and Q_g is the heat of reaction. Let us suppose also that the molecular weights of the unburnt propellant and the products differ very little, therefore the molecular weight is assumed to be constant. Similarly, the specific heats of the condensed and gas phases are assumed to be equal.

The problem is considered in the frame attached at the interface $x = 0$ between the phases.

The equations governing the combustion process in the model formulated above comprise the heat conduction equations in the condensed and gas phases ($-\infty < x \leq 0$)

$$\frac{d}{dx}\lambda\frac{dT^0}{dx} - m^0 c\frac{dT^0}{dx} = 0$$
$$\frac{d}{dx}\lambda_g\frac{dT_g^0}{dx} - m^0 c_p\frac{dT_g^0}{dx} + Q_g W(Y^0, T_g^0) = 0 \tag{1.85}$$

as well as the mass balance for the propellant products of gasification ($0 \leq x < \infty$)

$$\frac{d}{dx}D\rho_g\frac{dY^0}{dx} - m^0\frac{dY^0}{dx} - W(Y^0, T_g^0) = 0 \tag{1.86}$$

For the nth order reaction in the gas

$$W = \tilde{k}(\rho_g Y^0)^n \exp(-E_g/RT_g^0) \tag{1.87}$$

The required boundary conditions are

$$x \to -\infty, \quad T^0 = T_a,$$
$$x \to \infty, \quad \frac{dT_g^0}{dx} = 0, \quad \frac{dY^0}{dx} = 0, \tag{1.88}$$

The interface relations have the form

$$x = 0, \quad T^0 = T_g^0,$$
$$\lambda\left.\frac{dT^0}{dx}\right|_{0^-} = \lambda_g\left.\frac{dT_g^0}{dx}\right|_{0^+} - m^0 L$$
$$m^0 = m^0 Y^0 - D\rho_g\frac{dY^0}{dx}, \tag{1.89}$$

Finally, the condition of equilibrium evaporation relating concentration and temperature at the interface is

$$pY_s^0 = b\exp\left(-\frac{L\tilde{\mu}}{RT_s^0}\right) \tag{1.90}$$

where $\tilde{\mu}$ is molecular weight of the liquid. The constant parameter b may be excluded if the boiling temperature T_r at some given pressure p_r is known. In this case, the previous formula is rewritten as

$$pY_s^0 = p_r\exp\left[-\frac{L\tilde{\mu}}{R}\left(\frac{1}{T_s^0} - \frac{1}{T_r}\right)\right] \tag{1.91}$$

The mass burning rate can be found by the same considerations that apply to flame propagation in gas (Section 1.5). In particular, applying (1.39)–(1.49), one obtains the following expression identical to (1.53)

$$(m_n^0)^2 = 2n!\frac{\lambda_g(T_b^0)\tilde{k}_n[\rho_g(T_b^0)]^n}{c_p}\left(\frac{c_p R(T_b^0)^2}{Q_g E_g}\right)^{n+1}\exp\left(-\frac{E_g}{RT_b^0}\right) \tag{1.92}$$

except that, in contrast to (1.40), the combustion temperature is determined not only by the heat of reaction in gas, but also by the heat of evaporation:

$$T_b^0 = T_a + \frac{Q_g - L}{c_p} \tag{1.93}$$

The similarity of the concentration and temperature fields in the gas phase provides the relation between fuel temperature and concentration at the interface

$$Y_s^0 = \frac{c_p(T_b^0 - T_s^0)}{Q_g} \tag{1.94}$$

This relation, coupled with (1.91), allows Y_s^0 and T_s^0 to be found.

Finally, Eqs (1.85) and (1.86) with the conditions (1.88)–(1.91) provide temperature distributions in both phases, as well as the concentration profile in the gas phase.

Belyaev applied this model to the combustion of nitroglycol, while Allison and Faeth (1975) applied it to hydrazine combustion.

2

Equations of the Theory of Nonsteady Combustion

2.1 Major Assumptions

Let us list the major assumptions of the Zeldovich–Novozhilov (ZN) theory.

a) The theory is developed within the t_c approximation, i.e. the relaxation times of all chemical reaction zones and heating regions in the gas are considered negligible compared to the thermal inertia of the preheat zone in the condensed phase. It is straightforward to show that the thermal relaxation time of the latter preheat zone (zone I in Figure 1.1), $t_c \sim \kappa/(u^0)^2$, is much larger than the relaxation time of the reaction zone (zone II in Figure 1.1), t_{cr}. Indeed, the latter may be estimated as $t_{cr} \sim t_c \Delta T/(T_s^0 - T_a)$, where ΔT is the temperature range of the significant reaction rate. Because of the strong sensitivity of reaction rate to temperature $\Delta T << T_s^0 - T_a$ and therefore $t_{cr} < < t_c$, the range ΔT can be estimated for the Arrhenius reaction rate as $\Delta T \sim R(T_s^0)^2/E$. This gives, for example, $t_{cr}/t_c \propto 0.1$ for $E \sim 84$ kJ/mol. The reaction zone in the condensed phase is assumed therefore to be infinitely thin.
   Zeldovich (1942) estimates the gas phase inertia to be $t_g \propto l_g/u_g^0$, where l_g is the thickness of the preheat zone in gas and u_g^0 is the mean gas velocity in this zone. For ballistites, the relaxation time t_c varies from 0.5 to 10^{-3}s as the pressure changes from 1 to 100 atm. Therefore, $t_c/t_g \propto 7 \times 10^3$ at 1 atm and $t_c/t_g \propto 7$ at 100 atm.
   Detailed measurements of temperature profiles in propellants (Section 1.3) call for revision of the above estimations. Before proceeding to estimations based on real temperature profiles, let us note first that there are two major problems to be investigated by the nonsteady combustion theory.
   The first problem is the prediction of the propellant burning rate under given conditions, that is, pressure, tangential gas stream, and external radiation. This problem has been investigated extensively and is a major focus of this book. On the other hand, investigation of nonsteady combustion in real chambers requires a coupled problem to be solved where pressure has to be determined from the equations relating it to nonsteady propellant burning rate and combustion chamber temperature. The latter also changes with time under unsteady conditions.
   As discussed in Chapter 1, experimental data for ballistite N suggests that, within certain pressure and initial temperature limits, the reaction zone in the gas phase can be divided into two regions. The first, closest to the surface, controls the combustion rate.

Theory of Solid-Propellant Nonsteady Combustion, First Edition. Boris V. Novozhilov and Vasily B. Novozhilov.

Companion website: www.wiley.com/go/Novozhilov/solidpropellantnonsteadycombustion

The temperature within this zone is significantly lower than the combustion temperature; this zone is called the cold flame zone. The second zone is the gas flame zone, located at up to several millimetres away from the surface. Reaction proceeds here in the induction regime at a temperature equal to the combustion temperature. This zone has very little effect on the burning rate. This means that for the given time history of pressure, the presented theory will be valid if the cold flame zone inertia is small compared to the inertia of the condensed phase. Time lags in the second flame zone would not affect the results.

For a known temperature profile the gas phase inertia scale t_g should be obtained as

$$t_g = \int_0^{l_b} \frac{dx}{u_g^0(x)} = \frac{p\tilde{\mu}}{m^0 R} \int_0^{l_b} \frac{dx}{T_g^0(x)} \tag{2.1}$$

As these two flame zones merge at pressures above approximately 60 atm, the length scale l_b should be taken as the distance away from the surface where heat generation becomes negligible, for example at a pressure of 100 atm $l_b = 2\times10^{-2}$ cm (the temperature at this distance from the surface is about 0.9 of the combustion temperature) and $t_g = 1.8\times10^{-4}$ s. This value is several times less than $t_c = 10^{-3}$ s, therefore the assumption of negligible gas phase inertia is fulfilled.

In the case where the two flame zones are separated, the estimation of t_g should be based on the zone closest to the surface, as it is this zone that affects the burning rate. Experimental data (temperature profile at p = 50 atm) give $t_c = 3\times10^{-3}$ s, $t_{cr} \sim 10^{-4}$ s, and $t_c/t_{cr} = 30$. The same condition, $t_c >> t_{cr}$, is also fulfilled at lower pressures. Therefore, if the burning rate needs be predicted under known external conditions (pressure or tangential stream velocity), gas phase inertia may be neglected compared to the inertia of the preheat zone in the condensed phase.

b) The time scale of the unsteady process t_p (e.g. a period of pressure variation) is assumed to satisfy $t_p >> t_{cr}$ and $t_p >> t_g$. The theory would contradict experimental data if applied to sufficiently rapid unsteady processes with the frequency of the order of 10 kHz.
c) ZN theory (Zeldovich 1942; Novozhilov 1965a,b, 1973a,b) assumes that the steady-state dependencies $u^0(T_a, p)$, $T_s^0(T_a, p)$, and $T_b^0(T_a, p)$ for the burning rate, surface temperature, and combustion temperature, respectively, are known. Here T_a is the propellant initial temperature and p is the pressure or any other relevant external parameter. These dependencies should be obtained either from experimental data or from a specific propellant combustion model.
d) Propellant properties are assumed to be homogeneous and isotropic. This assumption is fulfilled if the nonuniformity scale is much smaller than the Michelson length. The latter is obtained considering steady-state combustion as $l = \kappa/u^0$, i.e. the ratio of thermal diffusivity of the condensed phase and the linear steady-state burning rate. This requirement is unconditionally true for ballistites. It is fulfilled for composite propellants in cases where fuel and oxidizer grain sizes are small compared to the Michelson length.
e) A one-dimensional problem is considered (with the exception of Section 3.3), i.e. plane flame fronts and phase separating interfaces. While a certain structure may develop at the burning surface of homogeneous propellants (Zenin 1992), the scales are small

compared to the Michelson length at pressures exceeding 10 atm. Composite propellants must have sufficiently small grain size to fulfil this requirement.

2.2 Zeldovich Theory: Constant Surface Temperature

The simplest form of nonsteady propellant combustion theory was developed by Zeldovich, who assumed that the surface temperature of the burning fuel is constant. It should be noted that such an assumption leads to remarkable contradiction with experimental data. However, it is instructive to consider this theory in order to appreciate the idea of using steady-state burning rate dependencies on pressure and initial temperature for predicting nonsteady burning rates. The principal findings of Zeldovich (1942, 1964) are as follows.

Let us assume that the propellant surface temperature is constant and the gas phase inertia is negligible. Remember that the superscript 0 refers to steady-state conditions; the subscript is omitted for the values referring to unsteady process.

Under steady-state combustion conditions, the burning rate is determined by the pressure and the initial temperature of the propellant:

$$u^0 = F_u(p, T_a) \tag{2.2}$$

Energy which is generated by combustion is spent partly on heating up the propellant from initial to surface temperature; the heat flux that is required for such heating is

$$\lambda f^0 = \rho c u^0 (T_s^0 - T_a) \tag{2.3}$$

Here f^0 is the temperature gradient at the interface (at the condensed phase side)

$$f^0 = \left. \frac{\mathrm{d}T^0}{\mathrm{d}x} \right|_{x=0} \tag{2.4}$$

The relation (2.3) can be obtained either from considering energy balance directly or from the Mickelson profile, which itself results from the conservation of energy.

The existence of relation (2.3) between the temperature gradient, initial temperature, and burning rate allows (2.2) to be rewritten in the form of burning rate dependence on initial temperature and gradient

$$u^0 = F_f(p, f^0) \tag{2.5}$$

Indeed, using (2.3) one can express initial temperature as a function of burning rate and gradient, and then substitute the result into (2.2). The use of the relation (2.5) instead of (2.2) for steady-state combustion is not convenient as experimental measurements easily provide initial propellant temperature. However, it is exactly this representation of the burning rate as a function of pressure and surface temperature gradient which allows a decisive step to the description of nonsteady combustion process to be made.

The processes occurring in the gas phase are controlled, due to its negligible inertia, by pressure and condensed phase temperature distribution in the vicinity of the interface. Temperature variations in the deeper fuel layers are irrelevant for gas phase behaviour.

Temperature distribution and rates of chemical transformations in the gas phase are determined not just by pressure but also by the condensed phase temperature gradient at the surface. The role of the latter is clearly seen from the fact that it controls heat transfer from the gas phase to the condensed phase. In other words, burning rate is affected only by the temperature gradient (2.4) at the surface, but not by the overall shape of the temperature distribution in the condensed phase.

The above reasoning means that the steady and instantaneous nonsteady burning rates corresponding to the same values (p, f^0) are equal. This conclusion means that the dependence (2.5) is universal, i.e. applicable to both steady and nonsteady combustion. Consequently, the superscript 0 may be omitted in (2.5).

The constant burning rate under steady-state conditions corresponds to the value f^0, which follows from the Michelson profile. The burning rate under nonsteady conditions is determined by instantaneous values of the pressure and gradient. As such, it is controlled by the time history of pressure and by the heat conduction process in the condensed phase, which determines the time history of the gradient. The latter is taken account of by the heat transfer equation

$$\rho c\frac{\partial T}{\partial t} = \frac{\partial}{\partial x}\lambda\frac{\partial T}{\partial x} - \rho c u\frac{\partial T}{\partial x}, \quad x \leq 0 \tag{2.6}$$

describing thermal inertia of the condensed phase.

The obvious boundary conditions

$$x = 0, \quad T = T_s; \quad x \to -\infty, \quad T = T_a \tag{2.7}$$

must be supplemented with the known burning rate and pressure dependencies

$$u = F_f(p, f), \quad p = p(t) \tag{2.8}$$

As discussed earlier, the burning rate dependence $u(p, f)$ should be obtained from the steady-state relation $u^0(p, T_a)$ taking into account the balance (2.3).

As an example, let us consider one of the simplest models for steady-state burning rates, namely power dependence on pressure and exponential dependence on initial temperature

$$u^0 = Ap^{\prime} \exp(\beta T_a) \tag{2.9}$$

This model was considered by Zeldovich (1942) in his first paper devoted to nonsteady combustion. Applying (2.3), one obtains the following dependence of the burning rate on pressure and temperature gradient

$$u = Ap^{\prime} \exp\left\{\beta\left(T_s - \frac{\kappa f}{u}\right)\right\} \tag{2.10}$$

As explained earlier, this dependence holds true for both steady-state and unsteady combustion regimes.

Let us plot the profile $u(f)$ at a fixed value of pressure. Considering steady-state burning first, plot the dependencies $u^0(T_a)$ (given by (2.9)) and $(T_s^0 - T_a)(T_a)$ as functions of initial temperature (curves 1 and 2 in Figure 2.1). The product of these dependencies, according to (2.3), gives the steady-state gradient (curve 3 in Figure 2.1). The important feature of this curve is the existence of the maximum value of the gradient. At low temperatures the function $f^0(T_a)$ increases due to the strong rise in the burning rate, while at high temperatures the function decreases as the difference $T_s^0 - T_a$ approaches zero. Curve 3 may now be used to obtain the dependence $u^0(f^0)$, which will be equally true in the unsteady regime. This

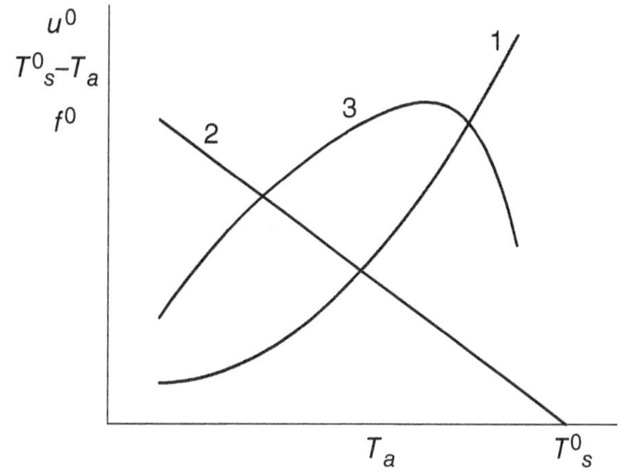

Figure 2.1 Procedure for obtaining the dependence of burning rate on surface temperature gradient.

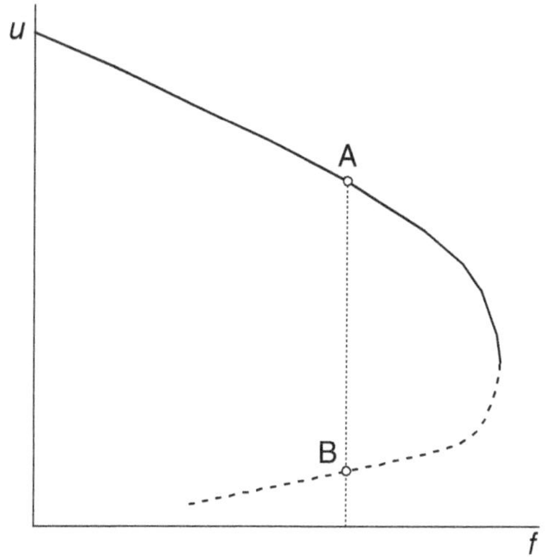

Figure 2.2 The dependence of burning rate on surface temperature gradient.

dependence is shown in Figure 2.2. Under steady-state conditions, each point on this curve corresponds to the constant burning rate at a fixed initial temperature. The maximum value of the burning rate corresponds to the initial temperature equal to the surface temperature. The existence of the maximum gradient value on this curve implies that the burning rate cannot be determined uniquely.

Zeldovich (1942, 1964) shows, however, that physically stable steady-state solutions are only observed for the decreasing branch of the curve where

$$\left(\frac{\partial u}{\partial f}\right)_p < 0 \tag{2.11}$$

The rest of the curve corresponds to physically irrelevant unstable combustion regimes and is therefore shown as dashed. For example, the combustion regime corresponding to point A in Figure 2.2 is stable, i.e. the temperature distribution in the condensed phase does not change with time. Point B corresponds to the same value of the temperature gradient, but lower initial temperature. The combustion regime corresponding to this point would be stable if the burning rate was sufficiently small. However, since at the same gradient value the larger value of the burning rate is possible (i.e. the burning rate corresponding to the point A), then it turns out that in reality combustion proceeds with the latter higher burning rate value. As a result deep layers of the propellant do not heat up, the temperature profile narrows, and the surface temperature gradient rises. As the latter reaches its maximum possible value combustion ceases.

Stability of combustion requires that a decrease in the temperature gradient at the surface leads to an increase in the burning rate. This can be understood qualitatively from the following reasoning: higher burning rates lead to quicker burnout of the preheat layer and therefore an increase in the gradient and an approach to steady-state conditions. Thus, at constant pressure and under the assumption of constant surface temperature, the combustion regime is only stable if (2.11) is satisfied. It is easy to show (see Problem 8) that the condition (2.11) is translated for the combustion model (2.9) into the following stability criterion

$$k < 1; \quad k = \beta(T_s^0 - T_a) \tag{2.12}$$

At the same time, the lowest value of the initial temperature that allows the steady-state combustion to proceed is given by

$$T_a^* = T_s^0 - \frac{1}{\beta} \tag{2.13}$$

Zeldovich proposed to use the plane (f, u) for graphical representation of unsteady combustion regimes. Since the burning rate and surface temperature gradient are related to each other at a particular instant of time, any unsteady process can be presented as a trajectory on such a plane.

Consider, for example, the burning rate model (2.9). Figure 2.3 shows the dependencies $u(f)$ for the two different values of pressure, $p_2 > p_1$. They correspond to steady-state

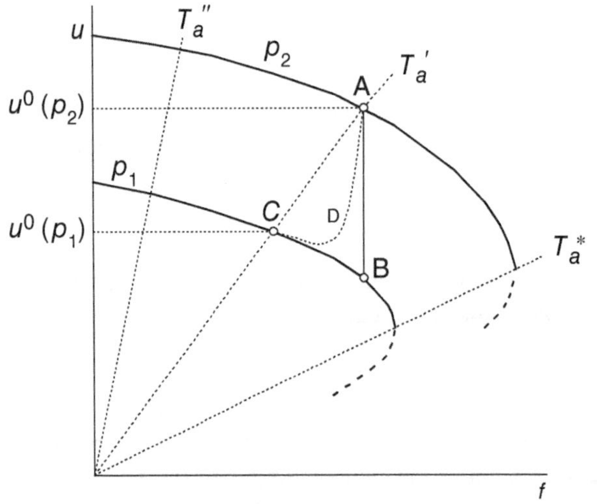

Figure 2.3 Transitional regimes in the plane (f, u).

combustion regimes at different initial temperatures. Note that initial temperature is constant along each of the lines intersecting at the origin of the coordinate frame ($T_a'' > T_a'$). Indeed, (2.3) implies that for fixed surface temperature, the steady-state burning rate and surface temperature gradient are proportional

$$u^0 = \frac{\kappa f^0}{T_s - T_a} \tag{2.14}$$

Smaller line slopes correspond to lower initial temperatures. The line T_a^* is the stability boundary. The parameter k assumes its maximum value $k = 1$ along this line.

If the time scale of pressure variation is large compared to the relaxation time of the preheat layer t_c, the process is quasi-steady. The trajectory of such a regime is represented by the line emerging from the origin and corresponding to the actual propellant initial temperature. The burning rate in this case follows steady-state dependence (2.9), $u^0 \sim p^\iota$. Similarly, the curves $p = \text{constant}$ represent quasi-steady regimes obtained by slow variation of initial temperature. The burning rate in this case is found again from (2.9), this time as a function of T_a, $u^0 \sim \exp(\beta T_a)$.

In the opposite limit of rapid (compared to t_c) pressure decrease from p_2 to p_1 the temperature profile in the condensed phase remains essentially the same and so does the surface temperature gradient. This process is shown by the vertical line AB in Figure 2.3. The burning rate at point B is smaller than the corresponding value for the steady-state combustion at the pressure p_1 (represented by point C). Heat conduction will therefore lead to a decrease in the surface temperature gradient and to a corresponding increase in the burning rate. The process will approach the steady-state regime along the curve segment BC. Pressure and burning rate are shown as a function of time in Figure 2.4. For a less rapid pressure drop the nonsteady process trajectory is represented by the dashed curve ADC in Figure 2.3. The pressure and burning rate for such a case are presented in Figure 2.5.

Any other nonsteady combustion process can be considered qualitatively in the same manner.

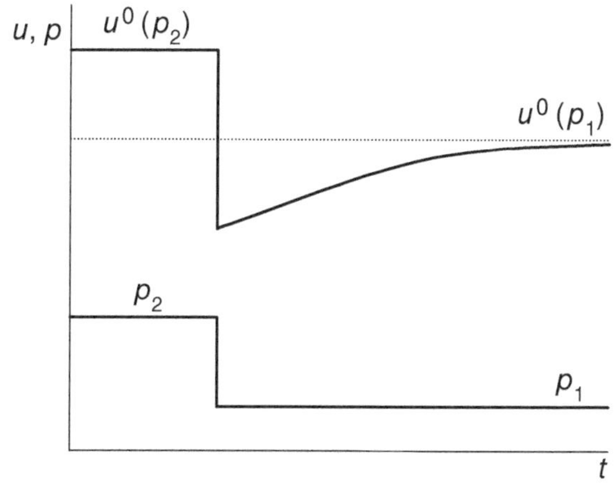

Figure 2.4 Burning rate variation under rapid pressure change.

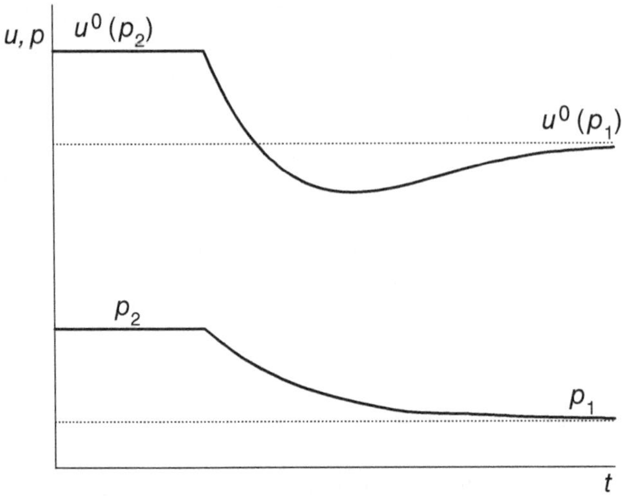

Figure 2.5 Burning rate variation under slow pressure change.

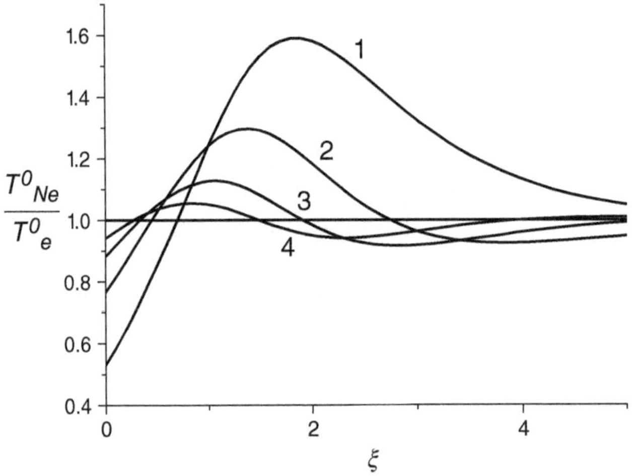

Figure 2.6 Ratio between approximate temperature profiles and the exact temperature profile $T_e^0(\xi)$. The numbers on each curve indicate the number of moments N used to calculate each of the approximate temperature profiles T_{Ne}^0.

2.3 Variable Surface Temperature

The present section extends the nonsteady combustion theory (Zeldovich 1942) to the case of variable surface temperature. Under the assumptions specified in Section 2.1 nonsteady propellant combustion processes can be predicted upon solution of the heat transfer equation in the condensed phase with the given initial temperature distribution. In order to impose appropriate boundary conditions, the dependencies of the burning rate, surface temperature, and combustion temperature on the pressure and surface temperature gradient must be known. These relations are obtained by transformation of the respective steady-state dependencies. Additionally, either the pressure time history or an equation that allows the latter to be determined needs be specified.

A major focus of nonsteady combustion theory has been on investigating the effect of pressure variation on the burning rate. It should be noted, however, that other factors affecting the burning rate can be accounted for in a similar manner. The most important of these is the velocity of the tangential gas stream near the burning surface. In the latter case condensed phase inertia results in the delay between the instantaneous value of the stream velocity and the corresponding value of the burning rate.

The developed theory also allows the effect of combustion product radiation on the burning rate to be taken into account. Some propellants are transparent, to a certain degree, to thermal radiation. This means that over the penetration depth of radiation l_r the temperature distribution in the propellant is affected by an additional heat source. Some problems concerning nonsteady combustion under radiation conditions are considered in Chapter 6.

From here on, within the present section, an external factor affecting the combustion zone and associated with the gas phase (i.e. having no inertia) is denoted as p and called 'pressure'. In reality, along with pressure, it may mean other external factors, such as tangential gas stream velocity.

The adopted t_c approximation implies that the condensed phase mass burning rate m and the associated heat of reaction Q_s are functions of the surface temperature T_s, pressure, heat supply rate from the gas phase, and heat transfer rate to the condensed phase. The heat supply rate from the gas is controlled by the temperature gradient on the gas side, f_g, while heat transfer to the condensed phase is controlled by the temperature gradient on the condensed phase side, f.

Therefore

$$m = m(T_s, p, f, f_g, Q_s) \tag{2.15}$$

$$Q_s = Q_s(T_s, p, f, f_g) \tag{2.16}$$

The energy balance at the interface ($x = 0$, propellant is at $x \leq 0$) takes the form

$$\lambda f = \lambda_g f_g + m Q_s \tag{2.17}$$

where λ and λ_g are the thermal conductivities of the solid and gas phases, respectively.

The burning rate in the gas phase depends on the pressure and combustion temperature

$$m_g = m_g(T_b, p) \tag{2.18}$$

and, due to negligible inertia of the combustion zone in the gas,

$$m = m_g \tag{2.19}$$

The energy balance for the gas phase reads

$$mcT_s + m(Q_s + Q_g) - \lambda f = mc_p T_b \tag{2.20}$$

while the total chemical energy

$$Q_s + Q_g = Q \tag{2.21}$$

is conserved.

Since pressure is considered to be known as a function of time, we have seven equations (2.15–2.21) containing eight unknowns: T_s, T_b, f, f_g, Q_s, Q_g, m, and m_g. This means that any of the unknowns can be considered as a function of the two parameters, that is pressure and temperature gradient f in the condensed phase at $x = 0$. For example, one can consider relations $m = m(p, f)$, $T_b = T_b(p, f)$, etc.

The presented reasoning is valid irrespective of whether a steady-state or unsteady regime is considered. Therefore, the obtained dependencies of the combustion process parameters on the two variables, p and f, are universally applicable in both the steady-state and nonsteady regimes.

Let us refer to steady-state dependencies on pressure and initial temperature as *steady-state burning laws*. The most important among these are the dependencies for the burning rate, surface temperature, and combustion temperature. They are written retaining the superscript 0 which refers to steady-state conditions

$$u^0 = F_u(p^0, T_a), \quad T_s^0 = F_s(p^0, T_a), \quad T_b^0 = F_b(p^0, T_a) \tag{2.22}$$

These dependencies may be obtained either from experimental data on steady-state combustion or from the particular propellant combustion model.

On the other hand, let us refer to dependencies of any variable on pressure and temperature gradient at the surface f as *nonsteady burning laws*. The examples of these are

$$u = u(p,f), \quad T_s = T_s(p,f), \quad T_b = T_b(p,f) \tag{2.23}$$

Steady-state burning laws (2.22) may be transformed into nonsteady ones using the Michelson distribution for the propellant temperature under a steady-state burning regime

$$T^0(x) = T_a + (T_s^0 - T_a)\exp\left({}^{u^0x}/_{\kappa}\right) \tag{2.24}$$

and the gradient value obtained directly from this distribution

$$f = \left.\frac{\partial T}{\partial x}\right|_{x=0}, \quad f^0 = \frac{u^0}{\kappa}(T_s^0 - T_a) \tag{2.25}$$

The latter expression allows (2.22) to be transformed to

$$\begin{aligned} u^0 &= F_u\left(p^0, T_s^0 - \kappa\frac{f^0}{u^0}\right) \\ T_s^0 &= F_s\left(p^0, T_s^0 - \kappa\frac{f^0}{u^0}\right) \\ T_b^0 &= F_b\left(p^0, T_s^0 - \kappa\frac{f^0}{u^0}\right) \end{aligned} \tag{2.26}$$

The major idea of the present theory is that the same dependencies are true for nonsteady combustion (i.e. as just discussed, the dependencies (2.26) are universal and therefore the superscript referring to steady-state may be omitted)

$$\begin{aligned} u &= F_u\left(p, T_s - \kappa\frac{f}{u}\right), \\ T_s &= F_s\left(p, T_s - \kappa\frac{f}{u}\right) \\ T_b &= F_b\left(p, T_s - \kappa\frac{f}{u}\right) \end{aligned} \tag{2.27}$$

The latter conclusion implies that the nonsteady burning rate can be obtained upon taking account of the inertia of the condensed phase. This can be done by solving the heat transfer equation

$$\rho c\frac{\partial T}{\partial t} = \frac{\partial}{\partial x}\lambda\frac{\partial T}{\partial x} - \rho c u\frac{\partial T}{\partial x} \tag{2.28}$$

with the boundary conditions

$$x \to -\infty, \quad T = T_a; \quad x = 0, \quad T = T_s \tag{2.29}$$

and given relations (2.27).

The initial temperature profile and pressure time history

$$t = 0, \quad T = T_i(x); \quad p = P(t) \tag{2.30}$$

must also be specified.

The following developments often make use of nondimensional variables. Let u^0 be a particular burning rate value (e.g. initial or mean) corresponding in the steady-state regime to pressure p^0. Nondimensional variables are introduced as

$$v = \frac{u}{u^0}, \quad \xi = \frac{u^0}{\kappa}x, \quad \tau = \frac{(u^0)^2}{\kappa}t, \quad \eta = \frac{p}{p^0} \tag{2.31}$$

where x and t are the dimensional cartesian coordinate and time, respectively. The spatial and time scales here are the respective scales of the preheat zone in the condensed phase.

Nondimensional propellant temperature, surface temperature gradient in the condensed phase, and surface temperature are introduced as

$$\theta = \frac{T - T_a}{T_s^0 - T_a}, \quad \varphi = \frac{f}{f^0}, \quad \vartheta = \frac{T_s - T_a}{T_s^0 - T_a} \tag{2.32}$$

where the surface temperature T_s^0 and the temperature gradient f^0 correspond to the steady-state regime with the burning rate u^0.

The problem of the nonsteady combustion theory is formulated now as the following: find burning rate $v(\tau)$ solving the heat transfer equation

$$\frac{\partial \theta}{\partial \tau} = \frac{\partial^2 \theta}{\partial \xi^2} - v\frac{\partial \theta}{\partial \xi} \quad (\xi \leq 0) \tag{2.33}$$

with the initial and boundary conditions

$$\theta(\xi, 0) = \theta_i(\xi), \quad \theta(-\infty, \tau) = 0, \quad \theta(0, \tau) = \vartheta(\tau) \tag{2.34}$$

given the relations

$$v = v(\eta, \varphi), \quad \vartheta = \vartheta(\eta, \varphi) \tag{2.35}$$

and the pressure dependence

$$\eta = \eta(\tau) \tag{2.36}$$

Nonsteady combustion temperature may be found, if needed, from (2.18).

Obviously, under steady-state conditions

$$\theta = e^{\xi}, \quad \eta = 1, \quad \vartheta = 1, \quad \varphi = 1, \quad v = 1 \tag{2.37}$$

2.4 Integral Formulation of the Theory

Solution of nonsteady combustion problems, obtained within the theory formulation discussed in the preceding section, produces, among other parameters, the time-dependent propellant temperature distribution $\theta(\xi, \tau)$. This field is really a by-product of the theory, since it is not usually used in ballistic studies (except in a few very special cases). Indeed, the major focus of nonsteady combustion theory is on prediction of the burning rate $u(\tau)$ based on given time dependencies of pressure and/or tangential gas stream velocity. It is natural, therefore, to re-formulate the theory in such a way (Novozhilov 1970) that the futile (in ballistic applications) field $\theta(\xi, \tau)$ is excluded.

Assuming $\theta = 0$ to the right from the propellant surface (i.e. at $\xi > 0$), let us subject the heat transfer Eq. (2.33)

$$\frac{\partial \theta}{\partial \tau} = \frac{\partial^2 \theta}{\partial \xi^2} - v\frac{\partial \theta}{\partial \xi} \quad (\xi \leq 0) \tag{2.38}$$

to the Fourier transform

$$F(k,\tau) = \int_{-\infty}^{0} \theta(\xi, \tau) e^{-ik\xi} \mathrm{d}\xi \tag{2.39}$$

This transforms the partial derivative over time into the full time derivative of the Fourier transform.

The first term on the right-hand side of the transformed equation may be integrated by parts twice to give

$$\int_{-\infty}^{0} \frac{\partial^2 \theta}{\partial \xi^2} e^{-ik\xi} \mathrm{d}\xi = \varphi + ik\vartheta - k^2 F \tag{2.40}$$

Finally, the convective term is transformed as follows

$$-v \int_{-\infty}^{0} \frac{\partial \theta}{\partial \xi} e^{-ik\xi} \mathrm{d}\xi = -v(\vartheta + ikF) \tag{2.41}$$

The evolution equation for the Fourier transform is therefore

$$\frac{dF}{d\tau} + (k^2 + ikv)F = \varphi - v\vartheta + ik\vartheta \tag{2.42}$$

and the initial condition is set as

$$F(k,0) = \int_{-\infty}^{0} \theta_i(\xi) e^{-ik\xi} \mathrm{d}\xi \tag{2.43}$$

The linear Eq. (2.42) (with the initial condition (2.43)) has the following solution

$$F(k,\tau) = \int_{0}^{\tau} [\varphi(\tau') - v(\tau')\vartheta(\tau') + ik\vartheta(\tau')] \exp[-k^2(\tau - \tau') - ikI]\mathrm{d}\tau' + F(k,0)\exp[-k^2\tau - ikJ] \tag{2.44}$$

where

$$I = \int_{\tau'}^{\tau} v(\tau'')\mathrm{d}\tau'', \quad J = \int_{0}^{\tau} v(\tau'')\mathrm{d}\tau'' \tag{2.45}$$

The solution (2.44) can now be subjected to the inverse transform

$$\theta(\xi,\tau) = \frac{1}{2\pi}\int_{-\infty}^{\infty} F(k,\tau)e^{ik\xi}\mathrm{d}k \tag{2.46}$$

The following integrals emerge upon this inverse transformation

$$\int_{-\infty}^{\infty} e^{-q^2k^2}\cos pk\mathrm{d}k = \frac{\sqrt{\pi}}{q}e^{-p^2/4q^2}, \quad \int_{-\infty}^{\infty} ke^{-q^2k^2}\sin pk\mathrm{d}k = \frac{\sqrt{\pi}p}{2q^3}e^{-p^2/4q^2} \tag{2.47}$$

The obtained integral equation for the temperature is

$$\theta(\xi,\tau) = \frac{1}{2\sqrt{\pi}}\left\{\int_0^{\tau}\left(\varphi(\tau') - \nu(\tau')\vartheta(\tau') + \frac{\vartheta(\tau')(I-\xi)}{2(\tau-\tau')}\right)\exp\frac{-(I-\xi)^2}{4(\tau-\tau')}\frac{\mathrm{d}\tau'}{\sqrt{\tau-\tau'}} + \right.$$
$$\left. + \frac{1}{\sqrt{\tau}}\int_{-\infty}^{0}\theta_i(z)\exp\frac{-(z+J-\xi)^2}{4\tau}\mathrm{d}z\right\} \tag{2.48}$$

There are three unknown functions of time present in (2.48): burning rate, temperature gradient at the surface, and surface temperature.

The two relations (2.35) are not sufficient for all the three functions to be determined. However, the third relation between ν, ϑ, and φ can be obtained by invoking (2.48) at the surface $\xi = 0$. It must be remembered though that the jump in the temperature profile occurs at $\xi = 0$ ($\theta(0,\tau) = \vartheta(\eta,\varphi)$ at $\xi = 0$ and $\theta(0,\tau) \equiv 0$ at $\xi > 0$). Thus, letting $\xi = 0$ in (2.48), one must multiply the right-hand side by a factor of 2. This leads to

$$\vartheta(\tau) = \frac{1}{\sqrt{\pi}}\left\{\int_0^{\tau}\left(\varphi(\tau') - \nu(\tau')\vartheta(\tau') + \frac{\vartheta(\tau')I}{2(\tau-\tau')}\right)\exp\left(\frac{-I^2}{4(\tau-\tau')}\right)\frac{\mathrm{d}\tau'}{\sqrt{\tau-\tau'}} + \right.$$
$$\left. + \frac{1}{\sqrt{\tau}}\int_{-\infty}^{0}\theta_i(z)\exp\frac{-(z+J)^2}{4\tau}\mathrm{d}z\right\} \tag{2.49}$$

Taking into account the instantaneous relations

$$\nu = \nu(\eta,\varphi), \quad \vartheta = \vartheta(\eta,\varphi) \tag{2.50}$$

one obtains the closed set of equations which allows any of the functions ν, ϑ, or φ to be determined upon specifying the time dependence $\eta(\tau)$ as an input.

The full propellant temperature distribution $\theta(\xi, \tau)$ can still be found, if needed, from (2.48).

The most important combustion process parameter is burning rate. If the relations (2.50) are explicitly specified, then the set of equations (2.49) and (2.50) may be reduced to a single equation for $\nu(\tau)$, and the latter value at any given time would be determined by the full pressure time history.

When solving internal ballistic problems, the use of the integral equation (2.49) has clear advantages over the original set of equations (2.33)–(2.36). First of all, the futile function $\theta(\xi, \tau)$ does not need be found. This leads to more efficient numerical solution methods for those problems that do not admit analytical solutions. Furthermore, a number of nonsteady combustion problems can be solved by perturbation methods, employing expansions with respect to some small parameter, for example to the amplitude of harmonically oscillating pressure. In this case, the use of an integral equation significantly simplifies calculations since various perturbations of steady-state temperature distribution would be excluded from consideration. Finally, the role of the obtained integral equation is that it closes the set of equations (containing among other variables pressure and burning rate) required for internal ballistic problems. For steady-state or quasi-steady combustion regimes this set is closed by the steady-state dependence $u^0 = u^0(p, T_a)$. For unsteady combustion, this steady-state dependence needs be replaced by the integral relation (2.49), together with the additional restrictions (2.50). Obviously, the partial differential equation formulation of the theory (2.33)–(2.36) may be used for the same purpose. In this case, however, due to necessity to consider propellant temperature distribution $\theta(\xi, \tau)$, the set of equations governing internal ballistics becomes significantly more complicated.

2.5 Theory Formulation through the Set of Ordinary Differential Equations

The set of equations (2.33)–(2.36), describing nonsteady propellant combustion, contains the heat transfer equation (2.33). The latter is a partial differential equation that can be replaced by the infinite set of ordinary differential equations (Novozhilov 2003, 2004).

In order to present such a set, we will make use of Laguerre polynomials, which are defined for $0 \le \xi < \infty$. Therefore, the sign of the spatial coordinate should be reversed in both Eq. (2.33) and the boundary conditions (2.34). The result is

$$\frac{\partial \theta}{\partial \tau} = \frac{\partial^2 \theta}{\partial \xi^2} + v\frac{\partial \theta}{\partial \xi} \quad (\xi \ge 0) \tag{2.51}$$

$$\theta(0, \tau) = \vartheta(\tau), \quad \theta(\infty, \tau) = 0 \tag{2.52}$$

The expression for the nondimensional temperature gradient also changes as

$$\varphi = -\left.\frac{\partial \theta}{\partial \xi}\right|_{\xi=0} \tag{2.53}$$

The solution is sought in the form of the following series

$$\theta(\xi, \tau) = e^{-\xi} \sum_{n=0}^{\infty} y_n(\tau) L_n(\xi) \tag{2.54}$$

where

$$L_n(\xi) = \frac{e^{\xi}}{n!} \frac{\mathrm{d}^n}{\mathrm{d}\xi^n} e^{-\xi} \xi^n \tag{2.55}$$

are the Laguerre polynomials. They are orthonormal (with weight $e^{-\xi}$) on $0 \le \xi < \infty$

$$\int_0^{\infty} e^{-\xi} L_m(\xi) L_n(\xi) \mathrm{d}\xi = \delta_{mn} \tag{2.56}$$

Using (2.54)–(2.56), the moments of the temperature distribution are represented as

$$y_n(\tau) = \int_0^{\infty} \theta(\xi, \tau) L_n(\xi) \mathrm{d}\xi \tag{2.57}$$

Let us multiply both sides of (2.51) by $L_n(\xi)$ and integrate over ξ from zero to infinity. The left-hand side will turn into a time derivative of y_n. The right-hand side can be integrated by parts, with the following identities taken into account

$$L_n(0) = 1, \quad \left.\frac{dL_n(\xi)}{d\xi}\right|_{\xi=0} = -n$$

$$\frac{dL_n(\xi)}{d\xi} = -\sum_{s=0}^{n-1} L_s(\xi), \quad \frac{d^2L_n(\xi)}{d\xi^2} = \sum_{s=0}^{n-2}(n-s-1)L_s(\xi) \tag{2.58}$$

The final result is the infinite set of ordinary differential equations written for the moments of temperature distribution in the condensed phase

$$y_n{}'(\tau) = \varphi(\tau) - n\vartheta(\tau) + \sum_{s=0}^{n-2}(n-s-1)y_s(\tau) + \\ + v(\tau)\left(\sum_{s=0}^{n-1} y_s(\tau) - \vartheta(\tau)\right), \quad n = 0, 1, 2, \ldots \tag{2.59}$$

The numerical solution of (2.59) requires that a finite number of equations, up to $n_{max} = N$, are retained. Obviously, the choice of n_{max} is dictated by the required accuracy, which improves as $n_{max} \to \infty$.

We obtain, therefore, the set of $N+1$ ordinary differential equations for the moments

$$y_n{}'(\tau) = \varphi(\tau) - n\vartheta(\tau) + \sum_{s=0}^{n-2}(n-s-1)y_s(\tau) + \\ + v(\tau)\left(\sum_{s=0}^{n-1} y_s(\tau) - \vartheta(\tau)\right), \quad n = 0, 1, 2, \ldots N \tag{2.60}$$

coupled with the burning rate and surface temperature dependencies (2.35)

$$v = v(\eta, \varphi), \quad \vartheta = \vartheta(\eta, \varphi) \tag{2.61}$$

Besides, either explicit pressure dependence on time

$$\eta = \eta(\tau) \tag{2.62}$$

or an equation which allows pressure to be determined must be given.

The set of equations (2.60)–(2.62) is written for $N+5$ unknowns, v, φ, ϑ, η, and $N+1$ distribution moment, while contains only $N+4$ equations. The lack of an extra equation which is required for the set to be closed is a consequence of the transformation from an infinite to a finite set of equations.

In the present case, the closure is rather simple: the expansion (2.54), being applied at $\xi = 0$, provides the required extra equation

$$\vartheta(\tau) = \sum_{s=0}^{N} y_s(\tau) \tag{2.63}$$

In particular, under steady-state conditions

$$\eta^0 = 1, \quad \varphi^0 = 1, \quad v^0 = 1, \quad \vartheta^0 = 1,$$
$$y_0^0 = 1, \quad y_n^0 = 0, \quad n = 1, 2, \ldots N$$

It should be noted that one can also obtain from (2.54)

$$\varphi(\tau) = \sum_{s=0}^{N} (s+1) y_s(\tau) \tag{2.64}$$

The latter relation can be used to control the accuracy of calculations, i.e. for the choice of an appropriate truncation number N. An example of application of the present method is illustrated in Figure 2.6. Please see the Problem 11 for details.

2.6 Linear Approximation

Burning propellant represents a nonlinear system. This is seen from the fact that the heat conduction equation, necessary to describe propellant combustion, includes a nonlinear convection term (equal to the product of the burning rate and the temperature gradient). Moreover, the nonsteady burning laws are generally nonlinear.

Analytical solutions to nonlinear problems are exceptional. In particular, in the nonsteady propellant combustion theory we can point out only two papers (Zeldovich 1964; Librovich and Novozhilov 1972) where the analytical solutions describing self-similar combustion waves were obtained.

There is, however, a number of nonsteady combustion problems where the burning regime is close to a steady-state one. These problems may be solved using perturbation methods. In particular, an original problem is linearized in the vicinity of the steady-state burning regime, and then an appropriate analytical solution is obtained. Linearization is an efficient method for the solving challenging and important problems of the nonsteady burning theory. A number of examples will be considered in later chapters.

The objective of the present section is to develop a linearized version of the theory of nonsteady propellant combustion.

When the burning regime is only weakly different from a steady state one, any time-dependent variable can be represented as a sum of its steady-state value and a small correction. For example, pressure can be written as $p(t) = p^0 + p_1(t)$, where $p_1 \ll p^0$ at any moment t. In all the transformations and final expressions of the linear approximation, only the values proportional to the first order of correction are taken into account.

Let us begin with linearization of nonsteady heat conduction equation

$$\frac{\partial T}{\partial t} = \kappa \frac{\partial^2 T}{\partial x^2} - u \frac{\partial T}{\partial x} \tag{2.65}$$

Let $T^0(x)$ be the steady-state solution satisfying the equation

$$\kappa \frac{d^2 T^0}{dx^2} - u^0 \frac{dT^0}{dx} = 0 \tag{2.66}$$

where u^0 is the steady-state burning rate.

Eq. (2.65) is linearized assuming

$$T(x,t) = T^0(x) + T_1(x,t), \quad u(t) = u^0 + u_1(t)$$
$$(T_1 \ll T^0;\ u_1 \ll u^0) \tag{2.67}$$

Substituting these relations into (2.65) and taking (2.66) into account yields the heat conduction equation in the linear approximation

$$\frac{\partial T_1}{\partial t} = \kappa \frac{\partial^2 T_1}{\partial x^2} - u^0 \frac{\partial T_1}{\partial x} - u_1 \frac{\mathrm{d}T^0}{\mathrm{d}x} \tag{2.68}$$

Linearization of the boundary and initial conditions

$$x \to -\infty, \quad T = T_a; \quad x = 0, \quad T(0,t) = T_s(t)$$
$$t = 0, \quad T(x,0) = T_i(x) \tag{2.69}$$

also presents no difficulties. The surface temperature and the initial temperature distributions are written in the form

$$T_s(t) = T_s^0 + T_{s1}(t), \quad T_i(x) = T_i^0(x) + T_{i1}(x) \tag{2.70}$$

and for the small corrections we obtain

$$x \to -\infty, \quad T_1 = 0; \quad x = 0, \quad T_1(t) = T_{s1}(t)$$
$$t = 0, \quad T_1(x) = T_{i1}(x) \tag{2.71}$$

The necessary components of the theory are steady-state dependencies of the burning rate, and the surface and combustion temperatures on the pressure and initial temperature. They are written (superscript 0 refers to steady-state as usual) as

$$u^0 = F_u(p^0, T_a), \quad T_s^0 = F_s(p^0, T_a), \quad T_b^0 = F_b(p^0, T_a) \tag{2.72}$$

and can be transformed into their nonsteady counterparts

$$u = F_u\left(p, T_s - \frac{\kappa f}{u}\right), \quad T_s = F_s\left(p, T_s - \frac{\kappa f}{u}\right), \quad T_b = F_b\left(p, T_s - \frac{\kappa f}{u}\right) \tag{2.73}$$

Let us proceed with their linearization. The first law takes, in the linear approximation, the form

$$u_1 = \left(\frac{\partial u^0}{\partial p^0}\right)_{T_a} p_1 + \left(\frac{\partial u^0}{\partial T_a}\right)_{p^0} \left(T_{s1} - \frac{\kappa f_1}{u^0} + \frac{\kappa f^0 u_1}{(u^0)^2}\right) \tag{2.74}$$

We shall use further the following nondimensional parameters

$$\iota = \left(\frac{\partial \ln u^0}{\partial \ln p^0}\right)_{T_a},$$

$$k = (T_s^0 - T_a)\left(\frac{\partial \ln u^0}{\partial T_a}\right)_{p^0},$$

$$\mu = \frac{1}{T_s^0 - T_a}\left(\frac{\partial T_s^0}{\partial \ln p^0}\right)_{T_a},$$

$$r = \left(\frac{\partial T_s^0}{\partial T_a}\right)_{p^0},$$

$$r_b = \left(\frac{\partial T_b^0}{\partial T_a}\right)_{p^0},$$

$$\mu_b = \frac{1}{T_s^0 - T_a}\left(\frac{\partial T_b^0}{\partial \ln p^0}\right)_{T_a}$$

$$\delta = \iota r - \mu k \tag{2.75}$$

In this notation, (2.74) can be written in the form

$$\frac{u_1}{u^0} = \iota \frac{p_1}{p^0} + k\left(\frac{T_{s1}}{T_s^0 - T_a} - \frac{f_1}{f^0} + \frac{u_1}{u^0}\right) \tag{2.76}$$

Let us introduce further the nondimensional variables

$$\tau = \frac{(u^0)^2 t}{\kappa}, \quad \xi = \frac{u^0 x}{\kappa}, \quad v_1 = \frac{u_1}{u^0}, \quad \eta_1 = \frac{p_1}{p^0}, \quad \varphi_1 = \frac{f_1}{f^0}$$
$$\theta_1 = \frac{T_1 - T_a}{T_s^0 - T_a}, \quad \theta_b = \frac{T_b - T_a}{T_s^0 - T_a}, \quad \vartheta = \frac{T_s - T_a}{T_s^0 - T_a} \tag{2.77}$$

Using these variables, (2.76) and the rest of the nonsteady burning laws are written as follows

$$v_1 = \iota\eta_1 + k(\vartheta_1 - \varphi_1 + v_1)$$
$$\vartheta_1 = \mu\eta_1 + r(\vartheta_1 - \varphi_1 + v_1)$$
$$\theta_{b1} = \mu_b\eta_1 + r_b(\vartheta_1 - \varphi_1 + v_1) \tag{2.78}$$

The latter equations can be resolved to deduce v_1, ϑ_1, and θ_{b1}

$$v_1 = \frac{1}{s}[k\varphi_1 - (\iota - \delta)\eta_1]$$
$$\vartheta_1 = \frac{1}{s}[r\varphi_1 - (\mu + \delta)\eta_1]$$
$$\theta_{b1} = \frac{1}{s}[r_b\varphi_1 + (s\mu_b - r_b(\iota + \mu))\eta_1], \quad s = k + r - 1 \tag{2.79}$$

Finally, we combine the results to write down (in nondimensional variables) the linearized heat transfer equation, together with the initial and boundary conditions

$$\frac{\partial\theta_1}{\partial\tau} = \frac{\partial\theta_1{}^2}{\partial\xi^2} - \frac{\partial\theta_1}{\partial\xi} - v_1\frac{d\theta^0}{d\xi}$$
$$\xi \to -\infty, \quad \theta_1 = 0; \quad \xi = 0, \quad \theta_1(0,\tau) = \vartheta_1(\tau)$$
$$\tau = 0, \quad \theta_1(\xi, 0) = \theta_{i1}(\xi) \tag{2.80}$$

The set of equations (2.79) and (2.80) represents the formulation of the nonsteady combustion problem (under given time dependence of pressure) in the linear approximation.

2.7 Formal Mathematical Justification of the Theory

The major idea of the presented theory of nonsteady combustion is the possibility of obtaining nonsteady burning laws from the steady-state ones. For the case of constant surface temperature, as was shown in Section 2.2, the procedure of transforming the steady-state dependence $u^0(p^0, T_a)$ into the nonsteady dependence $u(p, f)$ is straightforward.

The case of variable surface temperature was considered in Section 2.3. However, the consideration was simplified as only kinetic and thermal effects were taken into account.

A more rigorous proof of the existence, under the assumptions listed at the beginning of the present chapter, of nonsteady burning laws is presented below. Nonsteady burning laws refer to dependencies between instantaneous values of the temperature gradient, pressure, and any parameter describing inertia-free zones. The latter are the interface between the condensed and gas phases, and the region occupied by gas.

The following analysis uses the coordinate system in which a virgin condensed substance moves in the positive direction of the x axis with the speed coinciding with the linear burning rate $u(t)$. Therefore, the interface between the phases keeps its position at $x = 0$ irrespective of the combustion regime. The zone of chemical transformations in the condensed phase (zone II in Figure 1.1) is assumed to be infinitely thin.

Chemical transformations, complicated further by transport processes and gas movement, occur in the vicinity of the interface on both the condensed and gas phase sides. The space where initial substance is being transformed into the gaseous products of combustion comprises the three regions (Figure 1.1):

- Region I ($-\infty < x \leq 0$). This is the region where the condensed phase is being heated up. There are no any chemical transformation occurring.
- Region II ($x = 0$). Here, the condensed substance is being transformed into intermediate gaseous products. As mentioned previously, this region is considered as a plane ($x_c = 0$, Figure 1.1).
- Region III ($0 \leq x < \infty$). In this region, the intermediate products are transformed into the final products of combustion. This transformation is accompanied by gas movement and transport processes.

Consider the state of the system at any instantaneous moment in time. The combustion regime is either steady-state or unsteady. All the relationships derived below are related to inertia-free zones only and therefore apply to a regime of any type.

Gas pressure in region III is assumed to be constant along the coordinate. This is a standard assumption in combustion theory, since the velocity of gas motion is small compared to the speed of sound. An inertia-free condition demands all the partial derivatives with respect to time to be dropped from the equations describing combustion process, that is, in the t_c approximation only steady-state equations must be considered in inertia-free zones. The dependence of any variable on time may only appear due to a change in pressure or boundary conditions. The continuity equation implies that the mass velocity of the gas stream m_g, which coincides with the mass burning rate, does not depend on the coordinate. The gas density ρ_g, which explicitly enters the energy and reactants balances, is determined by the equation of state of the ideal gas

$$p = \rho RT \sum_{n=1}^{N} \frac{Y_n}{\tilde{\mu}_n} \tag{2.81}$$

where Y and $\tilde{\mu}$ are the mass fractions and molecular weights of components, respectively.

The number of independent components is $N - 1$, therefore $N - 1$ balance equations for components (diffusion equations), and furthermore the conservation of energy (heat transfer equation with chemical sources) must be considered for a complete description of region III. As each of these equations is of second order, there are $2N$ arbitrary constants emerging upon their integration. In addition, the problem eigenvalue, that is, the mass burning rate, has to be found from the same set of equations. Therefore, a complete problem formulation requires $2N + 1$ boundary conditions.

At $x \to \infty$, the boundary conditions must reflect the equilibrium conditions of the combustion products

$$\frac{dT}{dx} = 0, \quad \frac{dY_n}{dx} = 0 \tag{2.82}$$

The number of such boundary conditions is N. Additional $N + 1$ boundary conditions are set at the interface between the phases

$$x = 0, \quad T = T_s, \quad Y_n = Y_{ns} \quad \frac{dT}{dx} = f_{gs} \tag{2.83}$$

where the subscript gs refers to the derivative on the gas phase side. The values T_s, Y_{ns}, and f_{gs}, whose total number is $N + 1$, are not considered to be known. It is stated only that the conditions (2.82) and (2.83), combined with the known value of pressure p, are sufficient for the spatial distributions of the temperature and reactants mass fractions

$$T_g = T_g(x, T_s, f_{gs}, Y_{ns}, p), \quad Y_n = Y_n(x, T_s, f_{gs}, Y_{ns}, p) \tag{2.84}$$

to be determined.

Mass velocity m_g does not depend on the coordinate, and its value is also determined by pressure and the above parameters

$$m_g = m_g(T_s, f_{gs}, Y_{ns}, p) \tag{2.85}$$

The functions (2.84) and (2.85) fully determine the state of region III at the considered moment in time. In particular, setting $x \to \infty$ in (2.84) provides the composition and the temperature of the burning products.

The known functions Y_n allow the diffusion fluxes of components at the interface

$$d_{ns} = d_{ns}(T_s, f_{gs}, Y_{ns}, p) \tag{2.86}$$

to be found.

For that purpose, it is sufficient to take derivatives at $x = 0$. The set (2.86) contains $N - 1$ relationships.

Consider now region II, i.e. the plane $x = 0$. There are complicated physical and chemical processes occurring on this plane. The result of these processes is the transformation of condensed substance into gas. The mass velocity m_s of this transformation, the composition of the emerging products, and the rate of heat generation may depend on pressure, conditions at the boundaries of this zone, and its internal parameters. Since this zone is considered infinitely thin, there is only one internal parameter, namely, its temperature T_s. Adopting the notation f_{cs} for temperature gradient at the surface on the condensed phase side, the dependencies

$$\begin{aligned} m_s &= m_s(f_{cs}, T_s, f_{gs}, Y_{ns}, p), \quad m_{ns} = m_{ns}(f_{cs}, T_s, f_{gs}, Y_{ns}, p) \\ Q_s &= Q_s(f_{cs}, T_s, f_{gs}, Y_{ns}, p) \end{aligned} \tag{2.87}$$

may be introduced.

Here, m_{ns} are component mass fluxes generated at the plane $x = 0$ (some may be equal to zero) and Q_s is the combined heat of all the reactions in region II. The set (2.87) contains $N + 1$ relationships.

Regions I, II, and III are related by the balances of mass, energy, and components

$$m_s = m_g, \quad m_{ns} = m_g Y_{ns} + d_{ns}, \quad \lambda_{cs} f_{cs} = \lambda_{gs} f_{gs} + m_g Q_s \tag{2.88}$$

The number of the relationships (2.88) is $N + 1$.

Thus, the sets (2.85)–(2.88) contain $3N + 2$ relationships and exactly the same number of parameters which describe the instantaneous states of regions II and III, namely T_s, m_s, Q_s, m_{ns}, Y_{ns}, d_{ns}, f_{gs}, and m_g.

Additionally, the relationships (2.85)–(2.88) contain two more variables: the pressure p and the gradient f_{cs}. Obviously, these relationships allow any of the $3N + 2$ parameters named above to be expressed as a function of the latter two variables, for example, $T_s = T_s(p, f_{cs})$ or $m_s = m_s(p, f_{cs})$.

The mass burning rate may be denoted as m; note that $m_s = m_g = m$. Instead of f_{cs} the notation f will be used for temperature gradient at the surface, on the condensed phase side.

It has been proved that at any instance, there exist the dependencies

$$m = m(p, f), \quad T_s = T_s(p, f) \tag{2.89}$$

Besides the dependencies (2.89), the other parameters describing inertia-free zones may be of interest, for example the temperature of the combustion products $T_b = T_b(p, f)$. This function plays an important role in studying combustion in a semi-enclosed volume.

The relationships of the form (2.89) are called nonsteady burning laws. They may be put in correspondence with the steady-state dependencies

$$m^0 = m^0(p^0, T_a), \quad T_s^0 = T_s^0(p^0, T_a) \tag{2.90}$$

as was demonstrated in Section 2.3.

3

Combustion Under Constant Pressure

3.1 Stability Criterion for a Steady-state Combustion Regime

In the case of constant pressure, the steady-state Michelson solution (1.4) exists for any given pressure and initial temperature. However, this solution does not necessarily describe the real combustion process as it may be unstable. The real solution must remain stable under small perturbations of the temperature profile.

Therefore, the problem that emerges naturally is the one of the steady-state combustion regime stability to small deviation from the Michelson profile (under fixed pressure and initial temperature). This problem is essentially unsteady as it requires that time evolutions of small perturbations (e.g. of the burning rate or the temperature distribution) are considered. This is only possible within the framework of nonsteady combustion theory.

Zeldovich (1942) was the first to consider the steady-state combustion regime stability. He used the assumption of constant propellant surface temperature and did not succeed in reproducing experimental data. The apparent discrepancy between the model predictions and the observations led later to reformulation of the original Zeldovich theory and consideration of variable surface temperature.

The present chapter, following the results of Novozhilov (1965a), demonstrates that stability analysis taking account of variable surface temperature may be performed without detailed knowledge of the real causes of the latter variations.

The standard analysis of a given solution stability introduces small perturbations of such a solution and follows their time evolution in a linear approximation. A steady-state regime having small perturbations proportional to $\exp(\Omega t)$ (where Ω is a complex frequency) growing in time for at least one frequency value is unstable. In the opposite case ($\mathrm{Re}\Omega < 0$; all the perturbation amplitudes decrease with time) the steady-state regime is stable.

Following the above procedure, let us consider small perturbations of the steady-state burning rate and temperature distribution. Assuming these perturbations are proportional to $\exp(\Omega t)$, we shall find the characteristic equation for the frequency and the conditions ensuring $\mathrm{Re}\Omega < 0$.

Consider, first of all, the heat transfer process in the condensed phase (remember that the model preheat zone of the condensed phase is the only zone having inertia).

Theory of Solid-Propellant Nonsteady Combustion, First Edition. Boris V. Novozhilov and Vasily B. Novozhilov.

Companion website: www.wiley.com/go/Novozhilov/solidpropellantnonsteadycombustion

Let T_1 and u_1 be small perturbations of the temperature and the burning rate

$$T(x) = T^0(x) + T_1(x)\exp(\Omega t), u = u^0 + u_1 \exp(\Omega t) \tag{3.1}$$

i.e. their amplitudes are small compared to respective steady-state values

$$T_1(x) << T^0(x), u_1 << u^0 \tag{3.2}$$

The unperturbed regime corresponds to the steady-state heat transfer equation

$$\kappa\frac{d^2T^0}{dx^2} - u^0\frac{dT^0}{dx} = 0, -\infty < x \leq 0 \tag{3.3}$$

with the boundary conditions

$$x \to -\infty, \quad T = T_a; \quad x = 0, \quad T = T_s^0 \tag{3.4}$$

The solution of the latter boundary value problem is the Michelson profile

$$T^0(x) = T_a + (T_s^0 - T_a)\exp\left(\frac{u^0 x}{\kappa}\right) \tag{3.5}$$

The corresponding temperature gradient (at the condensed phase side) is

$$f = \left.\frac{\partial T}{\partial x}\right|_{x=0}, f^0 = \frac{u^0}{\kappa}(T_s^0 - T_a) \tag{3.6}$$

The perturbed temperature distribution satisfies the heat transfer equation

$$\frac{\partial T}{\partial t} = \kappa\frac{\partial^2 T}{\partial x^2} - u\frac{\partial T}{\partial x} \tag{3.7}$$

with the boundary conditions

$$x \to -\infty, T = T_a; x = 0, T = T_s \tag{3.8}$$

Let us linearize Eq. (3.7) by substituting expressions (3.1). Neglecting second-order terms and taking into account (3.5), we obtain in the linear approximation the equation for the amplitude T_1

$$\Omega T_1 = \kappa\frac{d^2T_1}{dx^2} - u^0\frac{dT_1}{dx} - \frac{u^0(T_s^0 - T_a)}{\kappa}u_1\exp\left(\frac{u^0x}{\kappa}\right) \tag{3.9}$$

The boundary conditions for this equation are

$$x \to -\infty, T_1 = 0; x = 0, T_1 = T_{s1} \tag{3.10}$$

By definition, at $x = 0$ we also have

$$\frac{dT_1}{dx} = f_1 \tag{3.11}$$

where f_1 is the amplitude of the small surface temperature gradient perturbation.

The solution of (3.9) has the form

$$T_1 = -\frac{u^0u_1(T_s^0 - T_a)}{\kappa\Omega}\exp\left(\frac{u^0x}{\kappa}\right) + C_1\exp\left(z_1\frac{u^0x}{\kappa}\right) + C_2\exp\left(z_2\frac{u^0x}{\kappa}\right)$$

$$z_{1,2} = \frac{1}{2}\left(1 \pm \sqrt{1 + 4\frac{\kappa\Omega}{(u^0)^2}}\right) \tag{3.12}$$

The first of the conditions in (3.10) requires that the perturbations must diminish at infinity, therefore $C_2 = 0$. The second boundary condition, along with the definition of f_1, yields

$$T_{s1} = -\frac{u^0 u_1 (T_s^0 - T_a)}{\kappa\Omega} + C_1, f_1 = -\frac{(u^0)^2 u_1 (T_s^0 - T_a)}{\kappa^2\Omega} + C_1 z_1 \frac{u^0}{\kappa} \tag{3.13}$$

Excluding the integration constant C_1 from the above two equations, we obtain the relation between the amplitudes of the burning rate, surface temperature, and surface temperature gradient

$$z_1 \frac{T_{s1}}{T_s^0 - T_a} - \frac{f_1}{f^0} + \frac{u_1}{z_1 u^0} = 0 \tag{3.14}$$

The extra two relations between T_{s1}, f_1 and u_1 can be obtained from nonsteady burning laws. As discussed in Chapter 2, in the nonsteady combustion regime (under constant pressure), the burning rate and surface temperature are functions of the temperature gradient only

$$u = F_u\left(T_s - \kappa\frac{f}{u}\right), T_s = F_s\left(T_s - \kappa\frac{f}{u}\right) \tag{3.15}$$

The Taylor expansions in the vicinity of

$$u^0 = F_u(T_a), T_s^0 = F_s(T_a) \tag{3.16}$$

are

$$\begin{aligned} u &= u^0 + \left(\frac{\partial F_u}{\partial T_a}\right)_p \left[T_s - T_s^0 - \kappa\left(\frac{f}{u} - \frac{f^0}{u^0}\right)\right] \\ T_s &= T_s^0 + \left(\frac{\partial F_s}{\partial T_a}\right)_p \left[T_s - T_s^0 - \kappa\left(\frac{f}{u} - \frac{f^0}{u^0}\right)\right] \end{aligned} \tag{3.17}$$

Furthermore, as in the linear approximation

$$\kappa\left(\frac{f}{u} - \frac{f^0}{u^0}\right) = (T_s^0 - T_a)\left(\frac{f_1}{f^0} - \frac{u_1}{u^0}\right) \tag{3.18}$$

Eqs. (3.17) take the form

$$\begin{aligned} \frac{u_1}{u^0} &= k\left(\frac{T_{s1}}{T_s^0 - T_a} - \frac{f_1}{f^0} + \frac{u_1}{u^0}\right) \\ \frac{T_{s1}}{T_s^0 - T_a} &= r\left(\frac{T_{s1}}{T_s^0 - T_a} - \frac{f_1}{f^0} + \frac{u_1}{u^0}\right) \end{aligned} \tag{3.19}$$

where

$$k = (T_s^0 - T_a)\left(\frac{\partial \ln u^0}{\partial T_a}\right)_p, \quad r = \left(\frac{\partial T_s^0}{\partial T_a}\right)_p \tag{3.20}$$

The latter parameters can be found from the steady-state combustion experiments.

As follows from (3.19), the perturbations of surface temperature and burning rate are related to the temperature gradient perturbation

$$\frac{u_1}{u^0} = \frac{k}{k + r - 1}\frac{f_1}{f^0}, \frac{T_{s1}}{T_s^0 - T_a} = \frac{r}{k + r - 1}\frac{f_1}{f^0} \tag{3.21}$$

Eqs. (3.14) and (3.21) are the set of the three linear homogeneous equations. A nontrivial solution only exists if

$$\begin{vmatrix} z_1 & -1 & \dfrac{1}{z_1} \\ 0 & -\dfrac{k}{k+r-1} & 1 \\ 1 & -\dfrac{r}{k+r-1} & 0 \end{vmatrix} = 0 \tag{3.22}$$

which leads to

$$rz_1^2 - (k + r - 1)z_1 + k = 0 \tag{3.23}$$

Taking the value of z_1

$$z_1 = \frac{1}{2}(1 + \sqrt{1 + 4\tilde{\omega}}), \quad \tilde{\omega} = \frac{\kappa\Omega}{(u^0)^2} \tag{3.24}$$

from the solution of (3.9) and substituting it into Eq. (3.23) leads to the algebraic equation for the complex nondimensional frequency $\tilde{\omega}$

$$r^2\tilde{\omega}^2 + [r(k + 1) - (k - 1)^2]\tilde{\omega} + k = 0 \tag{3.25}$$

Separating real and imaginary parts, the frequency can be written as

$$\tilde{\omega} = i\omega - \lambda \tag{3.26}$$

where λ is the oscillation damping decrement (by analogy with oscillations of a system with one degree of freedom). The requirement of the latter to be positive defines the region of stability. Separating real and imaginary parts in (3.25) results in the two equations

$$\begin{aligned} &r^2(\omega^2 - \lambda^2) + [r(k + 1) - (k - 1)^2]\lambda - k = 0 \\ &2\lambda r^2 - [r(k + 1) - (k - 1)^2] = 0 \end{aligned} \tag{3.27}$$

which yield

$$\lambda = \frac{r(k + 1) - (k - 1)^2}{2r^2}, \quad \omega = \sqrt{\frac{k}{r^2} - \lambda^2} \tag{3.28}$$

The latter two parameters are nondimensional. Their dimensional counterparts differ by a factor of $(u^0)^2/\kappa$.

The above analysis shows that within certain limits of the parameters k and r variation the propellant behaves as an oscillatory system. The frequency (3.28) can be considered as a natural frequency of its oscillations. The existence of the natural frequency of burning rate oscillations leads to a possibility of resonant interaction between a harmonically varying pressure and a burning propellant. This phenomenon is of great importance in studying low-frequency instability in a combustion chamber and the acoustic admittance of the burning propellant surface. These issues are considered in detail in Chapter 4.

In the linear approximation, the steady-state regime is stable if $\lambda > 0$, i.e. for

$$r \geq r^*, \quad r^* = \frac{(k - 1)^2}{(k + 1)} \tag{3.29}$$

or

$$k \leq 1 + \frac{r}{2} + \sqrt{2r + \frac{r^2}{4}} \tag{3.30}$$

The equality in the above relations determines the stability boundary. For parameter values that belong to this boundary small perturbations of the burning rate (or other variables) do not diminish with time. The natural frequency of these sustained oscillations is $\omega_1 = \sqrt{k}/r$.

For the case of constant surface temperature ($r = 0$) (Zeldovich 1942) expression (3.30) leads to the following stability criterion

$$k \leq 1 \tag{3.31}$$

In the vicinity of the stability boundary the asymptotic (at large times) behaviour of the burning rate is

$$v(\tau) = 1 + v_1 \exp(-\lambda\tau)\cos(\omega\tau + \psi) \tag{3.32}$$

where ψ is the phase which is determined by the initial condition.

The real part of the frequency Ω is equal to zero at the stability boundary, negative in the stability region, and positive in the unstable region. At the same time, the imaginary part is different from zero in the vicinity of the stability boundary (including the boundary itself). Therefore, in the linear approximation an oscillating regime with increasing amplitude occurs within the unstable region. Away from the stability boundary the imaginary part decreases until it reaches zero. As follows from (3.28), the zero imaginary part corresponds to $\lambda r_- = -\sqrt{k}$, or

$$r_- = (\sqrt{k} - 1)^2 \tag{3.33}$$

This means that perturbations exhibit exponential growth accompanied by oscillations for $r^* > r > r_-$, and grow exponentially without oscillations for $r < r_-$.

Figure 3.1 shows the boundaries of the oscillatory ($r = r^*$) and the exponential ($r = r_-$) instability regions. Both curves originate from the point $(1, 0)$, corresponding to the stability criterion for the case of constant surface temperature (3.31). The qualitative behaviour of perturbations (in the linear approximation) in time is also shown in Figure 3.1.

The existence of the analytical solution, discussed above, relies on the assumption of the thermal properties of the propellant being constant. In this case both the steady-state and the linearized nonsteady equations admit analytical solutions. There are other problems that can be considered that avoid the necessity of numerical integration. We can point out the investigations by Cozzi et al. (1999) and Gusachenko et al. (1999) which considered the problem of propellant combustion stability in the presence of phase transition in the condensed phase. There are two additional nondimensional parameters that need be considered in this case, namely $q/c(T_s^0 - T_a)$ and $T_p/(T_s^0 - T_a)$, where q and T_p are the latent heat and the temperature of the phase transition, respectively.

Let us emphasize, in conclusion, that the stability criterion in the model with variable surface temperature (and within the t_c approximation) is expressed via the just two parameters, k and r. The latter describe the sensitivity of the burning rate and the surface temperature to the changes in the initial temperature. It should be noted that the criterion (3.29) is obtained

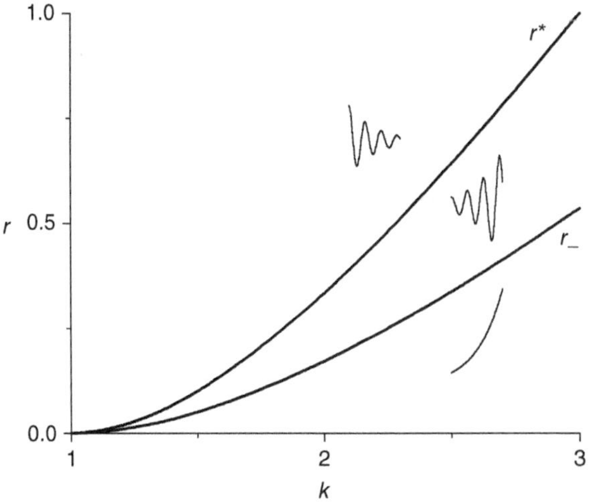

Figure 3.1 Stability boundary of the steady-state regime under constant pressure.

without any assumptions on the details of chemical reactions occurring in the condensed and gas phases.

A number of studies (De Luca 1992; Denison and Baum 1961; Istratov and Librovich 1964; Novikov and Ryazantsev 1965a,b) used different approaches to describe stability conditions. First, they proposed the model for steady-state combustion involving various assumptions on the mechanisms of chemical reactions, on the nature of transport processes, and the magnitudes of heat release in different combustion zones. Then unsteady governing equations were considered in the linear approximation. As a result, the conclusions of such analyses apply only to considered specific models that only roughly reflect real processes occurring in the course of propellant combustion. Naturally, the models involve a significant number of phenomenological parameters (activation energies and heats of combustion of chemical reactions, transport coefficients, specific heats of gas and propellant, etc.) that are in most cases unknown. The latter circumstance makes comparison with experiments quite difficult.

This flame model (FM) approach is much more complicated than the analysis presented in the present section. It requires laborious calculations of nonsteady processes in both the condensed and gas phases. Such calculations are unnecessary, as from the result of the present section it follows that consideration of *any* chemical reaction must lead to the criterion (3.29) where the parameters k and r would be related to kinetic and other parameters specific for a chosen model. Therefore, in order to investigate the stability of the steady-state combustion regime for any specific propellant combustion model, one only needs to calculate the relevant parameters k and r, and substitute them into the criterion (3.29).

We shall illustrate this point considering the Denison and Baum (1961) model. The authors used about 20 pages to discuss the following matters:

- the gas phase conservation equation
- the boundary condition
- the solution of the gas phase equation
- gas phase perturbation relations
- transient heat conduction in the solid, etc.

Finally, the authors came to the conclusion that the stability criterion is formulated exclusively in terms of the properties of the steady-state regime. This is exactly the idea embodied by Zeldovich into the Zeldovich–Novozhilov (ZN) method in 1942. Consequently, to obtain the stability criterion for the Denison and Baum model one only needs to find the parameters k and r and substitute them into (3.29).

The paper by Denison and Baum (1961) introduces the following nondimensional parameters (without loss of generality, specific heats of different phases are assumed to be equal).

$$\varepsilon = \frac{1}{2}\left(n + 2 + \frac{E_g}{RT_b^0}\right), \quad \alpha = \frac{T_b^0}{\left(T_s^0 - T_a\right)\varepsilon}$$
$$A = \left(T_s^0 - T_a\right)\frac{E_s}{R\left(T_s^0\right)^2}, \quad q = 1 + A(1 - \alpha) \tag{3.34}$$

Using these parameters, the stability conditions in the linear approximation are given by Denison and Baum (1961) as follows.

For

$$q > 1, q^2 - q - 2A > 0, q^2 < 4A \tag{3.35}$$

there is an unbounded oscillatory increase in T_s. The frequency of this oscillation is

$$\omega = \frac{1}{2}(q - 1)\sqrt{4A - q^2} \tag{3.36}$$

For

$$q > 1, q^2 - q - 2A > 0, q^2 > 4A \tag{3.37}$$

there is an unbounded nonoscillatory increase in T_s.

Let us demonstrate that these findings are identical to those obtained in the present section.

For the Denison and Baum model

$$m_s^0 = C_s \exp\left(-\frac{E_s}{RT_s^0}\right)$$
$$m_g^0 = C_g p^{n/2}(T_b^0)^{\left(n/2+1\right)} \exp\left(-\frac{E_g}{2RT_b^0}\right) \tag{3.38}$$

In these relationships C_s and C_g are constants. In the steady-state regime $m_g^0 = m_s^0$. Simple calculations show that

$$k = \frac{(T_s^0 - T_a)\varepsilon}{T_b^0}, r = \frac{R(T_s^0)^2\varepsilon}{2E_s T_b^0} \tag{3.39}$$

It is easy to see that

$$\alpha = \frac{1}{k}, A = \frac{k}{r}, q = 1 + \frac{k-1}{r} \tag{3.40}$$

The boundary of exponential instability is determined by the conditions

$$q > 1, q^2 = 4A \tag{3.41}$$

which give (3.33)

$$k > 1, r_- = (\sqrt{k} - 1)^2 \tag{3.42}$$

Substituting (3.40) into the conditions for the stability boundary (3.35) we obtain

$$k > 1, r < r^* \tag{3.43}$$

which conform with (3.29).

Finally, the substitution of (3.40) into (3.36) leads to the expression for the oscillation frequency (3.28).

The same, as discussed above, would be true for any available specific propellant combustion model (De Luca et al. 1995).

3.2 Asymptotical Perturbation Analysis

The analysis of the previous section assumed that small perturbations of burning rate and other variables depend exponentially on time. Such an analysis is not exhaustive since, strictly speaking, other time dependences of small perturbations are possible. This problem was addressed by Zeldovich et al. (1975) and Novikov and Ryazantsev (1966). However, the results of Zeldovich et al. (1975) are incorrect due to an error in the solution of the linear differential equation. The results of Novikov and Ryazantsev (1966) are limited to describing the boundary of stability only. The full analysis on the entire (k, r) plane is given by Kiskin and Novozhilov (1989), based on the use of a two-dimensional Laplace transform. The present section presents a simpler analysis based on a one-dimensional Laplace transform.

Let us investigate the evolution of small perturbations in the linear approximation.

We start with the heat transfer equation in nondimensional variables

$$\frac{\partial \theta}{\partial \tau} = \frac{\partial}{\partial \xi}\left(\frac{\partial \theta}{\partial \xi} - v\theta\right) \quad -\infty < \xi \le 0 \tag{3.44}$$

The initial and boundary conditions in these variables are

$$\begin{aligned} &\xi \to -\infty, \quad \theta = 0; \quad \xi = 0, \quad \theta = \vartheta(\tau) \\ &\tau = 0, \quad \theta(\xi, 0) = \theta_i(\xi) \end{aligned} \tag{3.45}$$

The steady-state solution is

$$\theta^0 = e^{\xi}, v^0 = 1, \vartheta^0 = 1, \varphi^0 = 1, \varphi = \left(\frac{\partial \theta}{\partial \xi}\right)_{\xi=0} \tag{3.46}$$

Let us linearize the unsteady solution in the vicinity of the steady-state regime

$$v = 1 + v_1, \vartheta = 1 + \vartheta_1, \varphi = 1 + \varphi_1, \theta_i = e^{\xi} + \theta_1 \tag{3.47}$$

bearing in mind that the perturbations with subscripts are small compared with the steady-state values.

The heat transfer equation, and the initial and boundary conditions in the linear approximation are written as

$$\begin{aligned} &\frac{\partial \theta_1}{\partial \tau} = \frac{\partial^2 \theta_1}{\partial \xi^2} - \frac{\partial \theta_1}{\partial \xi} - v_1 e^{\xi} \quad -\infty < \xi \le 0 \\ &\xi \to -\infty, \quad \theta_1 = 0; \quad \xi = 0, \quad \theta_1 = \vartheta_1(\tau) \\ &\tau = 0, \quad \theta_1(\xi, 0) = \theta_{1i}(\xi) \end{aligned} \tag{3.48}$$

The Laplace–Carson transform

$$\bar{f}(p) = p\int_{0}^{\infty} e^{-p\tau} f(\tau)\mathrm{d}\tau \tag{3.49}$$

is used to solve Eq. (3.48).

The transformed equation (corresponding to (3.48))

$$\bar{\theta}_1'' \varepsilon - \bar{\theta}_1' - p\bar{\theta}_1 = \bar{v}_1 e^{\xi} - p\theta_{1i} \tag{3.50}$$

with the conditions

$$\xi = 0, \quad \bar{\theta}_1 = \bar{\vartheta}_1(\tau); \xi \to -\infty, \quad \bar{\theta}_1 = 0 \tag{3.51}$$

similarly corresponds to (3.48) and has the following solution finite at $\xi \to -\infty$

$$\begin{aligned}\bar{\theta}_1(p,\xi) = -\frac{\bar{v}_1}{p}e^{\xi} + \left[C + \frac{p}{\sqrt{1+4p}}\int_{\xi}^{0} \theta_{1i}(u)e^{-zu}\mathrm{d}u\right]e^{z\xi} + \\ + \frac{p}{\sqrt{1+4p}}e^{(1-z)\xi}\int_{-\infty}^{\xi} \theta_{1i}(u)e^{(z-1)u}\mathrm{d}u\end{aligned} \tag{3.52}$$

where C is the integration constant and

$$z = \frac{1}{2} + \sqrt{p + \frac{1}{4}} \tag{3.53}$$

Recalling (3.21), we have

$$\bar{\vartheta}_1 = \frac{r}{k}\bar{v}_1, \bar{\varphi}_1 = \frac{k+r-1}{k}\bar{v}_1 \tag{3.54}$$

Using the solution of the transformed equation and the two latter relations, we find the transformation for the burning rate perturbation

$$\bar{v}_1 = \frac{kpV_1}{1 + \left(r - \frac{k}{z}\right)(z-1)}, V_1 = \int_{-\infty}^{0} \theta_{1i}(\xi)e^{(z-1)\xi}\mathrm{d}\xi \tag{3.55}$$

Consider first the case of constant surface temperature ($r = 0$). We have from (3.55)

$$\bar{v}_1 = \frac{k}{1-k}\int_{-\infty}^{0} \theta_{1i}(\xi)e^{-\xi/2}\frac{p\left(1/2 + \sqrt{p + 1/4}\right)e^{\sqrt{p+1/4}\,\xi}}{\sqrt{p + 1/4} - \tilde{k}}\mathrm{d}\xi, \tilde{k} = \frac{k+1}{2(k-1)} \tag{3.56}$$

Therefore to obtain the burning rate time history $v_1(\tau)$ the inverse transform of

$$\bar{\psi}(p) = \frac{p\left(1/2 + \sqrt{p + 1/4}\right)e^{-\sqrt{p+1/4}\,y}}{\sqrt{p + 1/4} - \tilde{k}} \tag{3.57}$$

is needed. Here, in order to emphasize that the coordinate ξ is negative, the positive variable $y = -\xi$ is introduced.

Using the Laplace–Carson transform tables (e.g. Table 3.1) one can find the inverse transform

$$v_1(\tau) = \frac{k}{2\sqrt{\pi}(k+1)}\frac{e^{-\tau/4}}{\tau^{3/2}}\int_{-\infty}^{0} \theta_{1i}(\xi)e^{-\xi/2}\left(J_1 - \frac{2k}{k-1}e^{\tilde{k}\xi}J_2\right)\mathrm{d}\xi \tag{3.58}$$

Table 3.1 Extraction from Laplace–Carson transform tables.

$\bar{f}(p) = p\int_0^\infty e^{-pt} f(t)dt$	$f(t)$
$\bar{f}(p)$	$\varphi(t)$
$\frac{p}{p-\beta}\bar{f}\left(\frac{p-\beta}{\alpha}\right)$	$e^{\beta t}\varphi(\alpha t)$
$e^{-\alpha p}\bar{f}(p)$	0 for $t < \alpha$ $\varphi(t-\alpha)$ for $t > \alpha$
$\sqrt{p}\bar{f}(\sqrt{p})$	$\frac{1}{2t\sqrt{\pi t}}\int_0^\infty \tau \exp\left(-\frac{\tau^2}{4t}\right)\varphi(\tau)\,d\tau$
$p^{n+1/2}\bar{f}(\sqrt{p})$	$\frac{1}{2^{\frac{1+n}{2}}\sqrt{\pi}t^{1+\frac{n}{2}}}\int_0^\infty \exp\left(-\frac{\tau^2}{4t}\right)\mathrm{He}_{n+1}\left(\frac{\tau}{\sqrt{2t}}\right)\varphi(\tau)d\tau,$ $\mathrm{He}_n(x) = (-1)^n e^{\frac{x^2}{2}}\frac{d^n}{dx^n}\left(e^{-\frac{x^2}{2}}\right)$
1	1
$\frac{1}{p-\alpha}$	$\frac{e^{\alpha t}-1}{\alpha}$
$\frac{pe^{-\alpha p}}{p+a}, \alpha > 0$	0 for $t < \alpha$ $e^{-a(t-\alpha)}$ for $t > \alpha$

where

$$J_1 = \int_y^\infty e^{-u^2/4\tau}\left(\frac{u^2}{2\tau}-1\right)du, \quad J_2 = \int_y^\infty e^{-u^2/4\tau + \tilde{k}u}\left(\frac{u^2}{2\tau}-1\right)du$$

$$\tilde{k} = \frac{k+1}{2(k-1)} \tag{3.59}$$

Integration produces the following results

$$J_1 = ye^{-y^2/4\tau}$$

$$J_2 = e^{-y^2/4\tau}\left[(y+2\tilde{k}\tau)e^{\tilde{k}y} + 2\sqrt{\pi}\tilde{k}^2\tau^{3/2}e^{y^2/4\tau+\tilde{k}^2\tau}\mathrm{erfc}\left(\frac{y-2\tilde{k}\tau}{2\sqrt{\tau}}\right)\right] \tag{3.60}$$

where

$$\mathrm{erfc}(\zeta) = \frac{2}{\sqrt{\pi}}\int_\zeta^\infty e^{-t^2}dt \tag{3.61}$$

Replacing y by $-\xi$ we get

$$J_1 = -\xi e^{-\xi^2/4\tau}$$

$$J_2 = e^{-\xi^2/4\tau}\left[(-\xi+2\tilde{k}\tau)e^{-\tilde{k}\xi} + 2\sqrt{\pi}\tilde{k}^2\tau^{3/2}e^{\xi^2/4\tau+\tilde{k}^2\tau}\mathrm{erfc}\left(-\frac{\xi+2\tilde{k}\tau}{2\sqrt{\tau}}\right)\right] \tag{3.62}$$

Substitution of these expressions into (3.58) leads to the final expression for the burning rate perturbation $v_1(\tau)$.

The results may be illustrated using an example of perturbation which has a form of δ function located at $\xi = \xi_0$

$$\theta_i(\xi) = e^{\xi}[1 + \Delta\delta(\xi - \xi_0)], \theta_{1i}(\xi) = \Delta e^{\xi}\delta(\xi - \xi_0) \tag{3.63}$$

In this case

$$v_1(\tau) = \frac{\Delta k}{2\sqrt{\pi}(k+1)} e^{\xi_0/2} \left(J_1(\xi_0, \tau) - \frac{2k}{k-1} e^{\tilde{k}\xi} J_2(\xi_0, \tau) \right) \frac{e^{-\tau/4}}{\tau^{3/2}} \tag{3.64}$$

Here

$$J_1(\xi_0, \tau) = -\xi_0 e^{-\xi_0^2/4\tau}$$

$$J_2(\xi_0, \tau) = e^{-\xi_0^2/4\tau} \left[(-\xi_0 + 2\tilde{k}\tau)e^{-\tilde{k}\xi_0} + 2\sqrt{\pi}\tilde{k}^2\tau^{3/2} e^{\xi_0^2/4\tau + \tilde{k}^2\tau} \operatorname{erfc}\left(-\frac{\xi_0 + 2\tilde{k}\tau}{2\sqrt{\tau}} \right) \right] \tag{3.65}$$

Let us consider the asymptotic behaviour of the burning rate at $\tau \to \infty$. Using the asymptotic properties of the complimentary error function (Abramowitz and Stegun 1972)

$$\zeta \to \infty, |\arg\zeta| < \frac{3\pi}{4}, \sqrt{\pi}\zeta e^{\zeta^2}\operatorname{erfc}(\zeta) \sim 1 - \frac{1}{2\zeta^2} \tag{3.66}$$

and the relation

$$\operatorname{erfc}(-\zeta) = 2 - \operatorname{erfc}(\zeta) \tag{3.67}$$

we obtain, for $\tau \gg 1$

$$J_1 \approx -\xi_0 \tag{3.68}$$

The value of J_2 depends on the sign of $\tilde{k}$, i.e.
for $\tilde{k} > 0$ $(k > 1)$

$$J_2 \approx 4\sqrt{\pi}\tilde{k}^2\tau^{3/2}e^{\tilde{k}^2\tau} \tag{3.69}$$

for $\tilde{k} < 0$ $(k < 1)$

$$J_2 \approx \frac{1}{\tilde{k}} e^{-\tilde{k}\xi_0} \tag{3.70}$$

Finally, at $\tau \to \infty$

$$k < 1, v_1(\tau) \approx -\frac{\Delta k}{2\sqrt{\pi}(k+1)} \left(\xi_0 + \frac{4k}{k+1} \right) e^{\xi_0/2} \frac{e^{-\tau/4}}{\tau^{3/2}} \tag{3.71}$$

$$k > 1, v_1(\tau) \approx -\frac{\Delta k^2(k+1)}{(k-1)^3} e^{k\xi_0/(k-1)} e^{k\tau/(k-1)^2} \tag{3.72}$$

Figures 3.2 and 3.3 illustrate burning rate behaviour in the case of constant surface temperature.

Let us turn now to the case of variable surface temperature ($r \neq 0$). In this case we have from (3.55)

$$\bar{v}_1 = \frac{k}{r} \int_{-\infty}^{0} \theta_{1i}(\xi) e^{-\xi/2} \frac{p\left(1/2 + \sqrt{p + 1/4}\right) e^{\sqrt{p+1/4}\xi}}{\left(\sqrt{p + 1/4} - a\right)\left(\sqrt{p + 1/4} - b\right)} d\xi$$

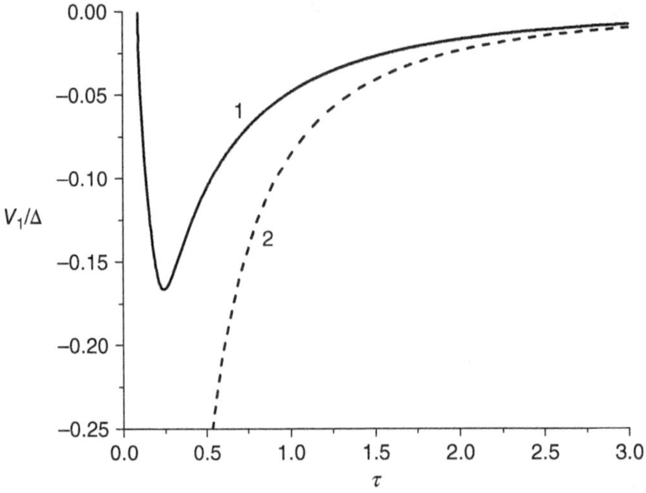

Figure 3.2 Burning rate (1) and its asymptotic (2) for $r = 0$, $k = 0.5$, and $\xi_0 = -0.1$. Constant surface temperature.

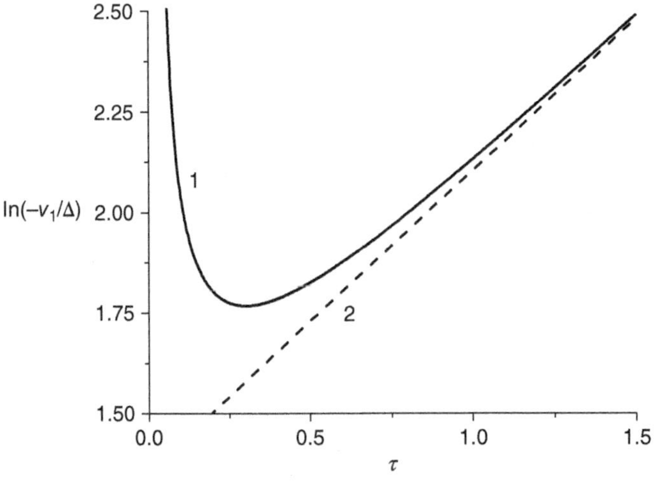

Figure 3.3 Burning rate (1) and its asymptotic (2) for $r = 0$, $k = 3$, and $\xi_0 = -0.1$. Constant surface temperature.

$$a = \frac{1}{2r}\left(k - 1 + \sqrt{(k-1)^2 - r[2(k+1) - r]}\right)$$
$$b = \frac{1}{2r}\left(k - 1 - \sqrt{(k-1)^2 - r[2(k+1) - r]}\right) \tag{3.73}$$

In order for the burning rate correction $v_1(\tau)$ to be found as a function of time, we need the inverse transform of

$$\overline{\psi}(p) = \frac{p\left(1/2 + \sqrt{p + 1/4}\right) e^{-\sqrt{p+1/4}\,y}}{\left(\sqrt{p + 1/4} - a\right)\left(\sqrt{p + 1/4} - b\right)} \tag{3.74}$$

(again, the negative coordinate ξ is replaced by the positive variable $y = -\xi$).

Using the Table 3.1 again

$$v_1(\tau) = \frac{k}{2\sqrt{\pi}r(a-b)} \frac{e^{-\tau/4}}{\tau^{3/2}} \int_{-\infty}^{0} \theta_{1i}(\xi) e^{-\xi/2} [W_a(\xi,\tau) - W_b(\xi,\tau)] d\xi$$

$$W_a(\xi,\tau) = \left(a + \frac{1}{2}\right) e^{a\xi} J_a(\tau), \quad J_a(\tau) = \int_{-\xi}^{\infty} u \exp\left(-\frac{u^2}{4\tau} + au\right) du \tag{3.75}$$

The integral is easily found

$$J_a(\tau) = 2\tau \left[\exp\left(-\frac{\xi(\xi + 4a\tau)}{4\tau}\right) + a\sqrt{\pi\tau} e^{a^2\tau} \text{erfc}\left(-\frac{\xi + 2a\tau}{2\sqrt{\tau}}\right)\right] \tag{3.76}$$

therefore

$$W_a(\xi,\tau) = \tau(2a+1) e^{a\xi} \times$$
$$\times \left[\exp\left(-\frac{\xi(\xi + 4a\tau)}{4\tau}\right) + a\sqrt{\pi\tau} e^{a^2\tau} \text{erfc}\left(-\frac{\xi + 2a\tau}{2\sqrt{\tau}}\right)\right] \tag{3.77}$$

In the case of the δ function (located at $\xi = \xi_0$) perturbation

$$\theta_{1i}(\xi) = \Delta e^{\xi} \delta(\xi - \xi_0) \tag{3.78}$$

the result is

$$v_1(\tau) = \frac{\Delta k e^{\xi_0/2}}{2\sqrt{\pi}r(a-b)} \frac{e^{-\tau/4}}{\tau^{3/2}} [W_a(\xi_0,\tau) - W_b(\xi_0,\tau)]$$

$$W_a(\xi_0,\tau) = \tau(2a+1) e^{a\xi_0} \times$$
$$\times \left[\exp\left(-\frac{\xi_0(\xi_0 + 4a\tau)}{4\tau}\right) + a\sqrt{\pi\tau} e^{a^2\tau} \text{erfc}\left(-\frac{\xi_0 + 2a\tau}{2\sqrt{\tau}}\right)\right] \tag{3.79}$$

To study the asymptotic behaviour of the burning rate at $\tau \to \infty$, let us start with $0 < k < 1$. In this region, the real parts of the values a and b are negative if $r < 2(k+1)$. Using the asymptotic properties (3.66), we find

$$W_a(\xi) = \left(1 + \frac{1}{2a}\right) \xi + \frac{1 + 2a}{2a^2} \tag{3.80}$$

and consequently the asymptotic expression at $\tau \to \infty$

$$v_1(\tau) = -\frac{\Delta k e^{\xi_0/2}}{\sqrt{\pi}[2(k+1) - r]} \frac{e^{-\tau/4}}{\tau^{3/2}} \left[\xi_0 + 2\frac{4k - r}{2(k+1) - r}\right] \tag{3.81}$$

Figure 3.4 illustrates burning rate behaviour for $0 < k < 1$.

The region $r > 2(k+1)$ is not considered as the existence of real systems having this parameter range is unlikely.

Let us turn now to the case $k > 1$. We note that if

$$0 \le r \le r_- \text{ or } r \ge r_+ \tag{3.82}$$

where

$$r_- = (\sqrt{k} - 1)^2, r_+ = (\sqrt{k} + 1)^2 \tag{3.83}$$

then a and b are real numbers. Outside this region they are complex (Figure 3.5).

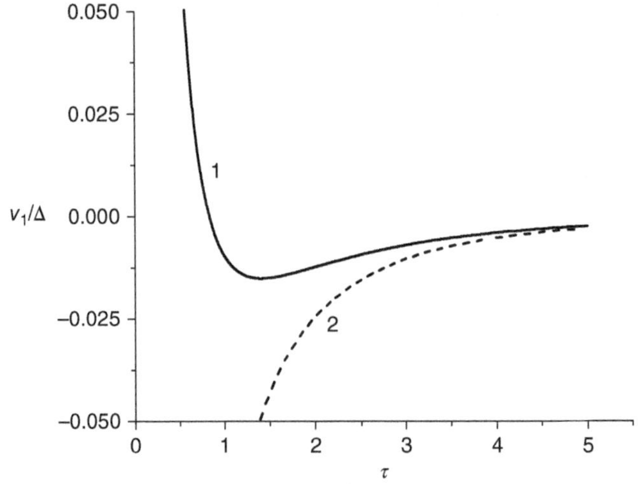

Figure 3.4 Burning rate (1) and its asymptotic (2) for $r = 0.25$, $k = 0.5$, and $\xi_0 = -0.1$. Variable surface temperature.

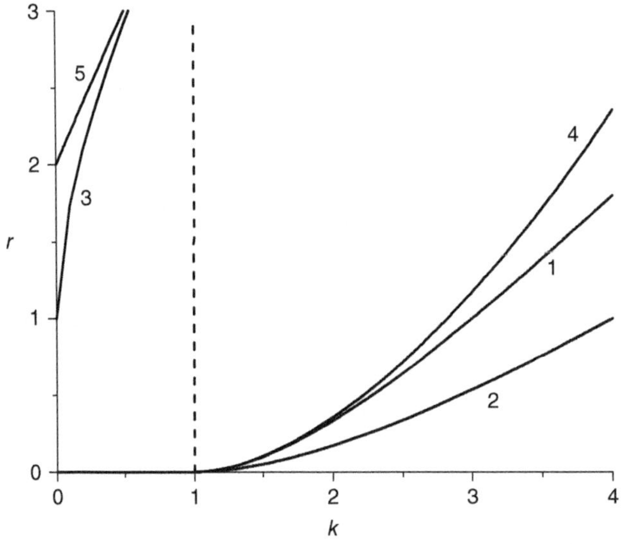

Figure 3.5 Regions of typical combustion regimes: 1, r^*; 2, r_-; 3, r_+; 4, $r_{d\omega}$ (3.94); 5, $r = 2(k+1)$. Variable surface temperature.

In the region $0 \leq r \leq (\sqrt{k} - 1)^2$ both a and b are real and positive. The asymptotic (at $\tau \to \infty$) can be found from (3.79) using the relation (3.67)

$$W_a(\xi, \tau) \approx 2a(2a+1)\sqrt{\pi} e^{a\xi} \tau^{3/2} e^{a^2\tau} \tag{3.84}$$

The burning rate, under the δ function initial perturbation, has the asymptotics

$$v_1(\tau) \approx \frac{\Delta k e^{\xi_0/2} e^{-\lambda\tau}}{r(a-b)}[U_a(\xi_0,\tau) - U_b(\xi_0,\tau)]$$

$$U_a(\xi_0,\tau) = a(2a+1)e^{a\xi_0}e^{\sqrt{\lambda^2-\omega_0^2}\tau}, \quad U_b(\xi_0,\tau) = b(2b+1)e^{b\xi_0}e^{-\sqrt{\lambda^2-\omega_0^2}\tau} \tag{3.85}$$

where λ and ω_0 are the oscillation damping decrement and the frequency corresponding to the stability boundary (3.28), respectively

$$\lambda = \frac{r(k+1)-(k-1)^2}{2r^2}, \omega_0 = \frac{\sqrt{k}}{r} \tag{3.86}$$

Note that in the considered region the decrement is negative.

Inside the considered region ($r < r_-$) the second term in the square brackets of (3.85) is negligible compared to the first one, therefore $\tau \to \infty$

$$v_1(\tau) \approx \frac{\Delta k a(2a+1)e^{\left(a+1/2\right)\xi_0}}{r(a-b)} e^{(-\lambda+\sqrt{\lambda^2-\omega_0^2})\tau} \tag{3.87}$$

Figure 3.6 shows the burning rate dependence for the parameters corresponding to this region.

At the boundary of the considered region ($r = r_-$) $a = b$, therefore both terms must be retained in (3.85), leading to

$$v_1(\tau) \approx 2k(\sqrt{k}+1)\exp\left[\frac{\sqrt{k}}{\sqrt{k}-1}\left(\xi_0 + \frac{\tau}{\sqrt{k}-1}\right)\right] \tag{3.88}$$

Consider now the case

$$k > 1, \quad r_- < r < r_+ \tag{3.89}$$

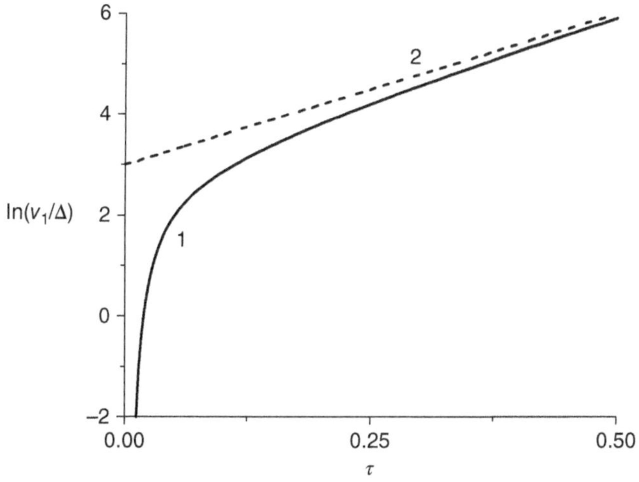

Figure 3.6 Burning rate (1) and its asymptotic (2) for $r = 0.5$, $k = 3$, and $\xi_0 = -0.5$. Variable surface temperature.

In this region both a and b are complex, with positive real parts. The asymptotic at $\tau \to \infty$ is found from (3.79), using (3.66) and (3.67)

$$W_a(\xi,\tau) \approx \left(1+\frac{1}{2a}\right)\xi + \frac{1+2a}{2a^2} + 2a(2a+1)\sqrt{\pi}e^{a\xi}\tau^{3/2}e^{a^2\tau} \tag{3.90}$$

For the δ function initial perturbation, asymptotically

$$v_1(\tau) \approx \frac{\Delta k e^{\xi_0/2}}{2\sqrt{\pi}r}\frac{e^{-\tau/4}}{\tau^{3/2}}\frac{\mathrm{Im}[W_a(\xi_0,\tau)]}{\mathrm{Im}[a]} \tag{3.91}$$

In the region $r_- < r < r^*$, where r^* corresponds to the stability boundary of the steady-state regime (3.29), the decrement is negative and the burning rate oscillates with increasing amplitude. The formulas (3.66) and (3.67) give

$$v_1(\tau) \approx \frac{\Delta k e^{\xi_0/2}e^{-\lambda\tau}}{r}\frac{\mathrm{Im}[a(2a+1)\exp(a\xi_0 + i\sqrt{\omega_0^2-\lambda^2})]}{\mathrm{Im}[a]} \tag{3.92}$$

Figure 3.7 shows burning rate dependence on time and its asymptotic.

Obviously, the condition $\lambda = 0$ provides a description of the stability boundary.

In the region $r^* < r < r_+$, i.e. in the region of stable combustion (decrement is positive), the formulas (3.90) and (3.91) give

$$\begin{aligned}
&v_1(\tau) \approx v_{1d}(\tau) + v_{1\omega}(\tau)\\
&v_{1d}(\tau) \approx \frac{\Delta k e^{\xi_0/2}}{2\sqrt{\pi}r}\frac{e^{-\tau/4}}{\tau^{3/2}}\frac{\mathrm{Im}\left[\left(1+\frac{1}{2a}\right)\xi_0 + \frac{1+2a}{2a^2}\right]}{\mathrm{Im}[a]}\\
&v_{1\omega}(\tau) \approx \frac{\Delta k e^{\xi_0/2}}{r}\frac{\mathrm{Im}[a(2a+1)e^{a\xi_0}\exp((-\lambda + i\sqrt{\omega_0^2-\lambda^2})\tau)]}{\mathrm{Im}[a]}
\end{aligned} \tag{3.93}$$

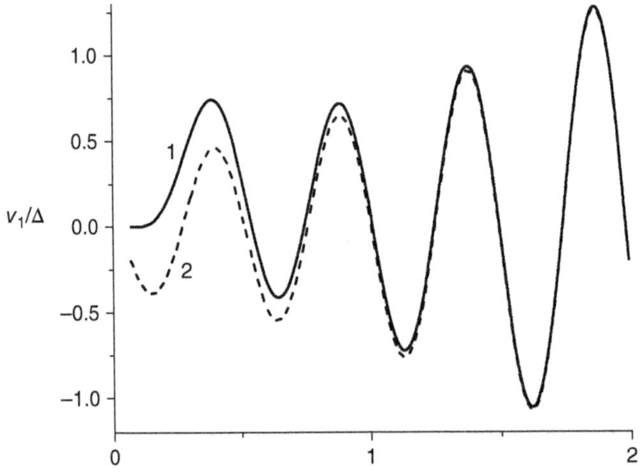

Figure 3.7 Burning rate (1) and its asymptotic (2) for $r = 0.095$, $k = 1.5$, and $\xi_0 = -2$. Variable surface temperature.

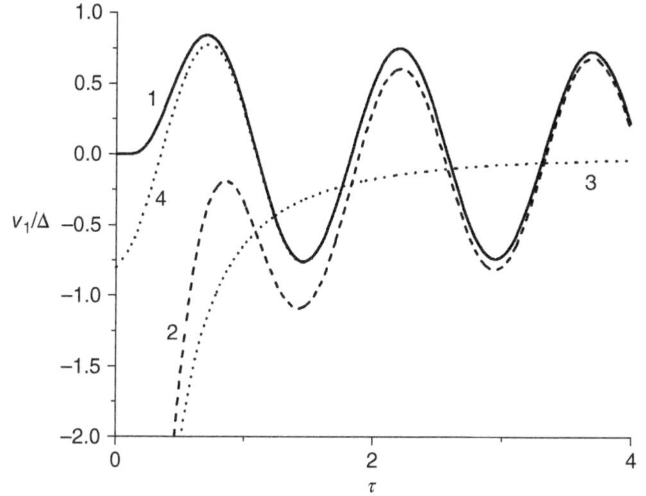

Figure 3.8 Burning rate (1), its asymptotic (2), and the terms v_{1d} (3) and $v_{1\omega}$ (4) for $r = 0.335$, $k = 2$, and $\xi_0 = -2$. Variable surface temperature.

It can be concluded that the asymptotic behaviour of the burning rate embraces two components, one diminishing monotonically with time and the other exponentially diminishing with oscillations.

To the logarithmic accuracy, the decrement of the first component is equal to $1/4$. Therefore, at $\lambda < 1/4$ the oscillating component decreases more slowly than monotonically decreasing. This case is presented in Figure 3.8. With the reversed inequality, the monotonically decreasing component prevails (Figure 3.9). The boundary between these two regimes is found from the restriction $\lambda = 1/4$, which leads to

$$r_{d\omega} = k + 1 - \sqrt{6k - k^2 - 1} \tag{3.94}$$

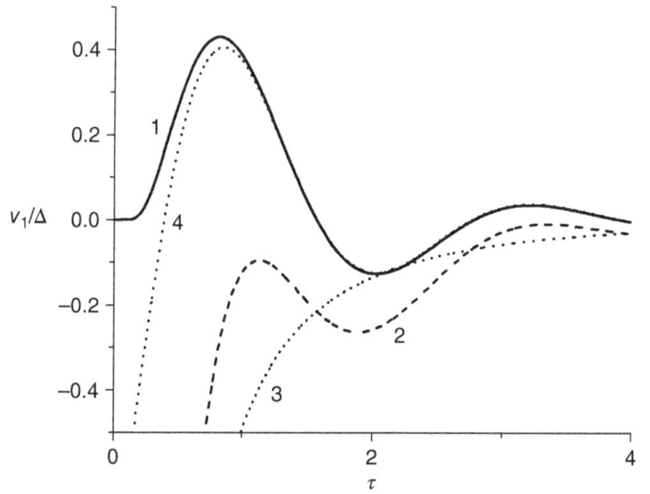

Figure 3.9 Burning rate (1), its asymptotic (2), and the terms v_{1d} (3) and $v_{1\omega}$ (4) for $r = 0.5$, $k = 2$, and $\xi_0 = -2$. Variable surface temperature.

The region $r > r_+$ may be considered in exactly the same manner as $0 \leq r < r_-$. However, it is unlikely that real systems corresponding to this region of parameters k and r exist.

3.3 Two-dimensional Combustion Stability of Gasless Systems

The stability criterion of the preceding section was obtained within a one-dimensional problem formulation. The surface of the propellant was assumed to be plain in the course of its derivation. One can pose, however, a problem of a steady-state combustion regime stability under constant pressure to perturbations of the surface shape. In contrast to the results of Section 3.1, the solution of such a problem in the same general form (i.e. independent of any specific combustion model) turns out to be impossible. The reason for this is as follows: one-dimensional case perturbations imposed at the surface result in conditions which can be observed under a steady-state combustion regime corresponding to different initial propellant temperatures. This allows the dependence $u(f)$, obtained from the steady-state relation $u^0(T_a)$, to be used in the analysis. Upon imposing two-dimensional perturbations with finite wavelength in both the condensed and the gas phases transversal heat and mass (in the gas phase) fluxes emerge. Consequently, conditions at the surface cannot be related to the characteristics of any steady-state regime. In order to solve the problem a specific combustion model has to be considered.

The stability criterion (3.29) is valid for disturbances with infinitely large wavelength. It is quite possible that in a particular specific combustion model formulation the least stable perturbation mode will have finite wavelength. In this case condition (3.29) would not be valid. The condition (3.29) would only apply if the least stable perturbation mode leaves the surface plain (i.e. has infinite wavelength). The problem is not currently solved for any model taking the gas phase reaction into account. In principle, both scenarios of the loss of stability are possible, i.e. the harmonics with either finite or infinite wavelength may be less stable. The result will be determined by the properties of a particular combustion model.

Usually propellant combustion is accompanied by a phase change from condensed to gaseous. There are systems, however, in which reaction products are left in the condensed state behind the exothermic reaction front. An example is the combustion of termites studied experimentally by Belyaev and Komkova (1950). A similar process is observed upon propagation of polymerization front through the condensed phase (Chechilo et al. 1972), as well as in the process of self-propagating high-temperature synthesis (Merzhanov 1994).

In the present section, the problem of stability of the steady-state combustion regime of gasless condensed systems under surface shape perturbations is investigated in the linear approximation.

Let us consider the combustion stability of a gasless system using the assumption of strong chemical reaction rate dependence on temperature (Makhviladze and Novozhilov 1971). This assumption allows the zone of chemical transformation of a virgin substance to be regarded as infinitely thin, compared to the preheat zone, and therefore it can be considered as a surface separating initial reactants (r zone) and reaction products (p zone). Relevant indices mark variables related to the respective zones.

Let us consider the laboratory frame where the undisturbed propellant's surface is stationary at $x = 0$, and virgin substance occupying the semi-space $x \leq 0$ moves from left to right with speed u^0 equal to the constant speed of propagation of the chemical reaction front.

Steady-state temperature distributions in both zones

$$T_r^0(x) = T_a + (T_s^0 - T_a)\exp\left(\frac{u^0 x}{\kappa}\right), T_p^0 = T_s^0 \tag{3.95}$$

satisfy the same heat transfer equation

$$\kappa\frac{d^2T_j^0}{dx^2} - u^0\frac{dT_j^0}{dx} = 0, j = r, p \tag{3.96}$$

and the boundary conditions

$$\begin{aligned}
&x \to -\infty, \quad T_r^0 = T_a; \quad x \to \infty, \quad \frac{dT_p^0}{dx} = 0\\
&x = 0, \quad T_r^0 = T_s^0, \quad T_p^0 = T_s^0\\
&T_s^0 = T_a + \frac{Q_s}{c}
\end{aligned} \tag{3.97}$$

Here T_a stands for initial temperature, T_s^0 for combustion temperature, and Q_s for heat of reaction.

Using the method of small perturbations, assume a disturbance of the surface in the form

$$\xi = D\exp(\Omega t + iKy) \tag{3.98}$$

where the y direction is along the unperturbed reaction front and Ω, K, and D are frequency, wavenumber, and amplitude of the perturbation, respectively.

Without any loss of generality, a two-dimensional problem may be considered.

Perturbed solutions are sought in the form

$$T_j = T_j^0 + \delta T_j(x)\exp(\Omega t + iKy), j = r, p \tag{3.99}$$

Substitution of (3.99) into the unsteady heat transfer equations

$$\frac{\partial T_j}{\partial x} = \kappa\left(\frac{\partial^2 T_j}{\partial x^2} + \frac{\partial^2 T_j}{\partial y^2}\right) - u^0\frac{\partial T_j}{\partial x}, j = r, p \tag{3.100}$$

leads to second-order ordinary differential equations with respect to $\delta T_r(x)$ and $\delta T_p(x)$ with the solutions

$$\begin{aligned}
&\delta T_r(x) = A\exp\left(z_r\frac{u^0 x}{2\kappa}\right), \quad z_r = 1 + \sqrt{1 + 4\omega + s^2}\\
&\delta T_p(x) = B\exp\left(z_p\frac{u^0 x}{2\kappa}\right), \quad z_p = 1 - \sqrt{1 + 4\omega + s^2}\\
&\omega = \frac{\kappa\Omega}{(u^0)^2}, \quad s = \frac{2\kappa K}{u^0}
\end{aligned} \tag{3.101}$$

Here ω is a nondimensional frequency, parameter s is the ratio of the width of the Michelson preheat zone κ/u^0 to the wavelength of the perturbation $2\pi/K$, and A and B are integration constants. The choice of signs in the z_r and z_p formulae is dictated by the requirement of the disappearance of perturbations at infinity.

The perturbed solutions are related to each other by the following conditions at the surface where chemical reaction occurs: the temperature distribution is continuous, while the heat flux experiences a jump as a result of energy release in the reaction zone. In the adopted approximation

$$x = \xi, \quad T_r = T_p, \; -\frac{\partial T_r}{\partial x} + \left(u^0 - \frac{\partial \xi}{\partial t}\right)\frac{Q_s}{\kappa c} = -\frac{\partial T_p}{\partial x} \tag{3.102}$$

Here use is made of the fact that the derivative normal to the surface differs from the derivative in the x direction by the value which is second-order infinitely small.

Assume further that reaction rate is a function of the temperature in the reaction zone only. Then, in the linear approximation, variation of the reaction rate with temperature is described by the parameter

$$k = (T_s^0 - T_a)\frac{\mathrm{d}\ln u^0}{\mathrm{d}T_a} \tag{3.103}$$

Its explicit form is determined by the steady-state burning law $u^0(T_s^0)$. In the case of Arrhenius dependence of reaction rate on temperature

$$u^0 \sim \exp\left(-\frac{E_s}{2RT_s^0}\right), k = \frac{E(T_s^0 - T_a)}{2RT_s^0} \tag{3.104}$$

Using the temperature sensitivity parameter k the condition of full reactant consumption may be written as

$$x = \xi, \; \frac{\partial \xi}{\partial t} = -\frac{ku(T_p - T_a)}{T_s^0 - T_a} \tag{3.105}$$

Linearizing (3.102) and (3.105) at $x = 0$

$$\begin{aligned} &\frac{\mathrm{d}T_r^0}{\mathrm{d}x}\xi + \delta T_r = \frac{\mathrm{d}T_p^0}{\mathrm{d}x}\xi + \delta T_p \\ &\frac{\mathrm{d}^2T_r^0}{\mathrm{d}x^2}\xi + \frac{\mathrm{d}\delta T_r}{\mathrm{d}x} + \frac{Q_s}{\kappa c}\frac{\partial \xi}{\partial t} = \frac{\mathrm{d}^2T_p^0}{\mathrm{d}x^2}\xi + \frac{\mathrm{d}\delta T_p}{\mathrm{d}x} \\ &\frac{\partial \xi}{\partial t} = -\frac{ku^0}{T_s^0 - T_a}\left(\frac{\mathrm{d}T_p^0}{\mathrm{d}x}\xi + \delta T_p\right) \end{aligned} \tag{3.106}$$

Substituting (3.98) here as well as solutions (3.95) and (3.101) provides a homogeneous set of three linear equations with respect to unknowns A, B, and D.

The solvability condition of this system is

$$16\omega^3 + 4\left(1 + 4k - k^2 + s^2\right)\omega^2 + 4k\left(1 + s^2\right)\omega + s^2k^2 = 0 \tag{3.107}$$

Letting $\omega = \pm i\Psi$ gives the stability boundary $k(s^2)$

$$k = \frac{4 + 3s^2 + \sqrt{\left(4 + 3s^2\right)^2 + 4\left(1 + s^2\right)^3}}{2\left(1 + s^2\right)} \tag{3.108}$$

Frequency is purely imaginary at the stability boundary, which means that the loss of stability is of oscillatory nature. Furthermore, the frequency may be expressed as a function of the wavelength of the perturbation

$$\Psi^2 = \frac{1}{8}\left(4 + 3s^2 + \sqrt{(4 + 3s^2)^2 + 4(1 + s^2)^3}\right) \tag{3.109}$$

Substitution of $s = 0$ in Eq. (3.108) leads to the conclusion that the region of instability of one-dimensional perturbations is

$$k_\infty > 2 + \sqrt{5} \approx 4.24 \tag{3.110}$$

On the other hand, the minimum value in (3.108) is attained at $s = 1$ and is $k_{min} = 4$. Therefore, the chemical reaction front is more stable to one-dimensional perturbations than to perturbations of higher dimensions.

Curves a in Figures 3.10 and 3.11 show the dependencies (3.108) and (3.109).

It is relevant to note that the undertaken analysis is similar in many respects to the investigation of thermodiffusive gaseous flame stability (Barenblatt et al. 1962). The latter study

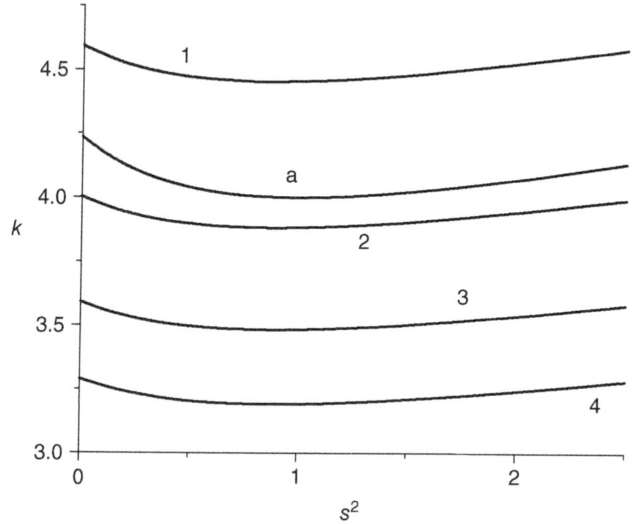

Figure 3.10 Stability boundary of the steady-state combustion regime: *a*, analytical result (3.108); 1–4, numerical results (3.115): 1, $\beta = 0$; 2, $\beta = 0.05$; 3, $\beta = 0.1$; 4, $\beta = 0.15$.

Figure 3.11 Frequency at the boundary of the steady-state combustion regime: *a*, analytical result (3.109); 1–2, numerical results (3.119): 1, $\beta = 0$; 2, $\beta = 0.15$.

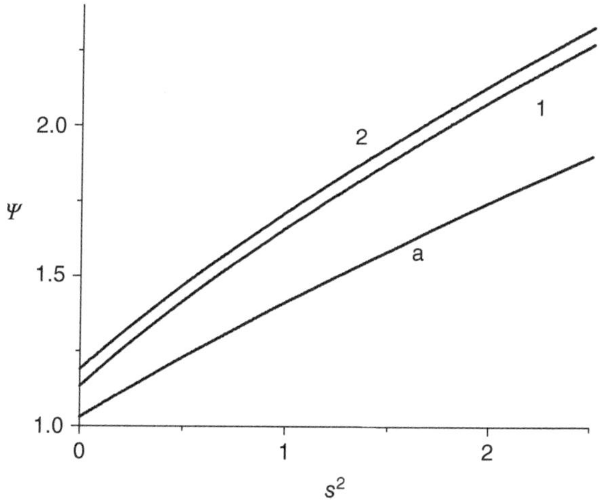

investigates the effect of the parameter D_g/κ (where D_g is the gas diffusion coefficient) on the gaseous flame stability to longwave disturbances. The problem was considered in the constant density approximation. Therefore, avoiding the longwave limitation, the limit $D_g/\kappa \to 0$ in the relevant relationships of the Barenblatt et al. (1962) paper leads to the same dispersion relation (3.107).

Apart from gasless combustion Makhviladze and Novozhilov (1971) also considered flameless combustion regime. In this case the condensed substance transforms into a gas where no chemical reactions take place. As was already pointed out such a regime may be observed in the combustion of ballistites under low pressure conditions. Here the stability of combustion is determined by the ratio of the thermophysical properties of the virgin propellant and the products of combustion

$$\alpha = \left(\frac{\rho_g D_g}{\rho \kappa}\right)^2 \tag{3.111}$$

A decrease in this parameter leads to expansion of the instability region. For $\alpha < 0.4$ the system becomes more stable to one-dimensional perturbations, compared to perturbations of higher dimensions.

As follows from (3.108), two-dimensional perturbations may turn out to be less stable than one-dimensional peturbations. However, within the framework of adopted approximation this conclusion is not fully justified. Indeed, the performed analysis neglects temperature variation in the reaction zone, with the order of magnitude $R(T_s^0)^2/E_s$, in comparison with the characteristic temperature interval $T_s^0 - T_a$. In other words

$$\frac{R(T_s^0)^2}{E(T_s^0 - T_a)} = \frac{1}{2k} \tag{3.112}$$

was assumed to be a small parameter, close to the stability boundary $k \approx 4$. Therefore, this assumption leads to an error of about 10% in the final answer. At the same time variation of the parameter k, while s changes from infinity to one, is only about 6%. Thus, the existence of the minimum on the curve $k(s)$ cannot be established for certainty as the magnitude of change of the parameter k is of the same order as an error of calculation.

For this reason, the finite thickness of the reaction zone was taken into account by Borisova et al. (1986). The problem was analysed in a linear approximation by means of numerical analysis. A first-order reaction along with constant propellant density and thermophysical properties was considered.

The governing equations for the process of gasless combustion, in the laboratory frame with virgin substance moving from left to right with the steady-state burning rate, are

$$\begin{aligned}
\frac{\partial T}{\partial x} &= \kappa\left(\frac{\partial^2 T}{\partial x^2} + \frac{\partial^2 T}{\partial y^2}\right) - u^0\frac{\partial T}{\partial x} + \frac{Q_s}{c}W(\eta, T)\\
\frac{\partial \eta}{\partial x} &= -u^0\frac{\partial \eta}{\partial x} + W(\eta, T)\\
W(\eta, T) &= Z(1-\eta)\exp\left(-\frac{E}{RT}\right)
\end{aligned} \tag{3.113}$$

where η is the progress variable of chemical transformation and Z is the pre-exponential factor.

Calculation results can be presented rather conveniently using curve fitting. Let

$$k = \Phi(\beta, s^2), \quad \beta = \frac{RT_s^0}{E} \tag{3.114}$$

be the relation between parameters at the stability boundary. Then the function $\Phi(\beta, s^2)$, to within 0.5% accuracy, may be represented for $0 \leq \beta \leq 0.15$ and $0 \leq s^2 \leq 2.4$ as

$$\Phi(\beta, s^2) = \frac{3s^2 + 3.39 + \sqrt{(3s^2 + 3.39)^2 + 4(s^2 + 2.35)^2(s^2 + 1)}}{2(s^2 + 1)(1 + 3.1\beta - 3.1\beta^2)} \tag{3.115}$$

This function, considered with a fixed value of the parameter β, has a minimum at $s^2 = 1$. The values $\Phi(\beta, 0)$ and $\Phi(\beta, 1)$ differ by 2–3% only. However, the existence of the minimum is established reliably. This implies a lower degree of stability with respect to two- or three-dimensional perturbations, compared to plain perturbations. Therefore, the qualitative result obtained analytically by Makhviladze and Novozhilov (1971) is ultimately valid.

For the plain perturbation ($s^2 = 0$) (3.115) gives

$$k_\infty = \frac{4.60}{1 + 3.1\beta - 3.1\beta^2} \tag{3.116}$$

This is practically identical to the curve-fitting formula

$$k_\infty = \frac{4.55}{1 + 2.5\beta} \tag{3.117}$$

obtained by Shkadinskii et al. (1971) using a numerical solution of the one-dimensional unsteady problem.

Within the assumption of the infinitely narrow reaction zone, the notion of reaction order is meaningless and the parameter β becomes irrelevant. As demonstrated above, based on the analysis of Makhviladze and Novozhilov (1971), in this case $k_\infty = 4.24$. For the most dangerous modes of perturbations ($s^2 = 1$) from (3.115)

$$k_{\min} = \frac{4.45}{1 + 3.1\beta - 3.1\beta^2} \tag{3.118}$$

while for infinitely narrow reaction zone $k_{\min} = 4$ (Makhviladze and Novozhilov 1971).

The frequency Ω at the stability boundary is purely imaginary $\Omega = i\Psi(\beta, s^2)$. Computational results, within 2% accuracy, may be approximated as

$$\Psi^2 = \left(0.28 + \beta + 0.34s^2 + \beta s^2\right)\Phi(\beta, s^2) \tag{3.119}$$

In particular, for $s = 0$ (plain perturbations)

$$\Psi_\infty^2 = \frac{4.60(0.28 + \beta)}{1 + 3.1\beta - 3.1\beta^2} \tag{3.120}$$

This is a very weak function of β and (to within the term of the order of β^2) is equal to $\Psi_\infty^2 = 1.29(1 + 0.5\beta)$. Under the narrow reaction zone assumption $\Psi_\infty^2 = 1.06$, i.e. an error of the approximate solution of Makhviladze and Novozhilov (1971) for the value β is about 10%.

Figures 3.10 and 3.11 show the dependencies $k = \Phi(\beta, s^2)$ and $\Psi(\beta, s^2)$ according to the formulae (3.115) and (3.119).

There have been attempts to improve the narrow reaction zone approximation analytically. In particular, Aldushin and Kasparyan (1979) proposed to use different, compared to Makhviladze and Novozhilov (1971) and Barenblatt et al. (1962), matching conditions in the chemical reaction zone. The reaction rate in the study of Aldushin and Kasparyan (1979) was considered as a function of not only the temperature (the usual approximation for an infinitely narrow zone approach), but also of the temperature gradients at the matching point. The claim made by Aldushin and Kasparyan (1979) that such an approach takes into account the inertia of the reaction zone is not valid as mass and energy conservation equations in this zone were not considered. Consequently, the parameter β does not appear in the final result. For comparison, the analysis of Aldushin and Kasparyan (1979) gives $k_\infty = 4.0$ and $k_{min} = 3.91$. The value of $\Psi_\infty^2 = 3$, obtained by their analysis, is also quite significantly different from that of Makhviladze and Novozhilov (1971) and Borisova et al. (1986).

Similar problems were considered by Borisova et al. (1990) assuming zero-order reactions with wide reaction zones. The results of this study suggest that the conclusion of weaker stability of gasless systems to two-dimensional, compared to plain, perturbations is probably quite general.

Investigation of two-dimensional combustion stability for simple condensed systems shows that stability boundaries for plain (infinite wavelength) and two-dimensional (with wavelength of the order of Michelson thickness) perturbations do not differ significantly. It may be expected, therefore, that in the case where a harmonic with the finite wavelength is least stable the relationship $r(k)$ at the stability boundary of the steady-state regime would not be significantly different from (3.29) corresponding to the case of infinite wavelength perturbation.

This consideration provides good reason to believe that the one-dimensional approach adopted in the ZN theory does not alter the final results significantly. However, the importance of further studies on the behaviour of two-dimensional perturbations within specific propellant combustion models should be emphasized.

3.4 Combustion Beyond the Stability Region

The preceding sections of this chapter investigated nonsteady combustion regimes in a linear approximation. It is natural to expect that deeper into the instability region the propellant combustion process would exhibit properties typically observed in nonlinear dynamical systems. The latter have been studied extensively over the last few decades (e.g. Landau and Lifshitz 1987; Mikhailov 2011). A number of studies have been devoted to analytical investigation of propellant combustion beyond the stability region. Nonlinearity of the problem forced considerations only in the vicinity of the stability boundary. Auto-oscillating regimes were found and investigated, both in stable and unstable regions of steady-state combustion, by Novozhilov (1966) and Novozhilov et al. (2002a). It was also shown that resonances may occur not only at the natural frequency of the burning propellant, but at its integer multiples as well (Novozhilov 1992b).

In the present section nonsteady modes of propellant combustion beyond the stability region are investigated by means of numerical analysis. The problem is considered within the framework of the ZN theory using a simple propellant model having just two parameters. Such a system is one of the simplest dynamical systems with distributed parameters.

The problem is solved by reducing the heat transfer partial differential equation to the infinite set of ordinary differential equations. This method was proposed by Novozhilov (2003, 2004) and is described in Section 2.5.

As shown in Section 2.5, the heat transfer equation

$$\frac{\partial \theta}{\partial \tau} = \frac{\partial^2 \theta}{\partial \xi^2} + v\frac{\partial \theta}{\partial \xi} \ (\xi \geq 0) \tag{3.121}$$

$$\theta(0, \tau) = \vartheta(\tau), \theta(\infty, \tau) = 0 \tag{3.122}$$

with relevant boundary conditions may be reduced to an infinite set of ordinary differential equations for the moments of temperature distribution in the condensed phase

$$y_n(\tau) = \int_0^\infty \theta(\xi, \tau) L_n(\xi) \mathrm{d}\xi \tag{3.123}$$

where

$$L_n(\xi) = \frac{e^\xi}{n!} \frac{\mathrm{d}^n}{\mathrm{d}\xi^n} e^{-\xi} \xi^n \tag{3.124}$$

are the Laguerre polynomials, orthonormal on the interval $0 \leq \xi < \infty$ with weight $e^{-\xi}$.

The set of ordinary differential equations has the form

$$\begin{aligned} y_n'(\tau) = \varphi(\tau) - n\vartheta(\tau) + \sum_{s=0}^{n-2} (n - s - 1) y_s(\tau) + \\ + v(\tau) \left(\sum_{s=0}^{n-1} y_s(\tau) - \vartheta(\tau) \right), \quad n = 0, 1, 2, \ldots \end{aligned} \tag{3.125}$$

where the temperature gradient is

$$\varphi = -\left. \frac{\partial \theta}{\partial \xi} \right|_{\xi = 0} \tag{3.126}$$

For numerical modelling, the set of equations (3.125) has to be truncated at some maximum value $n_{\max} = N$, which must be large enough to provide required accuracy.

The set of equations (3.125) has to be solved with the following restrictions (nonsteady burning laws) relating burning velocity, surface temperature, and temperature gradient

$$v = v(\varphi), \vartheta = \vartheta(\varphi) \tag{3.127}$$

The sets (3.125) and (3.127) contain all together $N + 3$ equations and $N + 4$ unknowns, that is, v, φ, and ϑ as well as $N + 1$ temperature distribution moments. The closure problem which inevitably arises upon transition from the infinite set of equations to the truncated one is not very difficult in the present case. We apply expansion (2.54) at $\xi = 0$ to get the required additional equation

$$\vartheta(\tau) = \sum_{s=0}^{N} y_s(\tau) \tag{3.128}$$

At the steady-state

$$\varphi^0 = 1, \quad \nu^0 = 1, \quad \vartheta^0 = 1$$
$$y_0^0 = 1, \quad y_n^0 = 0, \quad n = 1, 2, \dots N \tag{3.129}$$

Let us consider a specific propellant model, i.e. the set of steady-state burning laws. These are chosen in such a way that the problem contains the minimal number of parameters. The simplest choice of model (we assume that pressure remains constant) is given by the following steady-state dependencies

$$u^0 = A\exp(\beta T_a), u^0 = B\exp(\beta_s T_s^0) \tag{3.130}$$

where A, B, β, and β_s are constants. The parameters characterizing the sensitivity of the burning velocity and the surface temperature to variations of the initial temperature take the form

$$k = \beta\left(T_s^0 - T_a\right), r = \frac{\beta}{\beta_s} \tag{3.131}$$

Using the standard procedure of transformation to nonsteady dependencies (Section 2.3), the following nondimensional nonsteady burning laws are obtained

$$\nu = \exp\left[k\left(\vartheta - \frac{\varphi}{\nu}\right)\right], \nu = \exp\left[\frac{k}{r}(\vartheta - 1)\right] \tag{3.132}$$

The system under investigation contains the two parameters, namely k and r. Computations are made at the fixed value of $r = 1/3$. Transition from stable steady-state regimes to unstable ones occurs with the change of the parameter k, which serves therefore as the bifurcation parameter.

It follows from (3.28) and (3.29) that in the linear approximation the stability boundary is given by $k = 2$ with the natural frequency $\omega = 4.243$ and the period of oscillations $T = 2\pi/\omega$ equal to 1.481. It turns out that upon numerical modelling, the first bifurcation occurs at $k_1 = 1.9610$. At this value steady-state burning regime changes to an oscillating regime via the Andronov–Hopf bifurcation. The period of oscillations at the bifurcation point is equal to 1.505. Both the bifurcation value of k_1 and the period of oscillations at bifurcation are close to the values which are presented above and obtained from the theoretical linear stability analysis. This confirms good accuracy of the computations.

Figures 3.12–3.17 illustrate the system behaviour as the bifurcation parameter changes. Each figure contains two parts, (a) and (b). Part (a) shows the time history of the burning velocity, while part (b) is the two-dimensional projection of the phase trajectory on the plane (ν, y_0). We remember that the zeroth moment is a heat content of the condensed phase. The figure captions list the quantitative parameters, that is the values of the parameter k and the period of oscillations.

Figure 3.12 refers to the system which is within the region of stable steady-state combustion for values of k less than the one corresponding to the first bifurcation.

A cascade of bifurcations develops as the parameter k increases, and the period of oscillations doubles at each bifurcation. While errors associated with numerical analysis prevent quantitative description of the full infinite series of bifurcations, the following are the first several values of the parameter k (to within an absolute error of $\Delta k \approx 10^{-4}$) where successive

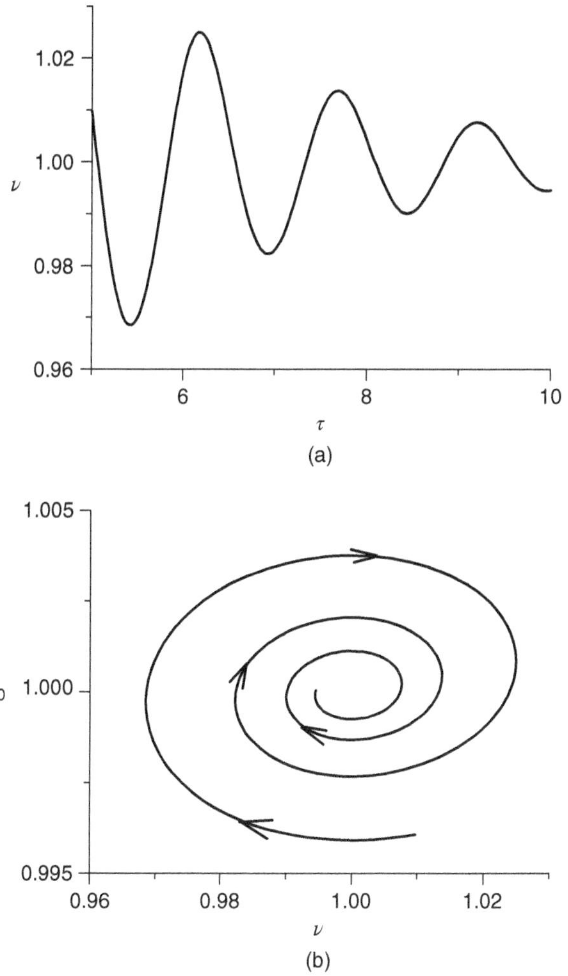

Figure 3.12 Stable steady-state regime: $k < k_1$, $k = 1.95$, and $T = 1.51$.

doubling of oscillation period occurs

$$k_1 = 1.9610, k_2 = 2.0764, k_3 = 2.0892, k_4 = 2.0919 \tag{3.133}$$

After the fourth bifurcation, as the bifurcation parameter increases slightly ($k = 2.095$), a chaotic combustion regime can be observed.

Figures 3.13–3.16 demonstrate the evolution of the system at successive monotonically increasing values of the bifurcation parameter k. The latter are chosen in such a way that the process of period doubling may be clearly seen. It is conventional (Landau and Lifshitz 1987) to refer to the obtained regimes as the T regime, the $2T$ regime and so on. Exact period doubling occurs at the bifurcation point. The period changes slightly as the bifurcation parameter increases between the two successive bifurcation points. Because of this, the factor of two difference between the periods shown in Figures 3.13–3.16 holds only approximately. Figure 3.17 shows a chaotic burning regime. It is impossible to uncover any

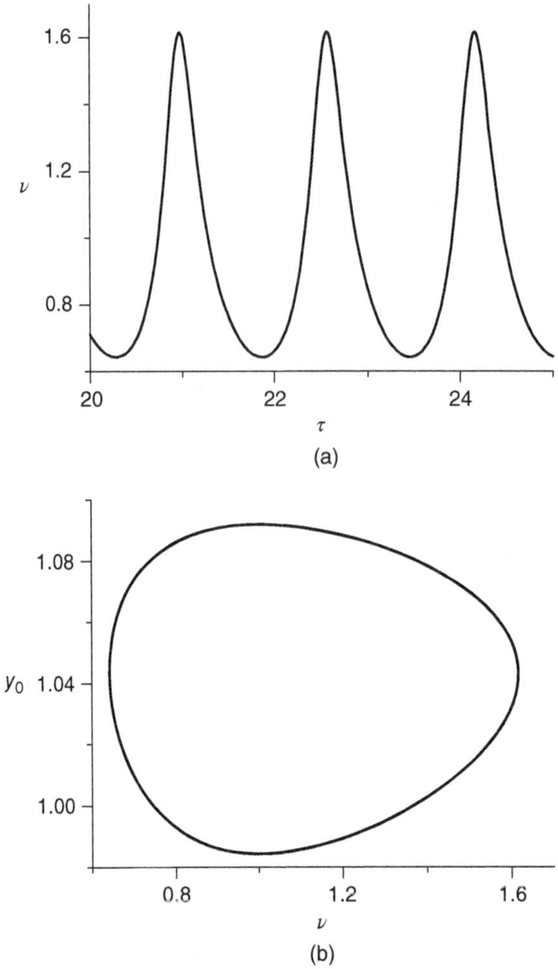

Figure 3.13 T regime: $k_1 < k < k_2$, $k = 2.05$, and $T = 1.59$.

periodicity in the burning velocity time history plot, and phase trajectories fill almost uniformly certain region in the phase space.

Successive establishment of the $2^{m-1}T$ regimes and transition to chaos is illustrated also in Figure 3.18 in a different way. Construction of the plot presented in Figure 3.18 is based on the idea, due to Lorenz, that in the chaotic regime there exists an approximate dependence

$$\mathrm{M}_{s+1} = \mathrm{P}(\mathrm{M}_s) \tag{3.134}$$

where M_s are the maximums of some function of time, related to the process. Figure 3.18 shows the maximum values of the burning velocity as a function of the preceding maximum. Thus, for the T regime there is only one maximum value, which is indicated by just one point on the plot. In the $2T$ regime there are two maxima, and correspondingly two points on the plot. In the course of the period doubling process and transition to chaos the number of points increases, and in the limit they collapse on a curve (3.134). The mapping (3.134) is approximate as the set it describes is known as a Cantor set. The other reason for the mapping (3.134) being approximate is an inevitable inaccuracy of numerical simulation.

Figure 3.14 2T regime: $k_2 < k < k_3$, $k = 2.085$, and $T = 3.35$.

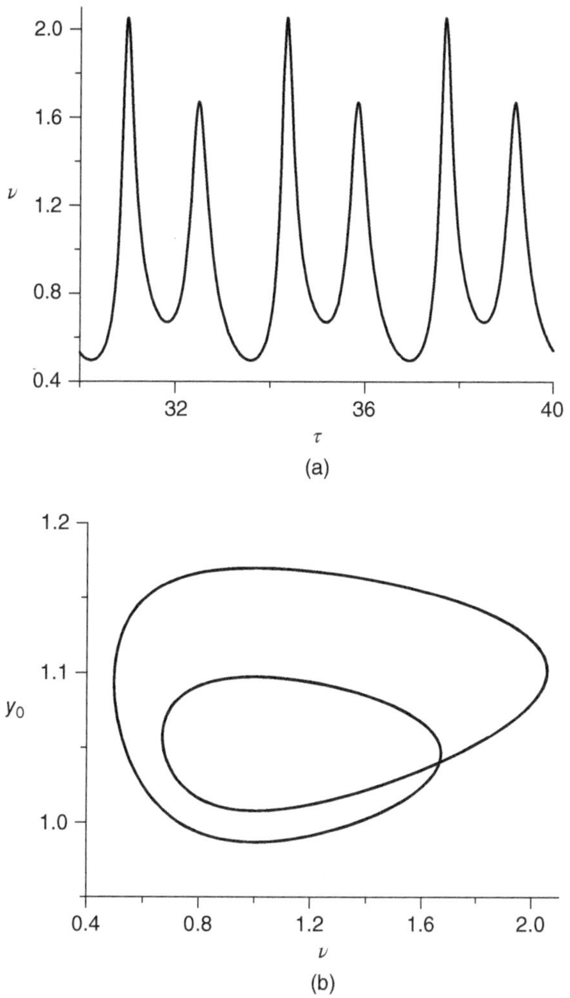

The computations clearly confirm that the transition from the stable steady-state regime to the chaotic one occurs via the classical Feigenbaum scenario of cascading bifurcations (Landau and Lifshitz 1987; Arnold 1988). In this scenario the successive bifurcation values k_m satisfy the simple law

$$\lim_{m\to\infty} \delta_m = \delta, \delta_m = \frac{k_m - k_{m-1}}{k_{m+1} - k_m} \tag{3.135}$$

where $\delta = 4.669\ldots$ is the universal Feigenbaum constant. From the bifurcation values of the parameter k provided earlier (3.133) one can estimate

$$\delta_2 = 9.0 \pm 0.2, \delta_3 = 4.7 \pm 0.4 \tag{3.136}$$

Therefore the sequence δ_m, as has been observed on many occasions, converges very quickly.

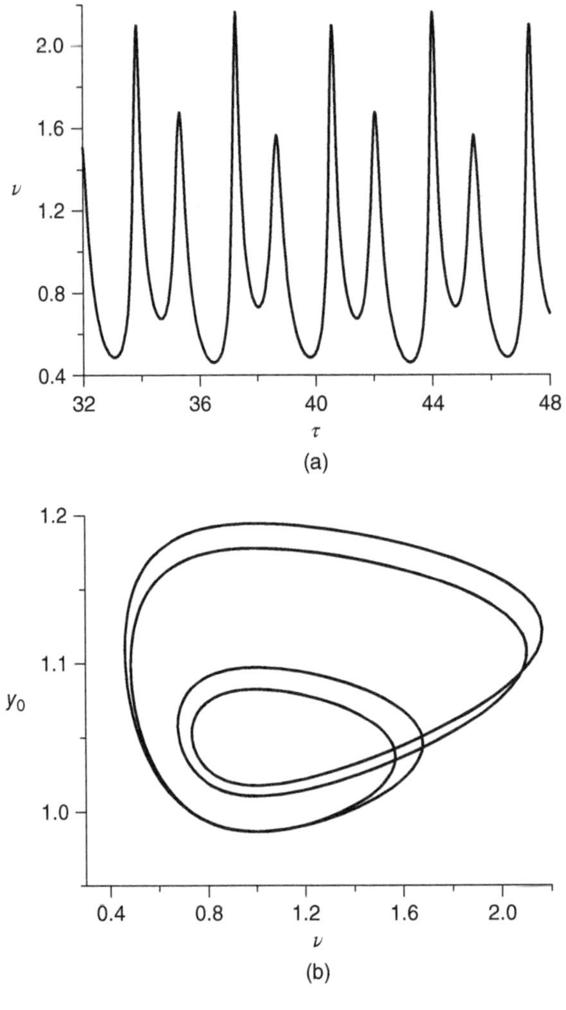

Figure 3.15 4*T* regime: $k_3 < k < k_4$, $k = 2.0905$, and $T = 6.74$.

3.5 Comparison with Experimental Data

Let us consider experimental data that can be compared to the outcomes of the theory of stability of condensed system combustion.

It should be noted, first of all, that it is impossible within the framework of the described ZN theory approach to determine whether combustion of a particular propellant would be stable or unstable at given conditions. Indeed, phenomenological relations obtained at steady-state burning conditions are an essential part of the theory. Therefore one can only judge whether the fact of stable propellant burning and the values of parameters, measured at such a stable regime and involved in the stability criterion, are mutually consistent.

For example, in the case of constant surface temperature of the propellant the theory would be consistent with the experiment if the measured value $k = \beta(T_s^0 - T_a)$ satisfies the condition $k < 1$.

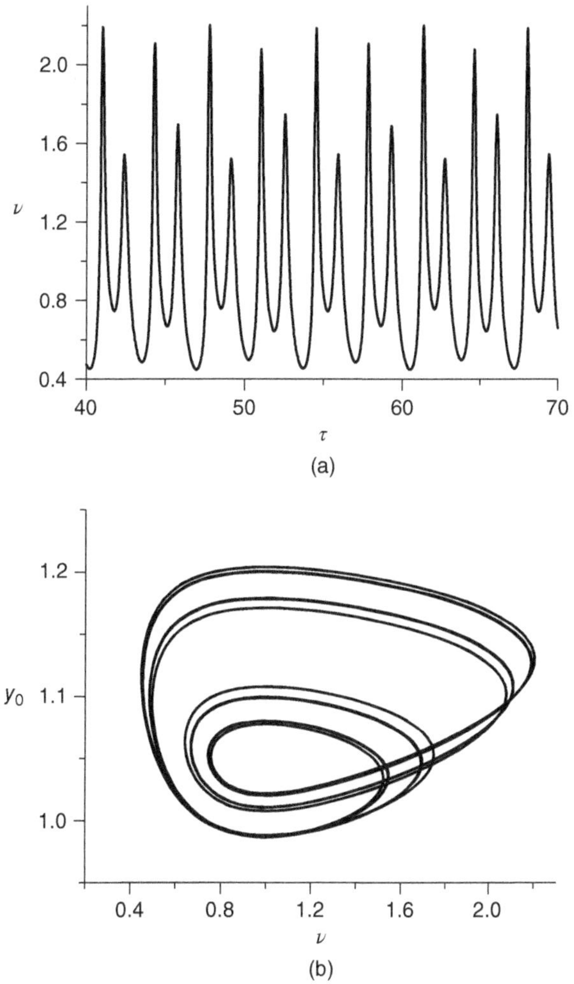

Figure 3.16 8T regime: $k > k_4$, $k = 2.0923$, and $T = 13.51$.

Strictly speaking, testing the theory with regard to its consistency with observed data can only be carried out using stable regimes of combustion. Taking again an example of combustion at constant surface temperature, the values of the parameter k measured at unstable combustion regimes should not be necessarily greater than unity. The theory does relate the value of the parameter k and the fact of stable combustion, but does not relate observable instability with the values of this parameter measured at such an unstable regime.

The same is true for the case of variable surface temperature. One can agree with the theory and the experiment only if at some steady-state combustion regime the condition

$$r > \frac{(k-1)^2}{k+1} \tag{3.137}$$

is fulfilled where the values of k and r are obtained by measurements at the very same regime.

In the unstable regime where burning velocity, surface temperature, and temperature distribution in the condensed phase all change in time the measured parameters k and r would

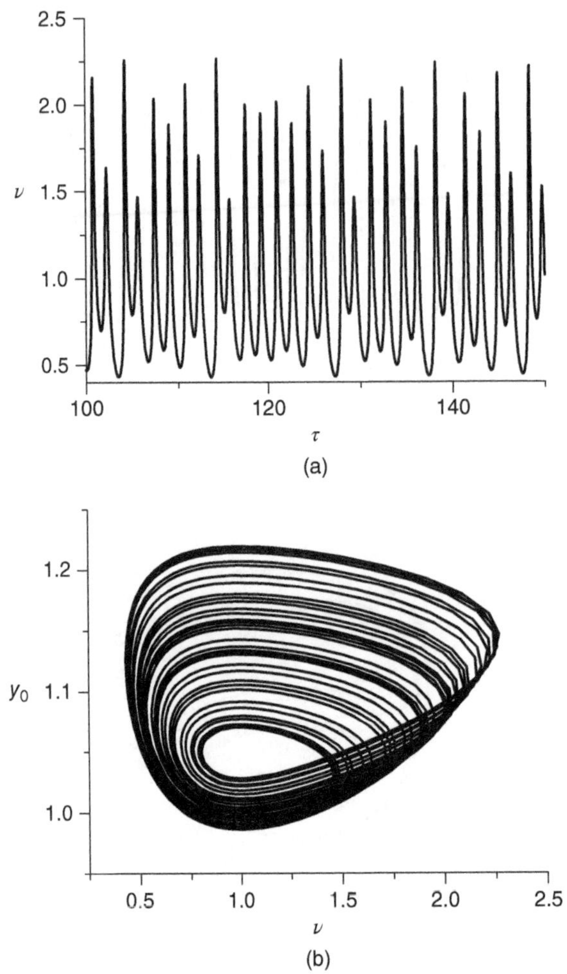

Figure 3.17 Chaotic combustion regime: $k = 2.0952$.

be related to this particular observed regime and not to the unstable steady-state regime the observed regime has originated from. Consequently, the parameters k and r measured in the unstable combustion regime are not obliged to satisfy the inequality similar to (3.137), but with the inequality sign reversed, i.e. $r < (k-1)^2/(k+1)$.

Kondrikov (1969) summarizes the combustion data of some nitrate esters (where surface temperature is constant and close to boiling temperature) in order to determine the value of the parameter k and relate this to the type of regime, either stable or unstable, which is observed in the experiment. Data are presented on the temperature coefficients of the burning rate β and the surface temperatures of methyl nitrate, nitroglycol, nitroglycerin, diethylene glycol dinitrate, and pentaerythritol tetranitrate (PETN) at atmospheric pressure. This data are used to derive the value of the parameter k for each of these substances. The k values for methyl nitrate and nitroglycol, at the initial temperature of 20 °C, turned out to be 0.34 and 0.90, respectively. The experiment demonstrates stable burning of these two nitrate esters at such conditions. For the other three nitrate esters, that is, nitroglycerin,

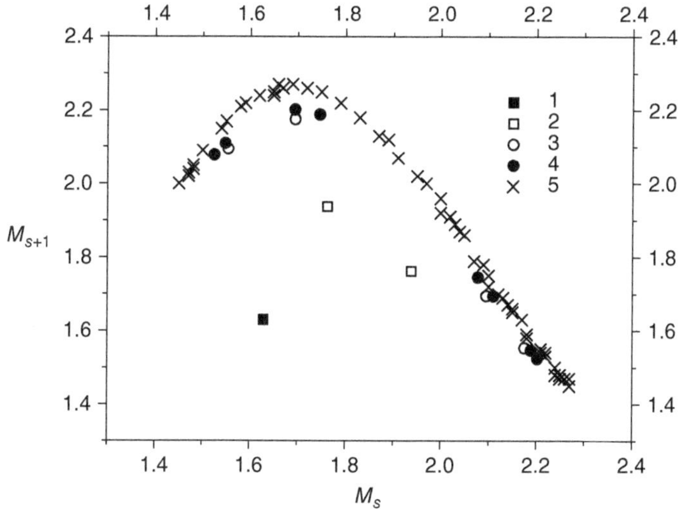

Figure 3.18 Succession of period doubling and transition to chaos upon increase of the parameter k. 1, T regime, $k = 2.05$; 2, $2T$ regime, $k = 2.08$; 3, $4T$ regime, $k = 2.091$; 4, $8T$ regime, $k = 2.0922$; 5, chaotic regime, $k = 2.095$.

diethylene glycol dinitrate, and PETN, the values of the parameter k exceed unity (1.29, 1.13, and 1.5, respectively). The paper by Kondrikov (1969) discusses the methodology of experimental burning of the latter substances and concludes that their combustion at atmospheric pressure and room temperature is quite unstable. It would probably only be sustained in the presence of an additional heat source.

Thus, for methyl nitrate and nitroglycol, the values of the parameter k and observed steady-state stable burning regimes conform with the conclusions of the theory.

In the case of a variable surface temperature, a comparison of the stability criterion (3.137) with experimental data is only possible if the dependence $T_s^0(T_a)$ is known as the criterion involves derivative of such a function. Such a comparison was carried out after the development of an appropriate thermocouple technique, which allowed data on the effects of the initial temperature on the surface temperature of burning propellant to be obtained (Zenin and Nefedova 1967; Zenin et al. 1966). Obviously, scattering of the measured data $T_s^0(T_a)$ prevents a very accurate estimation of the derivative r (1.14) of this function. Moreover, the obtained value of the derivative depends on the specific data processing procedure. It has been shown by Zenin and Nefedova (1967) that scattering in the surface temperature data allows the function $T_s^0(T_a)$ to be interpolated by a linear dependence. This leads to the derivative r being constant. Such a procedure, however, results in the estimated activation energy of the chemical reaction in the condensed phase being variable. For this reason, another approach to estimating the derivative r, based on the assumption of the activation energy being constant, has been proposed (Zenin and Nefedova 1967; Zenin et al. 1966). The idea of the method is to plot experimental data in the coordinates $(1/T_s^0, \ln u^0)$. Straight line approximating experimental points correspond to the Arrhenius dependence of the reaction rate on the surface temperature. In this case the derivative r turns out to be variable and increases with increasing initial temperature. Such a data processing technique used

Table 3.2 The parameters *k*, *r* and the critical values r^* for ballistite N at $p = 20$ atm.

Parameter	Unit	Values						
T_a	°C	−150	−100	−50	0	50	100	140
k		0.2	0.75	1.2	1.6	2.0	2.3	2.7
r		0.2	0.2	0.2	0.3	0.4	0.6	0.9
r^*		0.0	0.0	0.018	0.14	0.33	0.51	0.78

the formula (1.80), which corresponds to the zero-order chemical reaction with the heat flux from the gaseous to the condensed phase.

As was previously mentioned, at atmospheric pressure ballistite N burns in a stable regime only at elevated initial temperatures. Even at $T_a = 50\,°\text{C}$ oscillations of the temperature profile can be observed. The value of r obtained at $T_a = 100\,°\text{C}$ was 0.9. The maximum value of the parameter k (3.137) for such a value of r is 2.85. The experimental value $k = 2.7$ is somewhat less but close to the latter. Thus the stability of combustion at $p = 1$ atm and $T_a = 100\,°\text{C}$ conforms with the criterion (3.137). At such conditions combustion is likely to occur in the vicinity of the stability boundary.

At a pressure of 20 atm where stable combustion occurs at all the investigated initial temperatures the criterion (3.137) is fulfilled. Table 3.2 lists the relevant parameters k and r as well as the critical values r^*. The inequality $r > r^*$, which corresponds to stable combustion, holds in the entire temperature interval.

It is instructive to follow the relation between burning rate and the surface temperature gradient (at the condensed phase side) f. In the theory with constant surface temperature, the stability of the combustion regime corresponds to the condition $(\partial u/\partial f)_p < 0$.

The situation is different if the surface temperature varies. In this case, stable combustion is possible with both positive and negative signs of the burning rate derivative with respect to gradient (Novozhilov 1967a). Indeed,

$$\left(\frac{\partial u}{\partial f}\right)_p = \frac{k}{k+r-1}\frac{u}{f} \tag{3.138}$$

therefore the sign of the derivative at $k > 0$ coincides with the sign of the expression $k + r - 1$ which, if the stability criterion (3.137) is fulfilled, may be either positive or negative. A similar conclusion is true for the derivative of the surface temperature with respect to gradient

$$\left(\frac{\partial T_s}{\partial f}\right)_p = \frac{r}{k+r-1}\frac{T_s - T_a}{f} \tag{3.139}$$

as the sign of this value at $r > 0$ is also determined by the sign of the same expression $k + r - 1$.

The experimental data presented in Table 3.3 allow the dependence $u^0(f^0)$ for ballistite N to be plotted for a pressure of 20 atm. To fulfil this task, one has to use the relationship

$$f^0 = \frac{u^0}{\kappa}(T_s^0 - T_a) \tag{3.140}$$

Table 3.3 Burning rate and surface temperature of ballistite N as a function of initial temperature at $p = 20$ atm.

Parameter	Unit	Values						
T_a	°C	−150	−100	−50	0	50	100	140
u	cm/s	0.18	0.19	0.215	0.27	0.355	0.49	0.76
T_s	°C	275	275	290	310	340	375	410

Table 3.4 Parameters of the combustion zones of ballistite N as a function of initial temperature at $p = 20$ atm.

Parameter	Unit	Values						
T_a	°C	−150	−100	−50	0	50	100	140
Q_s	J/g	501.6	426.4	384.6	359.5	359.5	351.1	363.7
q	J/g	121.2	121.2	112.9	92.0	71.1	50.2	33.4
Q_b'	J/g	1295.8	1337.6	1379.4	1400.3	1442.1	1483.9	1504.8
T_b'	°C	1160	1200	1260	1320	1360	1400	1440
T_b	°C	1500	1550	1600	1650	1690	1740	1780

An alternative method is to use the data in Table 3.4 and then find the dependence $u^0(f^0)$ from the relationship

$$f^0 = \frac{u^0}{\kappa c}(Q_s + q) \tag{3.141}$$

Naturally, both approaches deliver identical results.

Figure 3.19 demonstrates that at initial temperatures $T_a > -100\,°\mathrm{C}$ the burning rate increases as the gradient increases. At $T_a \approx -100\,°\mathrm{C}$ the gradient probably achieves its minimum value. Further decrease of the initial temperature causes the derivative of the burning rate with respect to gradient to become negative. This conclusion should be considered, of course, with caution, as there is only one experimental measurement in the region with a negative temperature ($T_a = -150\,°\mathrm{C}$). In addition, in this region changes of the gradient are comparable with the errors in its calculation. However, under the quite plausible assumption of monotonic dependence of the parameters k and r on the initial temperature, the existence of the minimum of the gradient is inevitable. The value $k + r - 1$ changes sign as the initial temperature decreases.

Note that in the model with a constant surface temperature and a constant temperature coefficient of the burning rate (Section 2.2), the parameter $k = \beta(T_s^0 - T_a)$ increases as the initial temperature decreases. The stability limit is achieved at a rather low value of the initial temperature.

On the contrary, the experimental data for ballistite N suggest that the temperature coefficient of the burning rate increases as does the initial temperature, and specifically in such a way that the product of the temperature coefficient with the difference $T_s^0 - T_a$ turns out

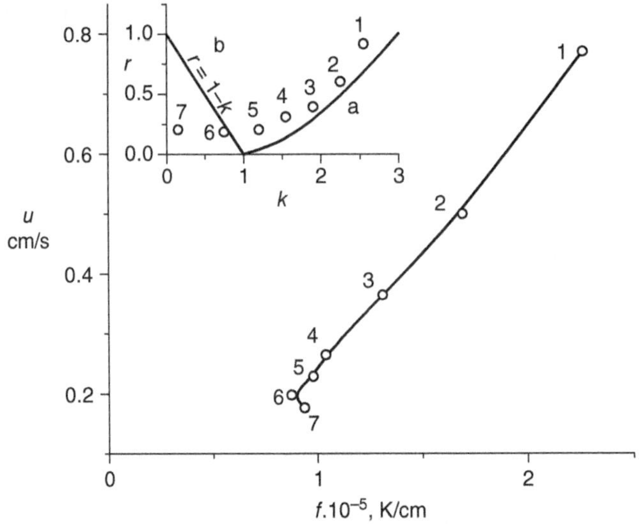

Figure 3.19 Burning rate dependence on the surface temperature gradient for ballistite N at different values of T_a: 1, 140 °C; 2, 100 °C; 3, 50 °C; 4, 0 °C; 5, −50 °C; 6, −100 °C; 7, −150 °C. Region a, unstable combustion regimes; region b, stable combustion regimes.

to be an increasing function of the initial temperature. Therefore, low initial temperatures correspond to small values of k and the question of how the stability boundary for ballistite propellants can be achieved remains open. As has been demonstrated, an increase in the initial temperature changes the parameters k and r in such a way that propellant is forced into the region of stable combustion. The initial temperature can only be increased up to a certain limit as at sufficiently high values volumetric thermal decomposition of the propellant occurs. On the other hand, the propellant also burns in a stable mode at quite low temperatures (down to −150 °C). If the stability criterion is correct then a further decrease of T_a would not result in instability (as in this region $k < 1$). It may well turn out that combustion will be stable even at a temperature of absolute zero. However, it may become impossible due to reasons not related to stability. At sufficiently low initial temperatures the temperatures of the combustion zones in the condensed and gaseous phases would not be high enough to sustain chemical reactions. Experimental investigation of propellant combustion at low temperatures is of considerable interest due to observed combustion quenching at rapid pressure drops. The existence of a minimal, critical initial temperature for combustion would possibly assist in understanding of such a phenomenon.

Apart from ballistite N, the stability criterion for the steady-state regime was also verified using ammonium perchlorate. Currently, this compound is quite often used as an oxidizer in composite propellants and its combustion behaviour has been studied extensively. In particular, the anomalous dependence of the burning rate on pressure was discovered by Friedman et al. (1957) and Glazkova (1963). The burning rate grows with pressure at low pressures, while dropping to some minimum afterwards, and then growing again.

Figure 3.20 shows the dependence of the burning rate on pressure (Glazkova 1963) at room temperature. It is evident that at a pressure of ~150 atm burning rate growth changes to rapid decay. Temperature distribution in the preheat and chemical reaction zones

Figure 3.20 Burning rate of ammonium perchlorate as a function of pressure.

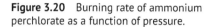

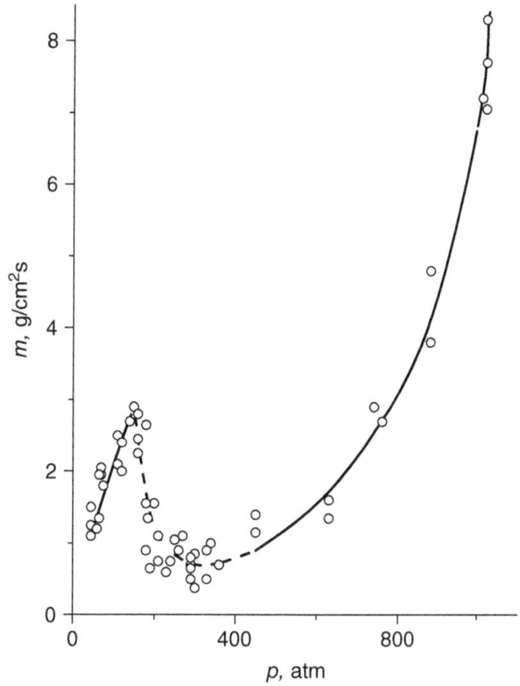

during combustion of ammonium perchlorate was studied by Bobolev et al. (1964). The temperature profile obtained by thin thermocouples confirmed the conclusion of Glazkova (1963) on the existence of the two combustion regimes. The first is realized at pressures of 40–150 atm, where the burning rate grows with pressure. Temperature profiles in this case look rather usual and similar, for example, to those obtained for ballistite N. Combustion within this pressure range is quite stable as the burning rate and temperature distribution are time-independent. The second regime occurs at higher pressures of 160–450 atm and features temperature pulsations within a flame zone, which is an indication of unstable combustion. The study of Bobolev et al. (1964) also discovered anomalous behaviour of the surface temperature as a function of pressure. In contrast to ballistite N, the surface temperature of ammonium perchlorate drops as the pressure rises.

The transition from stable to unstable combustion can be explained from the point of view of the stability criterion (3.137). In this regard the influence of the initial temperature on the parameters of the reaction zones of ammonium perchlorate was studied by Glazkova et al. (1970). Burning rate and surface temperature were measured at various pressures and initial temperatures. This data allowed derivatives involved in the stability criterion to be estimated and the validity of the criterion to be judged. Table 3.5 lists the experimental data and estimated values of the parameters k and r (Glazkova et al. 1970).

The value $r^* = (k-1)^2/(k+1)$ is a critical value, at given k, of the parameter r. At this value the system is at the stability boundary. Stable regimes correspond to $r > r^*$. For $k < 1$ the critical value is $r^* = 0$. Measurements were conducted for three different values of the initial temperature (20, 50 and 100 °C) and three values of pressure (50, 100, and 150 atm).

Table 3.5 Parameters k, r, r^* and other experimental data for ammonium perchlorate.

p (atm)	T_s^0(°C)	$\beta \times 10^3$ (°C)$^{-1}$	k	r	r^*
$T_a = 20\,°C$					
50	440	2.7	1.1	0.8	0.005
100	340	7.5	2.4	0.6	0.58
150	320	9.8	2.9	0.4	0.93
$T_a = 50\,°C$					
50	460	2.7	1.2	0.8	0.018
100	360	2.5	0.85	0.6	0.0
150	330	0.88	0.27	0.4	0.0
$T_a = 100\,°C$					
50	500	2.7	1.3	0.8	0.039
100	390	2.5	0.92	0.6	0.0
150	355	0.88	0.28	0.4	0.0

Unstable combustion was only observed at $T_a = 20\,°C$. An increase in the initial temperature widens the region of stable combustion of ammonium perchlorate substantially.

At initial temperature $T_a = 20\,°C$ the parameter k grows with increasing pressure, while the derivative of the surface temperature with respect to the initial temperature drops. At pressures greater than 100 atm the parameter r reaches its critical value and combustion instability develops, in agreement with the experiment. At high initial temperatures, due to a significant decrease in the temperature coefficient of the burning rate, the system moves into the region $k < 1$, where combustion is stable at any value of the parameter r.

The mechanism of ammonium perchlorate combustion consists of two processes at the phase interface, exothermic decomposition reaction and evaporation, as proposed by Manelis and Strunin (1971). Another study by the same authors (Strunin and Manelis 1971) investigated the stability of the steady-state combustion regime under the condition that combustion is controlled by the reaction in the condensed phase. It is assumed that the reaction is zero order and the surface temperature is determined by the volatility of the substance. Such a model results in specific expressions for the parameters involved in the stability criterion. It turns out that the parameters k and r are related to kinetic as well as physicochemical properties, such as activation energy, latent heat of vaporization, heat of combustion, and initial temperature. Substitution of these parameters into the stability criterion (3.137) leads to the conclusion that there are three possibilities, depending on the parameter magnitudes: (i) combustion is stable at all values of pressure; (ii) combustion is stable at low and high pressures and unstable in the intermediate region; and (iii) combustion is stable at low pressures and unstable at high pressures. It should be pointed out, however, that such an approach is oversimplified. At high pressures, gas phase inertia starts to play a significant role and the criterion (3.137) is not applicable.

One of the major conclusions of the present chapter is a possibility of auto-oscillations of propellant combustion under constant pressure. Such regimes were indeed observed experimentally, the earliest results being reported by Eisel et al. (1964), Maksimov (1964), Svetlichnyi et al. (1971), and Simonenko et al. (1980). Auto-oscillating regimes were identified by detecting different phenomena, such as pulsations of luminosity, temperature, flame, pressure or electrical conductivity in the combustion zone.

The authors of the above-mentioned papers interpreted their findings in different ways. In the study by Eisel et al. (1964) the cause of auto-oscillations was related to a high degree of fuel metallization. Maksimov (1964) studied combustion of ballistite N and explained the phenomenon by the growth and subsequent shedding of soot agglomerates from the propellant surface. On the other hand, the studies by Svetlichnyi et al. (1971) and Simonenko et al. (1980) correctly attributed auto-oscillations to the existence of the natural frequency

$$f \sim \frac{1}{2\pi}\frac{\sqrt{k}}{r}\frac{u^2}{\kappa} \tag{3.142}$$

of propellant combustion. This formula, however, may only qualitatively conform with experimental measurements. First, the parameters k and r are known with rather low accuracy. Second, the auto-oscillation spectrum is very complicated. The presented theory, as can be seen in Section 3.4, leads to the consideration of nT regimes (with various integers n), while in the experiments complicated spectrums are observed with ambiguity of oscillation frequency in some ranges of pressure variation.

There is little doubt that there is a need to study auto-oscillation combustion regimes more precisely, both theoretically and experimentally.

4

Combustion Under Harmonically Oscillating Pressure

4.1 Linear Burning Rate Response to Harmonically Oscillating Pressure

This chapter considers propellant combustion under the conditions of periodically varying pressure. The importance of such processes is warranted by the necessity to understand the reasons behind deviations of actual burning regimes in solid rocket engines from designed steady-state regimes. Among observed nonsteady effects, the are soft or hard (triggering) excitations of burning rate and pressure oscillations. Widespread usage of solid rocket engines proves that there are practical empirical methods of eliminating these unwelcome effects which normally appear early in the design stage. Several papers in the book by DeLuca et al. (1992) are devoted exactly to this aspect of solid rocket motor instability. The same reference suggests that theoretical understanding and the ability to explain various nonsteady processes observed during laboratory tests and practical applications of propellants are much more modest. The main difficulty, as pointed out by Novozhilov (2005), is to provide the correct description of interaction between combustion and an acoustic field. This is the problem that is discussed in the present chapter. It is natural to start with the linear approximation.

The set of Eqs (2.33–2.36) describing nonsteady combustion is, in general, nonlinear. The heat transfer equation involves nonlinear heat flux, which is a product of the burning rate and the temperature gradient. Similarly, both the steady-state and the nonsteady burning laws are in the general case nonlinear. For small deviations from the steady-state regime burning rate response to oscillating pressure can be found analytically in the linear approximation of the Zeldovich–Novozhilov (ZN) theory.

We need, first of all, a definition of the linear burning rate response.

If pressure near the burning surface oscillates with a small amplitude

$$p = p^0 + p_1 \cos \Omega t, \quad p_1 << p^0 \tag{4.1}$$

then the linear burning rate of the propellant would oscillate with the same frequency but with some phase shift, compared to pressure

$$u = u^0 + u_1 \cos(\Omega t + \psi), \quad u_1 << u^0 \tag{4.2}$$

Theory of Solid-Propellant Nonsteady Combustion, First Edition. Boris V. Novozhilov and Vasily B. Novozhilov.

Companion website: www.wiley.com/go/Novozhilov/solidpropellantnonsteadycombustion

The complex function

$$U(\Omega) = \frac{(u_1/u^0)}{(p_1/p^0)} \exp(i\psi) \tag{4.3}$$

is called the linear response function of the burning rate to oscillating pressure.

In the linear approximation it is convenient to use the method of nondimensional complex amplitudes. Within this approach

$$\eta = 1 + [\eta_1 \exp(i\omega\tau) + c.c.], \quad v = 1 + [v_1 \exp(i\omega\tau) + c.c.] \tag{4.4}$$

where

$$\omega = \frac{\Omega\kappa}{(u^0)^2}, \quad \eta_1 = \frac{p_1}{2p^0}, \quad v_1 = \frac{u_1}{2u^0}\exp(i\psi) \tag{4.5}$$

and *c.c.* stands for complex conjugate. In these variables the response function takes the form

$$U(\omega) = \frac{v_1}{\eta_1} \tag{4.6}$$

In the linear approximation the response function under constant pressure conditions was found by Novozhilov (1965b).

The following analysis make use of nondimensional variables

$$\tau = \frac{(u^0)^2 t}{\kappa}, \quad \xi = \frac{u^0 x}{\kappa}, \quad v = \frac{u}{u^0}, \quad \eta = \frac{p}{p^0} \tag{4.7}$$

$$\theta = \frac{T - T_a}{T_s^0 - T_a}, \quad \vartheta = \frac{T_s - T_a}{T_s^0 - T_a}, \quad \varphi = \frac{f}{f^0} \tag{4.8}$$

Within the framework of the ZN theory the nonsteady propellant combustion regime is investigated by solving the heat transfer equation with the appropriate boundary conditions

$$\frac{\partial\theta}{\partial\tau} = \frac{\partial}{\partial\xi}\left(\frac{\partial\theta}{\partial\xi} - v\theta\right), \quad -\infty < \xi \le 0$$

$$\xi \to -\infty, \quad \theta = 0, \quad \xi = 0, \quad \theta = \vartheta(\tau) \tag{4.9}$$

We shall consider only stabilized combustion regimes, therefore consideration of the initial conditions is not necessary.

Under the steady-state regime, that is for $\eta^0 = 1$

$$\theta^0 = e^{\xi}, \quad \varphi^0 = 1, \quad v^0 = 1, \quad \vartheta^0 = 1 \tag{4.10}$$

All the time-dependent variables are expressed in the form analogous to (4.4), that is, as a sum of the steady-state value and a small harmonic perturbation. In this way

$$\theta(\xi, \tau) = e^{\xi} + [\theta_1(\xi)\exp(i\omega\tau) + c.c.] \tag{4.11}$$

Substitution of the burning rate and condensed phase temperature expansions into the heat transfer Eq. (4.9) leads, on neglecting quadratic terms, to the linear ordinary differential equation

$$\theta_1'' - \theta_1' - i\omega\theta_1 = v_1 e^{\xi} \tag{4.12}$$

with the solution

$$\theta_1 = Ce^{z_1\xi} - \frac{v_1}{i\omega}e^{\xi} \tag{4.13}$$

where

$$z_1 = \frac{1}{2}(1 + \sqrt{1 + 4i\omega}), \quad z_1(z_1 - 1) = i\omega \tag{4.14}$$

The solution (4.13) satisfies the boundary condition at $\xi \to -\infty$. The second of the boundary conditions (4.9) allows the linear corrections for the surface temperature and its gradient to be found

$$\vartheta_1 = C - \frac{v_1}{i\omega}, \quad \varphi_1 = Cz_1 - \frac{v_1}{i\omega} \tag{4.15}$$

Elimination of the integration constant C results in

$$z_1\vartheta_1 - \varphi_1 + \frac{v_1}{z_1} = 0 \tag{4.16}$$

This is a relation between the linear corrections for the burning rate, surface temperature and its gradient, obtained from the linearized heat transfer equation.

Two additional relations between the above corrections are deducible from nonsteady burning laws. As has been already pointed out, steady-state dependencies of the burning rate and surface temperature on pressure and initial temperature are the key elements of the theory. These are written down in the form (retaining the superscript zero for steady-state regime)

$$u^0 = F_u(p^0, T_a), \quad T_s^0 = F_s(p^0, T_a) \tag{4.17}$$

and may be found either from steady-state combustion experiments or from some specific theoretical propellant combustion model.

The steady-state burning laws (4.17) may be transformed into the nonsteady ones

$$u = F_u\left(p, T_s - \kappa\frac{f}{u}\right), \quad T_s = F_s\left(p, T_s - \kappa\frac{f}{u}\right) \tag{4.18}$$

The method of transformation from steady-state to nonsteady burning laws was described in Section 2.3.

In principle, nonsteady burning laws (4.18) always allow explicit dependencies to be written down

$$v = v(\varphi, \eta), \quad \vartheta = \vartheta(\varphi, \eta) \tag{4.19}$$

in nondimensional variables, which, on expansion, provide the desired relations between linear corrections

$$v_1 = \left(\frac{\partial v}{\partial \varphi}\right)_\eta \varphi_1 + \left(\frac{\partial v}{\partial \eta}\right)_\varphi \eta_1, \quad \vartheta_1 = \left(\frac{\partial \vartheta}{\partial \varphi}\right)_\eta \varphi_1 + \left(\frac{\partial \vartheta}{\partial \eta}\right)_\varphi \eta_1 \tag{4.20}$$

The derivatives are taken at the steady-state burning regime. Let us express these derivatives through experimentally observable quantities.

Steady-state combustion experiments allow the burning law dependencies $u^0(p^0, T_a)$ and $T_s^0(p^0, T_a)$ to be measured and subsequently the derivatives

$$k = (T_s^0 - T_0)\left(\frac{\partial \ln u^0}{\partial T_a}\right)_{p^0}, \quad r = \left(\frac{\partial T_s^0}{\partial T_a}\right)_{p^0}$$
$$\iota = \left(\frac{\partial \ln u^0}{\partial \ln p^0}\right)_{T_a}, \quad \mu = \frac{1}{T_s^0 - T_0}\left(\frac{\partial T_s^0}{\partial \ln p^0}\right)_{T_a}$$
$$\delta = \iota r - \mu k \tag{4.21}$$

to be obtained. These derivatives describe, in the linear approximation, the sensitivity of the burning rate and surface temperature to pressure and initial temperature variations.

Transformation from derivatives (4.21) to those involved in (4.20) is readily done using Jacobians. From the Michelson relation for the steady-state combustion regime (Section 2.3)

$$\kappa f^0 = u^0(T_s^0 - T_a) \tag{4.22}$$

it follows that

$$(T_s^0 - T_a)\left(\frac{\partial \ln f^0}{\partial T_a}\right)_{p^0} = k + r - 1, \quad \left(\frac{\partial \ln f^0}{\partial p^0}\right)_{T_a} = \iota + \mu \tag{4.23}$$

Consequently, for example

$$\left(\frac{\partial \ln u^0}{\partial \ln p^0}\right)_{f^0} = \frac{\partial(\ln u^0, \ln f^0)}{\partial(\ln p^0, \ln f^0)} = \frac{\partial(\ln u^0, \ln f^0)/\partial(\ln p^0, T_a)}{\partial(\ln p^0, \ln f^0)/\partial(\ln p^0, T_a)} =$$
$$= \frac{\partial(\ln u^0, \ln f^0)/\partial(\ln p^0, T_a)}{(\partial \ln f^0/\partial T_a)_{p^0}} = \frac{\iota(r-1) - \mu k}{k + r - 1} \tag{4.24}$$

In the same way, the following relations may be obtained

$$\left(\frac{\partial \ln u^0}{\partial \ln p^0}\right)_{f^0} = \frac{\iota(r-1) - \mu k}{k + r - 1}, \quad \frac{1}{T_s^0 - T_a}\left(\frac{\partial T_s^0}{\partial \ln p^0}\right)_{f^0} = \frac{\mu(k-1) - \iota r}{k + r - 1}$$
$$\left(\frac{\partial \ln u^0}{\partial \ln f^0}\right)_{p^0} = \frac{k}{k + r - 1}, \quad \frac{1}{T_s^0 - T_a}\left(\frac{\partial T_s^0}{\partial \ln f^0}\right)_{p^0} = \frac{r}{k + r - 1} \tag{4.25}$$

The relations (4.20) take now the form

$$v_1 = \frac{k\varphi_1 + (\delta - \iota)\eta_1}{k + r - 1}, \quad \vartheta_1 = \frac{r\varphi_1 - (\delta + \mu)\eta_1}{k + r - 1} \tag{4.26}$$

The three relations (4.16) and (4.26) between the linear corrections lead to the following expression for the linear response function of the burning rate to oscillating pressure $U = v_1/\eta_1$

$$U = \frac{\iota + \delta(z_1 - 1)}{1 + (z_1 - 1)(r - k/z_1)} \tag{4.27}$$

It is important to note that the response function involves only the parameters of the steady-state combustion regime. These parameters can be found through experimentation. This circumstance allows the response of propellant to varying pressure to be calculated without consideration of the details of chemical transformations and transport processes in the flame front.

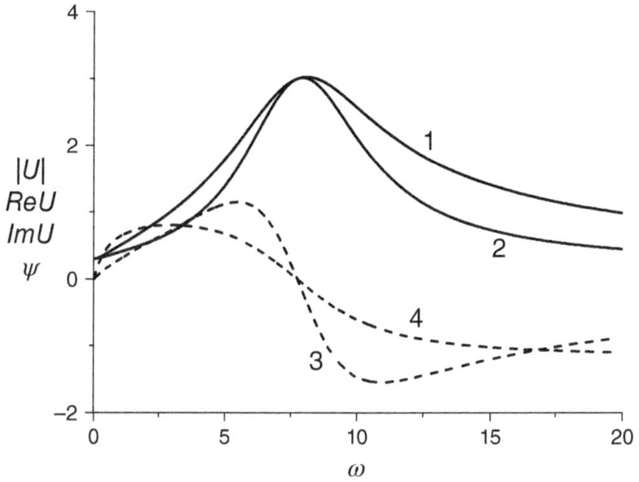

Figure 4.1 Linear response function of burning rate. $k = 1.5$, $r = 0.15$, $\iota = 0.3$, $\delta = 0.$ 1, modulus; 2, real part; 3 imaginary part; 4, argument.

Figure 4.1 shows, for a realistic set of steady-state regime parameters, the real and imaginary parts of the response function, as well as its modulus and argument. First of all, the resonance-type behaviour of the response function needs be pointed out. The maximum value of the real part is achieved at a frequency that is close to the natural propellant frequency ω_1 (3.28), and the resonance value is several times larger than the steady-state response $U(0)$.

The resonance behaviour of the function $U(\omega)$ is determined by its denominator, depending only on the parameters k and r, which describe the stability boundary for the steady-state regime (3.29). For some reasons which are not yet completely understood, a steady-state regime for the majority of volatile condensed systems occurs near its stability boundary. Therefore, for typical values of the parameters k and r the modulus of the denominator of the response function is small compared to unity, which leads to resonance.

Let us consider now an important question of the influence of experimental errors in measuring the propellant steady-state parameters on the accuracy of calculation of the response function (Novozhilov 2000). The resonance shape of the latter results in high sensitivity to the parameters k and r, that is, small changes in these parameters lead to large variations of the denominator and consequently of the response function itself. This may be illustrated if the response function is written in the form

$$U = \frac{N}{D}, \quad N = \iota + \delta(z_1 - 1), \quad D = 1 + \left(r - \frac{k}{z_1}\right)(z_1 - 1) \tag{4.28}$$

The relative error in calculation of this function

$$\frac{\Delta U}{U} = \frac{\Delta N}{N} + \frac{\Delta D}{D} \tag{4.29}$$

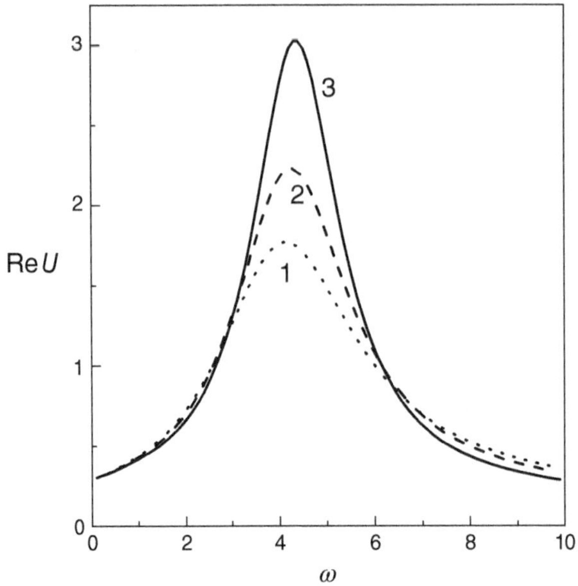

Figure 4.2 Effect of variation of parameter k on the real part of the burning rate response function. $\imath = 0.3$, $r = 0.3$, $\delta = 0.1$, $k = 1.70$; 2, $k = 1.75$; 3, $k = 1.80$.

is determined largely (as the denominator is small) by the second term. For the purpose of illustration, let us take $\Delta k = 0$, then

$$\frac{\Delta U}{U} \approx \frac{r(z_1 - 1)}{D} \frac{\Delta r}{r} \tag{4.30}$$

and a relatively small error in the measurement of the parameter r, for example, $\Delta r/r = 10\%$, would lead to the relative error in the response function being several times larger (near the resonance) than the latter. This point is illustrated in Figures 4.2 and 4.3. Figure 4.4 also shows the strong influence of the parameter δ on the response function.

Let us discuss briefly the magnitudes of experimental errors encountered typically in steady-state propellant combustion experiments. Unfortunately, existing publications do not pay due attention to this important issue. The error in estimation of the parameter $k = \beta(T_s^0 - T_a)$ is largely determined by the measurement error of the temperature sensitivity coefficient of the burning rate

$$\beta = \left(\frac{\partial \ln u^0}{\partial T_a}\right)_{p^0} \tag{4.31}$$

By the order of magnitude $\beta \approx 3 \text{ x } 10^{-3}\,\mathrm{K}^{-1}$. Its experimental estimation is based on the measurements of propellant burning rate at different initial temperatures $T_{a,1}$ and $T_{a,2}$

$$\beta = \frac{\ln u_2^0 - \ln u_1^0}{T_{a,2} - T_{a,1}} \tag{4.32}$$

An absolute error of this method is essentially related to the relative error occurring in the measurement of temperature sensitivity of the burning rate

$$\Delta\beta = \frac{2}{T_{a,2} - T_{a,1}} \frac{\Delta u^0}{u^0} \tag{4.33}$$

Figure 4.3 Effect of variation of parameter r on the real part of the burning rate response function. $\iota = 0.3$, $k = 1.8$, $\delta = 0.\ 1$, $r = 0.33$; 2, $r = 0.30$; 3, $r = 0.27$.

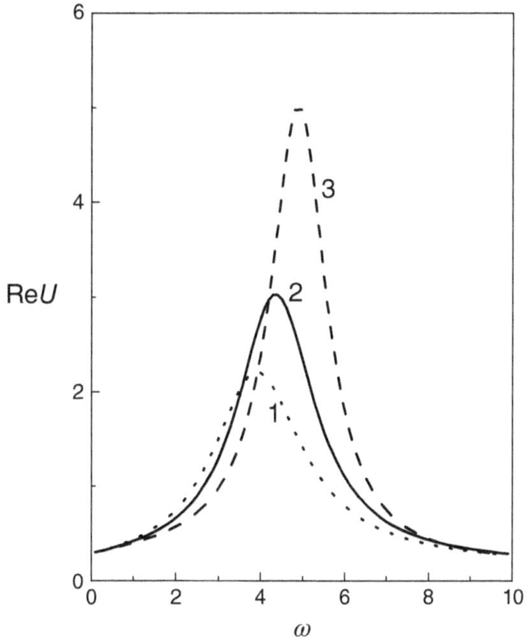

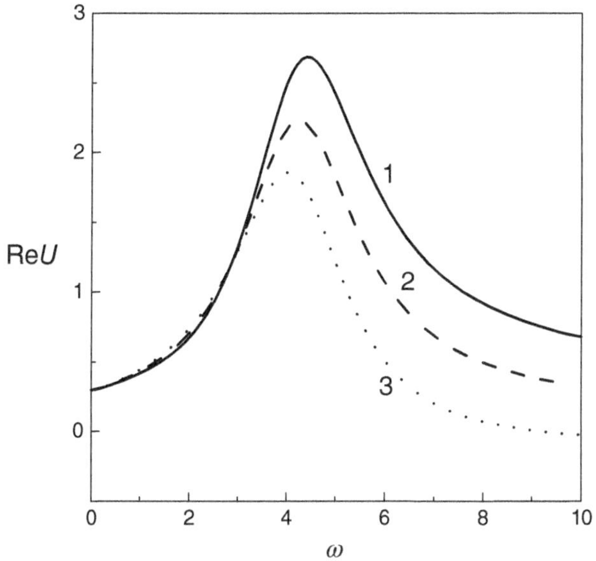

Figure 4.4 Effect of variation of parameter δ on the real part of the burning rate response function. $\iota = 0.3$, $k = 1.75$, $r = 0.3$. 1, $\delta = 0.05$; 2, $\delta = 0.0$; 3, $\delta = -0.05$.

Under typical measurement accuracy, $\Delta u^0/u^0 = 5\%$ and $T_{a,2} - T_{a,1} = 100$ K resulting in $\Delta\beta = 10^{-3}$ K^{-1}, which is comparable, by the order of magnitude, with the value of the burning rate temperature sensitivity coefficient itself. The relative error $\Delta k/k$ turns out to be up to 20–30%. With significant improvement in the accuracy of the burning rate measurement, the relative error in estimation of the parameter k is still significant. For example, even with $\Delta u^0/u^0 = 1\%$ the error is $\Delta k/k = 5$–7%.

The second important parameter is the surface temperature sensitivity to initial temperature at steady-state conditions r, which is usually estimated using thermocouple measurements of the surface temperature and further differentiating the experimental dependence $T_s^0(T_a)$. For a thermocouple of thickness h and for the surface temperature gradient $f^0 = (u^0/\kappa)(T_s^0 - T_a)$ the error in measuring the surface temperature is $\Delta T_s^0 = hf$. At high pressures where $u^0 \sim 10^{-2}$m/s even thin 5 μm thermocouples would produce relative surface temperature measurement error of the order of 20%. This error is reduced at lower pressures, but additional inaccuracy comes with differentiation of the experimental data $T_s^0(T_a)$. Overall, it is believed that the relative error in the estimation of the parameter r is at best not less than 10%.

The above analysis suggests that presently the response function may be only determined with relatively large, of the order of tens of percentage error.

Obviously the error in estimation of the response function is carried further into inaccuracies of those conclusions of the nonsteady combustion theory that are based on application of such functions.

Any periodical pressure variation can be written, in the linear approximation, as a superposition of infinite number of harmonics

$$\eta = 1 + \sum_{m=1}^{\infty} [\eta_m \exp(im\omega\tau) + c.c.] \tag{4.34}$$

Then the burning rate and the temperature distribution in the linear approximation may be written as

$$\begin{aligned} v &= 1 + \sum_{m=1}^{\infty} [v_m \exp(im\omega\tau) + c.c.] \\ \theta(\xi,\tau) &= e^{\xi} + \sum_{m=1}^{\infty} [\theta_m(\xi) e^{im\omega\tau} + c.c.] \end{aligned} \tag{4.35}$$

Substitution of these functions into the heat transfer Eq. (4.9) and its solution leads to a relation similar to (4.16)

$$z_m \vartheta_m - \varphi_m + \frac{v_m}{z_m} = 0 \tag{4.36}$$

where

$$z_m = \frac{1}{2}(1 + \sqrt{1 + 4im\omega}), \quad z_m(z_m - 1) = im\omega \tag{4.37}$$

Linearization of nonsteady burning laws (4.18) gives

$$v_m = \frac{k\varphi_m + (\delta - \iota)\eta_m}{k + r - 1}, \quad \vartheta_m = \frac{r\varphi_m - (\delta + \mu)\eta_m}{k + r - 1} \tag{4.38}$$

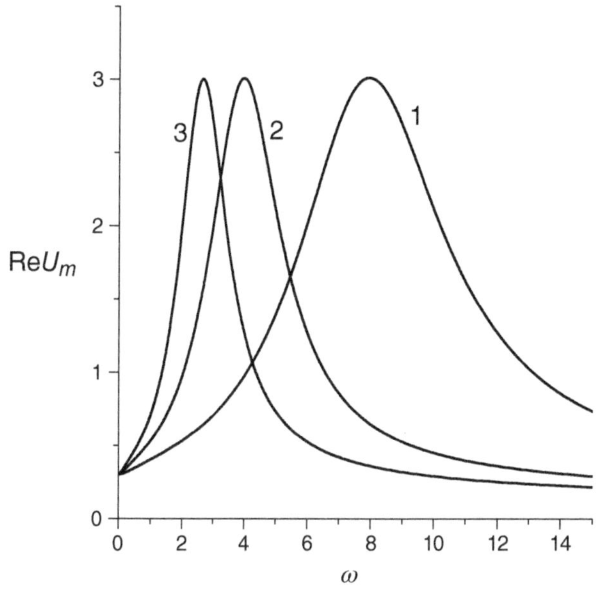

Figure 4.5 Real part of the linear response function of burning rate. $k = 1.5$, $r = 0.15$, $\iota = 0.3$, $\delta = 0.1$, $m = 1$; 2, $m = 2$; 3, $m = 3$.

The relations (4.36) and (4.38) allow the linear response function for the mth harmonic to be found

$$U_m(\omega) = \frac{\iota + \delta(z_m - 1)}{1 + (z_m - 1)(r - k/z_m)} \tag{4.39}$$

It is easy to verify that the response function for any harmonic can be expressed in terms of the response function of the primary harmonic

$$U_m(\omega) = U_1(m\omega), \quad U_m(\omega/m) = U_1(\omega) \tag{4.40}$$

where the latter $U_1 \equiv U$ is given by the formula (4.27).

The real parts of response functions of the first three harmonics are shown in Figure 4.5. It is evident from this figure that the resonances of various harmonics occur at the frequencies $\omega_m = \omega_1/m$.

In conclusion of this section, we discuss the alternative approaches to the problem, the vast majority of which are within the same framework of the linear approximation. Development of a consistent theory of nonsteady processes within a chamber of solid rocket engines has been attempted for over 50 years. Most publications are the result of investigations in the United States. Unfortunately, all these attempts proved to be unsatisfactory as the interaction between combustion and acoustic field is described by the phenomenological formula by Culick (1968) with$A$, B, and n being model parameters:

$$U = \frac{nAB}{z + \frac{A}{z} - (1 - A) + AB} \tag{4.41}$$

This formula was published three years later than the paper by Novozhilov (1965b). It is not based on any theory but rather aggregates the data obtained by a number of authors who considered specific propellant combustion models (the most famous of these is the study by Denison and Baum (1961)).

Culick (1968) did not unveil the physical meaning of the model parameters. It is easy to see that (4.41) is a particular case of formula (4.39) if one assigns

$$n = \iota, \quad A = \frac{k}{r}, \quad B = \frac{1}{k}, \quad \delta = 0 \tag{4.42}$$

The approach proposed by Culick (1968) will not succeed for the following reasons.

The constants A, B, and n cannot be calculated even if a specific propellant combustion model is formulated. Their dependencies on pressure and propellant initial temperature are not determined.

Boundary conditions would require the nonsteady temperature of the gases leaving the fuel surface to be specified. It is impossible to obtain this temperature from (4.41).

Finally, the discussed approach is strictly limited to consideration of linear interaction between pressure and propellant burning rate.

ZN theory overcomes the above problems. In contrast to the approach by Culick (1968) it offers the possibility to extend analysis beyond the linear approximation.

The following chapters of the book are devoted to systematic studies of nonlinear phenomena.

4.2 Acoustic Admittance of Propellant Surface

One of the important problems in the theory of nonsteady combustion of condensed systems is the possibility of amplification of pressure waves upon their reflection from the surface of the propellant. A solution to this problem is related to the practically important issue of suppressing pressure oscillations in the chambers of rocket engines. If a chamber is capable of augmenting pressure waves at frequencies which coincide with one of the chamber's own acoustic frequencies, then the propellant combustion process in such a chamber would be unstable and both the pressure and burning rate would change in time. As the frequency of the pressure and burning rate variations would be determined by the acoustic properties of the chamber, such instability is called an acoustic or high-frequency instability. The linear theory developed in the previous section cannot predict an amplitude of oscillations developing in a chamber. Such a prediction can only be done on the basis of nonlinear analysis, taking into account the finite value of the amplitude and the magnitude of energy dissipation that it determines. In the linear approximation one can only determine such conditions at which the amplitude of small oscillations would either grow or decrease.

Acoustic instability, as well as low-frequency instability, develops due to interaction between the two objects: the combustion chamber and the surface of the burning propellant. The acoustic properties of chambers of various configurations have been studied extensively. Therefore, the major difficulty in studying high-frequency instability is to determine the relevant properties of the fuel that describe its reaction to varying pressure. The reaction of the surface to an incident acoustic wave is normally described by the magnitude of acoustic admittance. Let us discuss the definition of this quantity first.

Let the acoustic wave propagate towards the flat surface $x = 0$ from the right direction ($x > 0$). The reflected wave having the same frequency propagates from left to right and its amplitude, in the linear approximation, is proportional to the amplitude of the incident wave. The proportionality coefficient is determined by the frequency and the acoustic properties of the surface.

It is convenient to describe the plane acoustic wave using the velocity potential ψ_a which satisfies the wave equation

$$\frac{\partial^2 \psi_a}{\partial t^2} - a^2 \frac{\partial^2 \psi_a}{\partial x^2} = 0 \tag{4.43}$$

where a stands for the speed of sound. Gas pressure and velocity variations are related to the potential via the simple relationships

$$p_1 = -\rho_g \frac{\partial \psi_a}{\partial t}, \quad u_{g1} = \frac{\partial \psi_a}{\partial x} \tag{4.44}$$

Let us find the conditions leading to the amplification of the wave on reflection. The acoustic field corresponding to superposition of the incident and reflected waves has the potential

$$\psi_a = A_+ \exp(iKx + i\Omega t) + A_- \exp(-iKx + i\Omega t) \tag{4.45}$$

where A_+ and A_- are the amplitudes of the incident and reflected waves, respectively, and K is a wave vector. The amplitudes ratio $\sigma_a = A_-/A_+$ can be expressed via the complex amplitudes of pressure and velocity at the surface. From (4.44) and (4.45)

$$\begin{aligned} u_{g1} &= iK[A_+ \exp(iKx + i\Omega t) - A_- \exp(-iKx + i\Omega t)] \\ p_1 &= -i\rho_g \Omega [A_+ \exp(iKx + i\Omega t) + A_- \exp(-iKx + i\Omega t)] \end{aligned} \tag{4.46}$$

At the surface ($x = 0$) these expressions give

$$-\frac{u_{g1}(0)}{p_1(0)} = \frac{K(A_+ - A_-)}{\rho_g \Omega (A_+ + A_-)} \tag{4.47}$$

The nondimensional value

$$\zeta = -\rho_g a \frac{u_{g1}(0)}{p_1(0)} \tag{4.48}$$

is called the acoustic admittance of the surface.

The ratio of the complex amplitudes of the incident and reflected waves σ_a is related to the acoustic admittance via very simple relationships

$$\zeta = \frac{1 - \sigma_a}{1 + \sigma_a}, \quad \sigma_a = \frac{1 - \zeta}{1 + \zeta} \tag{4.49}$$

which follow from (4.47) and (4.48) using the relation between the wave vector and the frequency $aK = \Omega$.

The wave is amplified upon reflection only if the absolute value of the ratio σ_a is larger than unity. From the second of the relations (4.49)

$$|\sigma_a|^2 = 1 - \frac{4\text{Re}\zeta}{|1 + \zeta|^2} \tag{4.50}$$

This means that amplification occurs if, and only if, the real part of the surface acoustic admittance is negative.

In the process of combustion of gases or volatile condensed systems, an acoustic wave always spreads in a medium moving with some characteristic speed u_g^0. In that case, acoustic admittance can be expressed in the form

$$\zeta = -\gamma \mathrm{M} G \tag{4.51}$$

where γ is the specific heat ratio, M is the Mach number, and G is the gas velocity response to oscillating pressure

$$\mathrm{M} = \frac{u_g^0}{a}, \quad G = \frac{g_1}{\eta_1} \tag{4.52}$$

The latter property is similar to the response function of the burning rate to oscillating pressure. It is called a linear response function of the gas velocity to oscillating pressure and is equal to the ratio of nondimensional complex amplitudes of the gas velocity and pressure

$$g_1 = \frac{u_{g1}}{u_g^0}, \quad \eta_1 = \frac{p_1}{p^0} \tag{4.53}$$

On concluding these preliminaries, let us proceed to the calculation of the acoustic admittance of the propellant burning surface. It is sufficient, according to (4.51), to find the response function G of the velocity of combustion products leaving the burning surface to harmonically oscillating pressure.

One has to be careful with definition of the surface of the burning condensed combustion system. Let us consider the unperturbed state of the gas. In the absence of acoustic wave, pressure is essentially uniform over the space (gas velocity is small compared to the speed of sound) while temperature changes considerably in the vicinity of the surface, from surface temperature to combustion temperature. Therefore, the gas parameters (pressure and temperature) are only constant, in the absence of acoustic waves, beyond the cross-section where the combustion temperature T_b^0 is achieved and where all the chemical reactions are fully completed. It is natural to consider this cross-section as the surface whose acoustic admittance determines parameters of solid propellant with respect to sound effects. To emphasize the fact that all the values are considered precisely at this surface, from here on the argument in expression (4.48) needs be assumed to be the coordinate x_b corresponding to the just defined surface. It should be kept in mind that the value of x_b is small compared to the length of the acoustic wave. Within the framework of the ZN theory, acoustic admittance is easily related to the response function of the burning rate and the response function describing the temperature change of the gases leaving the burning surface (Novozhilov 1965b, 1968; Gostintsev et al. 1970). The latter linear response function for the temperature of combustion products is defined as

$$\Theta(\omega) = \frac{b_1}{\eta_1}, \quad b_1 = \frac{T_{b1}}{T_b^0} \tag{4.54}$$

where T_{b1} is the complex amplitude of combustion temperature variation.

Upon neglecting the inertia of the gas phase, an assumption adopted within the ZN theory, the combustion temperature is fully determined by initial temperature and burning rate.

In order to calculate acoustic admittance use is made of the mass conservation equation and the equation of state for an ideal gas

$$\rho u = \rho_g u_g, \quad \rho_g = \frac{p\mu_g}{RT_b} \tag{4.55}$$

These two equations lead to

$$g = \frac{vb}{\eta} \tag{4.56}$$

where $\eta = p/p^0$ and $b = T_b/T_b^0$. From here

$$g_1 = v_1 + b_1 - \eta_1 \tag{4.57}$$

or

$$G = U + \Theta - 1 \tag{4.58}$$

To calculate the combustion temperature response function, let us recall that the state of the gas phase (including the temperature of combustion) is fully determined by the pressure and temperature gradient at the surface. Since the burning rate is determined by the same parameters then

$$T_b = T_b(u, p) \text{ or } u = u(p, T_b) \tag{4.59}$$

which further leads to

$$v_1 = \left(\frac{\partial \ln u^0}{\partial \ln p^0}\right)_{T_b^0} \eta_1 + \left(\frac{\partial \ln u^0}{\partial T_b^0}\right)_{p^0} T_{b1} \tag{4.60}$$

But since for complete combustion and equal specific heats of the phases

$$\left(\frac{\partial \ln u^0}{\partial \ln p^0}\right)_{T_b^0} = \left(\frac{\partial \ln u^0}{\partial \ln p^0}\right)_{T_a} = \iota$$

$$\left(\frac{\partial \ln u^0}{\partial T_b^0}\right)_{p^0} = \left(\frac{\partial \ln u^0}{\partial T_a}\right)_{p^0} = \frac{k}{T_s^0 - T_a} \tag{4.61}$$

then

$$\Theta = \frac{\Delta}{k}(U - \iota), \quad \Delta = \frac{T_s^0 - T_a}{T_b^0} \tag{4.62}$$

Not that the function Θ is much less than the corresponding function U (since Δ is a small parameter). Therefore

$$G \approx U - 1 \tag{4.63}$$

This conclusion is illustrated in Figure 4.6.

Taking into account (4.62), the following expression for the acoustic admittance (4.51) is obtained

$$\zeta = \gamma \mathrm{M}\left[1 + \frac{\Delta\iota}{k} - \left(1 + \frac{\Delta}{k}\right)U\right] \tag{4.64}$$

It must be emphasized that the formula (4.62) is derived based on the assumption that the heat generation rate is constant during combustion, which implies that combustion is

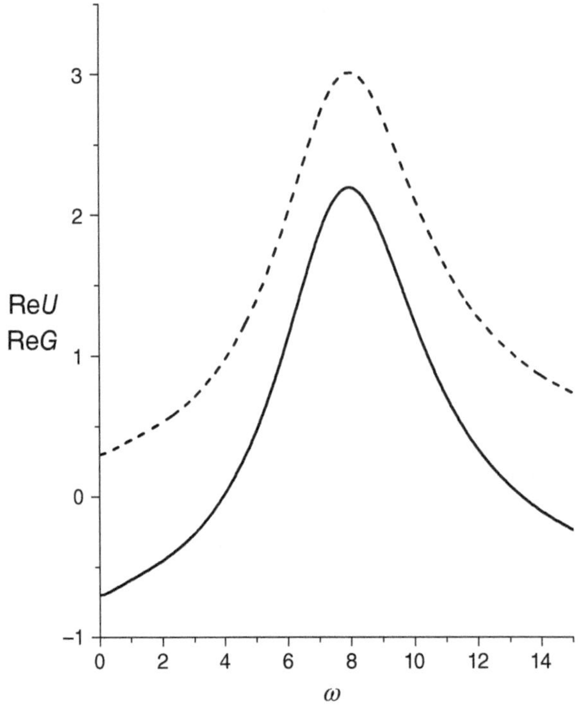

Figure 4.6 Real parts of linear response functions of burning rate U and G.

complete. Therefore, application of the formula (4.64) to ballistite propellants is restricted by conditions of rather high pressures. For pressures below 50–60 atm, the heat generation rate changes with pressure and its initial temperature. At such conditions, (4.62) must be replaced with an expression involving derivatives of the combustion temperature with respect to ambient parameters (Gostintsev et al. 1970). Nevertheless, such a correction may not be consistent as in some cases neglecting the thermal inertia of the gas phase is not justifiable within the relevant pressure interval.

For the acoustic wave to be amplified, it is necessary that the real part of expression (4.64) is negative, that is

$$\mathrm{Re}U > \frac{1 + {}^{\Delta_1}/_k}{1 + {}^{\Delta}/_k} \tag{4.65}$$

It is obvious that the most favourable conditions for this inequality to be fulfilled occur near resonance, where the response function of the burning rate is at a maximum and the phase shift between pressure and burning rate changes significantly.

To conclude this section, it is worthwhile noting that in periodic nonsteady combustion regimes (e.g. during combustion under harmonically oscillating pressure), variable combustion temperatures lead to the generation of temperature waves in the gas with wavelengths much shorter than the acoustic wavelength. Indeed, temperature perturbations leave the surface, along with combustion products, with the velocity u_g^0. Therefore, a temperature wave with length $2\pi u_g^0/\Omega$, which is less than the acoustic wavelength by a factor of

M, will be superimposed onto the acoustic wave. If one neglects dissipation processes occurring in the products of combustion, then the gas flow is isentropic and the above-mentioned temperature waves may be called entropic waves. Generation of the temperature or entropic waves is related to the fact that the temperature and density perturbations arising from the interaction between sound and burning surface cannot be carried away by the acoustic wave alone.

Temperature variations are not only the result of adiabatic compression of the gas within the acoustic wave, but are also the consequence of the nonsteady nature of the combustion process, which implies variable combustion temperature.

4.3 Quadratic Response Functions

Interaction of propellant combustion with varying pressure is a focus point in studying nonsteady processes in the chamber of a solid fuel rocket engine. Linear analysis is the simplest method and, combined with the approximation of linear acoustics, may be used to find the stability conditions of a steady-state engine operation regime. However, consideration of processes involving finite pressure amplitudes (such as sustained oscillations or triggering) requires restrictions of the linear approximation to be removed.

Most of the studies on nonsteady effects in a combustion chamber (e.g. Culick and Yang (1992)) considered nonlinear behaviour for acoustics only. The response function of the burning rate to oscillating pressure was only involved in a linear approximation. It was not until the papers by Novozhilov (1965c, 2002) and Novozhilov et al. (1996) that an attempt was made to take into consideration the nonlinear interaction between the acoustic wave and the combustion process of solid propellants. The simplest case involving just the first two harmonics of an acoustic field was considered. This is surely not sufficient for full investigation of the multimode problem.

An analytical investigation of the nonlinear effects of interaction between acoustics and combustion is obviously only possible in lower order (quadratic or at best third order with respect to oscillation amplitude) approximations.

The aim of the present section is to consider, in quadratic with respect to pressure amplitude approximation, the interaction between various pressure harmonics and find the response functions of various propellant combustion parameters (burning rate, temperature, and velocity of combustion products) to oscillating pressure (Novozhilov 2006).

A periodic acoustic field in a combustion chamber normally contains many harmonics. Their interaction can be described by introducing nonlinear response functions of propellant burning rate to harmonically oscillating pressure. Using the following analysis, these functions are derived in the lowest (quadratic) with respect to pressure amplitude approximation. The physical meaning of the quadratic response functions may be illustrated using an example of pressure in the vicinity of the burning propellant surface containing just two harmonics with small amplitudes (without loss of generality $n > m$)

$$\eta = 1 + [\eta_m \exp(im\omega\tau) + \eta_n \exp(in\omega\tau) + c.c.] \tag{4.66}$$

Propellant burning rate, in the quadratic with respect to pressure amplitude approximation, would contain the same harmonics as well as harmonics with combinational frequencies (including time-independent corrections to the steady-state burning rate)

$$\begin{aligned} v = 1 &+ v_{m,-m} + v_{n,-n} + \{v_m \exp(im\omega\tau) + v_n \exp(in\omega\tau) \\ &+ v_{m,m} \exp[2im\omega\tau] + v_{n,n} \exp[2in\omega\tau] \\ &+ v_{n,-m} \exp[i(n-m)\omega\tau] + v_{n,m} \exp[i(n+m)\omega\tau] + c.c.\} \end{aligned} \tag{4.67}$$

It is obvious that apart from the linear relationships

$$U_m(\omega) = \frac{v_m}{\eta_m}, \quad U_n(\omega) = \frac{v_n}{\eta_n} \tag{4.68}$$

one can also write

$$\begin{aligned} v_{m,-m} &= U_{m,-m}(\omega)\eta_m\bar{\eta}_m, \quad v_{n,-n} = U_{n,-n}(\omega)\eta_n\bar{\eta}_n \\ v_{m,m} &= U_{m,m}(\omega)\eta_m^2, \quad v_{n,n} = U_{n,n}(\omega)\eta_n^2 \\ v_{n,-m} &= U_{n,-m}(\omega)\eta_n\bar{\eta}_m, \quad v_{n,m} = U_{n,m}(\omega)\eta_n\eta_m \end{aligned} \tag{4.69}$$

The complex functions

$$\begin{aligned} U_{m,-m}(\omega) &= \frac{v_{m,-m}}{|\eta_m|^2}, \quad U_{m,m}(\omega) = \frac{v_{m,m}}{\eta_m^2} \\ U_{n,-m}(\omega) &= \frac{v_{n,-m}}{\eta_n\bar{\eta}_m}, \quad U_{n,m}(\omega) = \frac{v_{n,m}}{\eta_n\eta_m} \end{aligned} \tag{4.70}$$

are called the quadratic response functions of the burning rate to oscillating pressure. In the above notation, the two lower indexes show which of the harmonics are interacting while their sum gives the generated burning rate harmonic number. The quadratic response functions for the zero, first, and second harmonics

$$U_{1,1} = \frac{v_{1,1}}{\eta_1^2}, \quad U_{1,-1} = \frac{v_{1,-1}}{|\eta_1|^2}, \quad U_{2,-2} = \frac{v_{2,-2}}{|\eta_2|^2}, \quad U_{2,-1} = \frac{v_{2,-1}}{\eta_2\bar{\eta}_1} \tag{4.71}$$

were first obtained in analytical form by Novozhilov et al. (1996) and applied later to investigation of nonlinear effects developing in the combustion chambers of solid rocket motors (Novozhilov 2002).

Nonlinear response functions in the quadratic with respect to pressure amplitude approximation are derived below. In the general case their expressions assume a very complicated form. The reason for this is that the result in the linear approximation includes just four parameters describing the steady-state combustion regime, namely, the first derivatives of burning rate and surface temperature with respect to initial temperature and pressure. Quadratic response functions must involve second derivatives. Therefore, the number of parameters increases to 10 (two first and three second derivatives for each of the two steady-state burning laws). This significantly complicates the analysis as well as its final result. On the other hand, even the first derivatives are obtained from experiments with significant errors, not to mention the second derivatives. In view of this fact, it is preferable to derive nonlinear response functions for some specific propellant combustion model. In the present section, the model with the minimum number of parameters is considered. The simplest of such models is described by the following steady-state dependencies

$$u^0 = A(p^0)^\iota \exp(\beta T_a), \quad u^0 = B\exp(\beta_s T_s^0) \tag{4.72}$$

where A, B, β, and β_s are constants.

Using the standard ZN theory method of transformation to nonsteady dependencies, the following nondimensional nonsteady burning laws are obtained

$$v = \eta^{\iota} \exp\left[k\left(\vartheta - \frac{\varphi}{v}\right)\right], \quad v = \exp\left[\frac{k}{r}(\vartheta - 1)\right] \tag{4.73}$$

where

$$k = \beta(T_s^0 - T_a), \quad r = \beta/\beta_s \tag{4.74}$$

In the following analysis we will need expansions of the functions (4.73) up to second-order terms. Let us express all the variables (v, ϑ, φ и η) in the form

$$y = 1 + y_e \tag{4.75}$$

where the bottom index denotes the correction to the steady-state value.

Rewrite (4.73) in the form

$$\vartheta = 1 + \frac{r}{k}\ln v, \quad \varphi = v\left(1 + \frac{r-1}{k}\ln v + \frac{\iota}{k}\ln\eta\right) \tag{4.76}$$

Expansion of these relationships up to quadratic terms provides

$$\vartheta_e = \frac{r}{k}v_e - \frac{r}{2k}v_e^2, \quad \varphi_e = \frac{s}{k}v_e + \frac{\iota}{k}\eta_e + \frac{1}{2}\left(\frac{r-1}{k}v_e^2 + 2\frac{\iota}{k}v_e\eta_e - \frac{\iota}{k}\eta_e^2\right) \tag{4.77}$$

where $s = k + r - 1$.

First of all, let us calculate the quadratic correction to the steady-state value of the burning rate. In the quadratic approximation, it emerges as a result of self-interaction of any of the harmonics and therefore is a sum of all such contributions. It is sufficient to find a correction corresponding to any one arbitrary harmonic

$$U_{m,-m} = \frac{v_{m,-m}}{|\eta_m|^2} \tag{4.78}$$

Let us take pressure in the form

$$\eta(\tau) = 1 + (\eta_m e^{im\omega\tau} + \text{c.c.}) \tag{4.79}$$

and include two time-independent quadratic corrections into the expressions for the burning rate and temperature distribution

$$\begin{aligned} v(\tau) &= 1 + v_{m,-m} + (v_m e^{im\omega\tau} + \text{c.c.}) \\ \theta(\xi,\tau) &= e^{\xi} + \theta_{m,-m}(\xi) + [\theta_m(\xi)e^{im\omega\tau} + \text{c.c.}] \end{aligned} \tag{4.80}$$

The heat transfer Eq. (4.9) written to second-order accuracy

$$\theta''_{m,-m} - \theta'_{m,-m} = v_{m,-m}e^{\xi} + (v_m\overline{\theta}'_m + \text{c.c.}) \tag{4.81}$$

contains the linear solution (4.13).

Integrating Eq. (4.81) over the space ($-\infty < \xi \leq 0$) gives

$$\varphi_{m,-m} - \vartheta_{m,-m} - v_{m,-m} = v_m\overline{\vartheta}_m + \text{c.c.} \tag{4.82}$$

The right-hand side of this expression contains the product of linear corrections. It follows from (4.77) that in the linear approximation

$$\vartheta_m = \frac{r}{k}v_m \tag{4.83}$$

therefore

$$\varphi_{m,-m} - \vartheta_{m,-m} - v_{m,-m} = 2\frac{r}{k}|v_m|^2 \tag{4.84}$$

Consider now nonlinear relationships (4.77) following from nonsteady burning laws. In the present case

$$\begin{aligned} &\eta_e = \eta_m e^{im\omega\tau} + \text{c.c.}, \quad v_e = v_{m,-m} + (v_m e^{im\omega\tau} + \text{c.c.}) \\ &\vartheta_e = \vartheta_{m,-m} + (\vartheta_m e^{im\omega\tau} + \text{c.c.}), \quad \varphi_e = \varphi_{m,-m} + (\varphi_m e^{im\omega\tau} + \text{c.c.}) \end{aligned} \tag{4.85}$$

Substituting these expressions into (4.77) gives

$$\begin{aligned} \vartheta_{m,-m} &= \frac{r}{k}(v_{m,-m} - |v_m|^2) \\ \varphi_{m,-m} &= \frac{1}{k}[s v_{m,-m} + (r-1)|v_m|^2 + \imath(v_m\bar{\eta}_m + \bar{v}_m\eta_m - |\eta_m|^2)] \end{aligned} \tag{4.86}$$

The expressions (4.84) and (4.86) allow the response function corresponding to the quadratic correction to the steady-state burning rate to be found (for the model with constant surface temperature this problem was solved by Novozhilov (1965c))

$$U_{m,-m} = -|U_m|^2 + \imath(2\text{Re}U_m - 1) \tag{4.87}$$

In the same way as for the linear response functions, the obtained quadratic response function can be expressed through the response function of the primary harmonic

$$U_{m,-m}(\omega) = U_{1,-1}(m\omega), \quad U_{m,-m}(\omega/m) = U_{1,-1}(\omega) \tag{4.88}$$

and resonance occurs at frequencies close to ω_1/m. Frequency dependencies of the response functions $U_{m,-m}$ determined by self-interaction of the first three harmonics are shown in Figure. 4.7.

Apart from quadratic correction to the steady state value of the burning rate self-interaction of any harmonic leads, in the quadratic approximation, to generation of the mode with double frequency. Therefore, the problem now turns into the calculation of the response function

$$U_{m,m} = \frac{v_{m,m}}{\eta_m^2} \tag{4.89}$$

Using the same technique, take pressure in the form

$$\eta(\tau) = 1 + (\eta_m e^{im\omega\tau} + \text{c.c.}) \tag{4.90}$$

and expressions for the burning rate and the temperature distribution as

$$\begin{aligned} v(\tau) &= 1 + (v_m e^{im\omega\tau} + v_{m,m}e^{2im\omega\tau} + \text{c.c.}) \\ \theta(\xi,\tau) &= e^{\xi} + [\theta_m(\xi)e^{im\omega\tau} + \theta_{m,m}(\xi)e^{2im\omega\tau} + \text{c.c.}] \end{aligned} \tag{4.91}$$

A heat transfer equation written to second-order accuracy for the harmonic with double frequency

$$\theta''_{m,m} - \theta'_{m,m} - 2im\omega\theta_{m,m} = \left(v_{m,m} - \frac{v_m^2}{im\omega}\right)e^{\xi} + a_m z_m v_m^2 e^{z_m\xi} \tag{4.92}$$

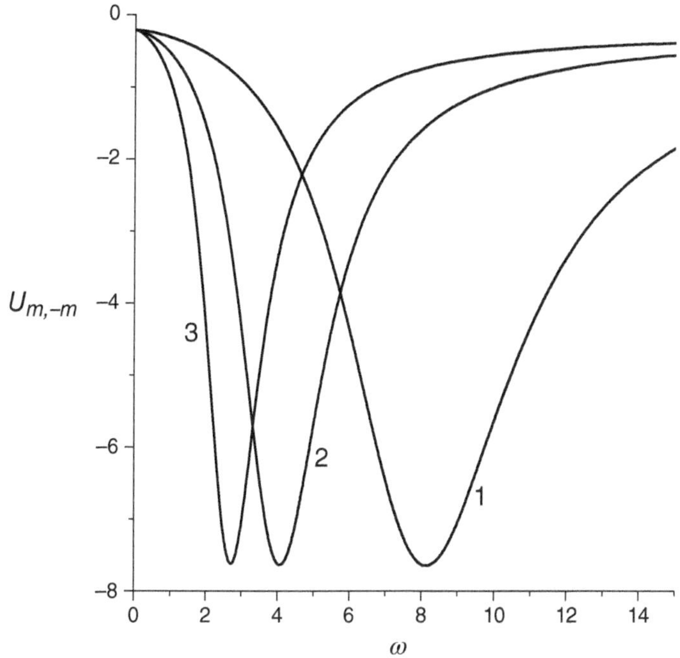

Figure 4.7 Quadratic correction to the constant component of burning rate. $k = 1.5$, $r = 0.15$, $\iota = 0.3$, $\delta = 0.1$, $m = 1$; 2, $m = 2$; 3, $m = 3$.

has the solution

$$\theta_{m,m} = C_{m,m}e^{z_{2m}\xi} + A_{m,m}e^{\xi} + B_{m,m}e^{z_m\xi}$$
$$A_{m,m} = -\frac{1}{2im\omega}\left(v_{m,m} - \frac{v_m^2}{im\omega}\right), \quad B_{m,m} = -\frac{a_m z_m v_m^2}{im\omega} \tag{4.93}$$

Stemming from this solution are quadratic corrections to the surface temperature and its gradient, and eliminating the integration constant $C_{m,m}$ we obtain

$$z_{2m}\vartheta_{m,m} - \varphi_{m,m} + \frac{v_{m,m}}{z_{2m}} = \frac{v_m^2}{im\omega}\left[\frac{1}{z_{2m}} - a_m z_m(z_{2m} - z_m)\right] \tag{4.94}$$

One the other hand, nonsteady burning laws (4.77) imply

$$\vartheta_{m,m} = \frac{r}{k}\left(v_{m,m} - \frac{v_m^2}{2}\right)$$
$$\varphi_{m,m} = \frac{1}{k}\left[s v_{m,m} + \frac{(r-1)v_m^2}{2} + \iota v_m \eta_m - \frac{\iota\eta_m^2}{2}\right] \tag{4.95}$$

The three relationships (4.94) and (4.95) allow the response function for the harmonic with double frequency to be found

$$U_{m,m} = \frac{U_m^2 U_{2m}}{2\iota}(-1 + rR_{m,m} + kK_{m,m}) + U_{2m}\left(U_m - \frac{1}{2}\right)$$
$$R_{m,m} = 3 + z_{2m} - \frac{4z_m}{z_{2m}}, \quad K_{m,m} = \frac{z_{2m} - 1}{(m\omega)^2}(2z_m - z_{2m} - 1) \tag{4.96}$$

The frequency dependencies of this response function are illustrated in Figure 4.8.

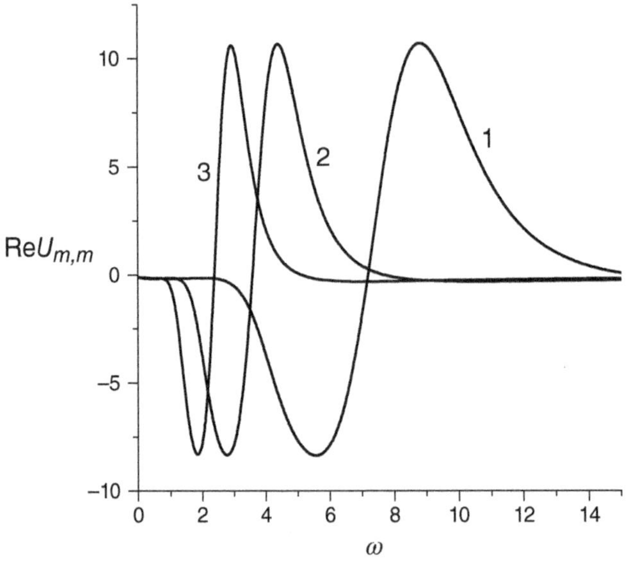

Figure 4.8 Real part of the response function of burning rate for double frequency. $k = 1.5$, $r = 0.15$, $\iota = 0.3$, $\delta = 0.1$, $m = 1$; 2, $m = 2$; 3, $m = 3$.

Burning velocity modes with combinational frequencies are being generated on quadratic interaction of two pressure harmonics. We shall derive here the corresponding response functions

$$U_{n,-m}(\omega) = \frac{v_{n,-m}}{\eta_n \bar{\eta}_m}, \quad U_{n,m}(\omega) = \frac{v_{n,m}}{\eta_n \eta_m} \tag{4.97}$$

In order to find the first of these functions take pressure in the form

$$\eta(\tau) = 1 + (\eta_m e^{im\omega\tau} + \eta_n e^{in\omega\tau} + \text{c.c.}) \tag{4.98}$$

and single out, apart from linear contributions, also the term corresponding to the combinational frequency $(n-m)\omega$ in the expressions for the burning rate and temperature distribution in the body of the propellant

$$v(\tau) = 1 + (v_m e^{im\omega\tau} + v_n e^{in\omega\tau} + v_{n,-m} e^{i(n-m)\omega\tau} + \text{c.c.})$$
$$\theta(\xi,\tau) = e^{\xi} + [\theta_m(\xi) e^{im\omega\tau} + \theta_n(\xi) e^{in\omega\tau} + \theta_{n,-m}(\xi) e^{i(n-m)\omega\tau} + \text{c.c.}] \tag{4.99}$$

Substitution of these expressions into the heat transfer Eq. (4.9) results in the equation

$$\theta''_{n,-m} - \theta'_{n,-m} - i(n-m)\omega\theta_{n,-m} = \left(v_{n,-m} + \frac{v_n \bar{v}_m}{i\omega} \frac{(n-m)}{nm} \right) e^{\xi} + $$
$$+ v_n \bar{v}_m (\bar{a}_m \bar{z}_m e^{\bar{z}_m \xi} + a_n z_n e^{z_n \xi}) \tag{4.100}$$

the solution of which, satisfying the boundary condition at $\xi \to -\infty$, has the form

$$\theta_{n,-m} = C_{n,-m} e^{z_{n-m}\xi} + A_{n,-m} e^{\xi} + M_{n,-m} e^{\bar{z}_m \xi} + N_{n,-m} e^{z_n \xi}$$
$$A_{n,-m} = -\frac{1}{i(n-m)\omega} \left(v_{n,-m} + \frac{v_n \bar{v}_m}{i\omega} \frac{(n-m)}{nm} \right)$$
$$M_{n,-m} = -\frac{\bar{a}_m \bar{z}_m v_n \bar{v}_m}{in\omega}, \quad N_{n,-m} = \frac{a_n z_n v_n \bar{v}_m}{im\omega} \tag{4.101}$$

The boundary condition at $\xi = 0$, on elimination of the integration constant C_{n-m}, leads to the relationship

$$z_{n-m}\vartheta_{n,-m} - \varphi_{n,-m} + \frac{v_{n,-m}}{z_{n-m}} = \\ v_n\bar{v}_m\left[\frac{z_{n-m}-1}{nm\omega^2} - \frac{z_{n-m}-\bar{z}_m}{in\omega}\bar{a}_m\bar{z}_m + \frac{z_{n-m}-z_n}{im\omega}a_n z_n\right] \tag{4.102}$$

Two additional relationships between quadratic corrections to the burning rate, surface temperature, and its gradient may be obtained from nonsteady burning laws

$$\begin{aligned}\vartheta_{n,-m} &= \frac{r}{k}(v_{n,-m} - v_n\bar{v}_m)\\ \varphi_{n,-m} &= \frac{1}{k}[sv_{n,-m} + (r-1)v_n\bar{v}_m + \imath(v_n\bar{\eta}_m + \bar{v}_m\eta_n) - \imath\eta_n\bar{\eta}_m]\end{aligned} \tag{4.103}$$

The last three relationships give the desired response function

$$\begin{aligned}U_{n,-m} &= \frac{U_{n-m}U_n\bar{U}_m}{\imath}(-1 + rR_{n,-m} + kK_{n,-m}) + U_{n-m}(U_n + \bar{U}_m - 1)\\ R_{n,-m} &= -\frac{(n-m)^2}{nm} + (z_{n-m}-1)\left(1 + \frac{nz_n - m\bar{z}_m}{inm\omega}\right)\\ K_{n,-m} &= \frac{(z_{n-m}-1)}{nm\omega^2}(1 - \bar{z}_m - z_n + z_{n-m})\end{aligned} \tag{4.104}$$

Substitution $m \to -m$ provides the response function corresponding to the addition of frequencies

$$\begin{aligned}U_{n,m} &= \frac{U_{n+m}U_nU_m}{\imath}(-1 + rR_{n,m} + kK_{n,m}) + U_{n+m}(U_n + U_m - 1)\\ R_{n,m} &= \frac{(n+m)^2}{nm} + (z_{n+m}-1)\left(1 - \frac{nz_n + mz_m}{inm\omega}\right)\\ K_{n,m} &= -\frac{(z_{n+m}-1)}{nm\omega^2}(1 - z_m - z_n + z_{n+m})\end{aligned} \tag{4.105}$$

Here the obvious relationships

$$\bar{z}_{-m} = z_m, \quad \bar{U}_{-m} = U_m \tag{4.106}$$

have been used.

Response functions corresponding to the generation of modes with combinational frequencies contain linear response functions of the original pressure variation frequencies and their combinational frequencies. It is expected therefore that the quadratic response functions would have extremums in the vicinities of ω_1/m, ω_1/n, $\omega_1/(n-m)$, and $\omega_1/(n+m)$. This is illustrated in Figures 4.9 and 4.10.

The results obtained above for the quadratic response functions corresponding to the constant correction (4.87), double frequency (4.96), and combinational frequencies (4.104) and (4.105) may be summarized in the universal form

$$\begin{aligned}U_{s,l} &= \left(1 - \frac{\delta_{s,l}}{2}\right)U_{s+l}\left[\frac{U_sU_l}{\imath}(kK_{s,l} + rR_{s,l} - 1) + U_s + U_l - 1\right]\\ K_{s,l} &= \frac{s+l}{z_{s+l}sl}\left(\frac{s+l}{z_{s+l}} - \frac{s}{z_s} - \frac{l}{z_l}\right)\\ R_{s,l} &= \frac{(s+l)^2}{sl} - \frac{s+l}{z_{s+l}sl}(sz_s + lz_l) + z_{s+l} - 1\end{aligned} \tag{4.107}$$

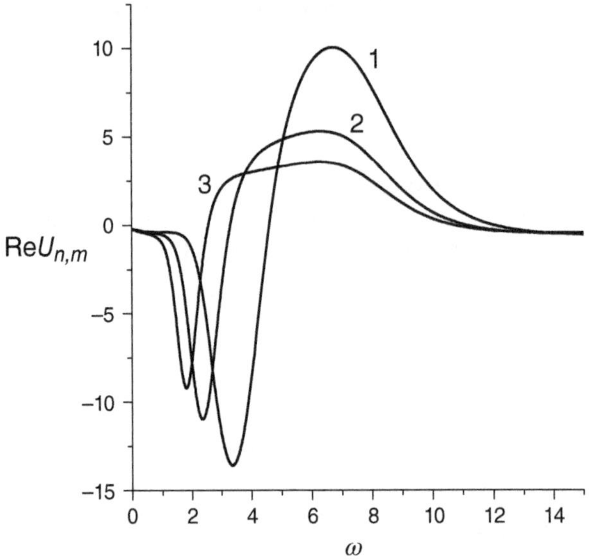

Figure 4.9 Quadratic response function of burning rate $U_{n,m}$. $k = 1.5$, $r = 0.15$, $\iota = 0.3$, $\delta = 0$. $m = 1$. 1, $n = 2$; 2, $n = 3$; 3, $n = 4$.

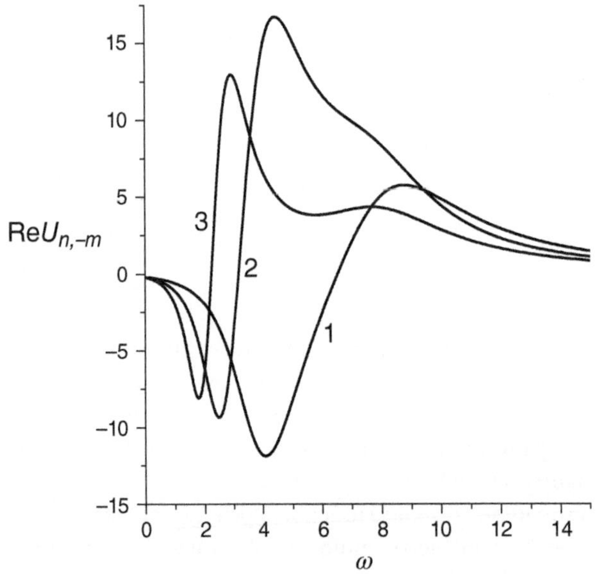

Figure 4.10 Quadratic response function of burning rate $U_{n,-m}$. $k = 1.5$, $r = 0.15$, $\iota = 0.3$, $\delta = 0$. $m = 1$. 1, $n = 2$; 2, $n = 3$; 3, $n = 4$.

where $\delta_{s,l}$ is the Kronecker delta, and the linear response function and the characteristic roots are expressed as

$$U_s = \frac{\iota}{1 + (z_s - 1)\left(r - {}^{k}/_{z_s}\right)}, \quad z_s = \frac{1}{2}(1 + \sqrt{1 + 4is\omega}) \tag{4.108}$$

4.4 Acoustic Admittance in the Second-order Approximation

For solution of acoustic problems, the response function of gas velocity to oscillating pressure must be specified on the surfaces enclosing the combustion chamber. This property is known as an acoustic admittance of the surface (the latter could be the surface of the propellant, the nozzle critical cross-section or any arbitrary surface).

Linear and quadratic expressions for acoustic admittance of the surface of burning propellant may be obtained from (4.68) and (4.70) where the linear burning rate v should be replaced with the nondimensional velocity $g = u_g/u_g^0$ of gases leaving the combustion zone. Here u_g is the dimensional gas velocity.

This results in the following definitions of the linear and nonlinear components of acoustic admittance

$$G_n(\omega) = \frac{g_n}{\eta_n}, \quad G_{m,-m}(\omega) = \frac{g_{m,-m}}{|\eta_m|^2}, \quad G_{m,m}(\omega) = \frac{g_{m,m}}{\eta_m^2}$$
$$G_{n,-m}(\omega) = \frac{g_{n,-m}}{\eta_n \bar{\eta}_m}, \quad G_{n,m}(\omega) = \frac{g_{n,m}}{\eta_n \eta_m} \tag{4.109}$$

In the framework of the ZN theory, acoustic admittance can be easily related to the response function of the burning rate and the response function describing the temperature change of gases leaving the burning surface. The linear and quadratic response functions of the temperature of combustion products are defined as

$$\Theta_n(\omega) = \frac{b_n}{\eta_n}, \quad \Theta_{m,-m}(\omega) = \frac{b_{m,-m}}{|\eta_m|^2}, \quad \Theta_{m,m}(\omega) = \frac{b_{m,m}}{\eta_m^2}$$
$$\Theta_{n,-m}(\omega) = \frac{b_{n,-m}}{\eta_n \bar{\eta}_m}, \quad \Theta_{n,m}(\omega) = \frac{b_{n,m}}{\eta_n \eta_m} \tag{4.110}$$

where $b = T_b/T_b^0$ is the nondimensional combustion temperature.

With negligible inertia of the gas phase, assumed by the ZN theory, combustion temperature is fully determined by the initial temperature and burning rate. In order to derive the response functions corresponding to the velocity of combustion products and the combustion temperature it is necessary to know the dependence of the combustion temperature in the steady-state regime on initial temperature and pressure

$$T_b^0 = F_b(T_a, p^0) \tag{4.111}$$

Let us assume, for simplicity's sake, that the combustion temperature does not depend on pressure and its variation is equal to that of initial temperature, that is

$$T_b^0 = T_a + \text{constant} \tag{4.112}$$

The standard procedure of transforming steady-state dependencies into nonsteady burning laws gives

$$b = 1 + \Delta\left(\vartheta - \frac{\varphi}{v}\right), \quad \Delta = \frac{T_s^0 - T_a}{T_b^0} \tag{4.113}$$

or, taking note of (4.73),

$$b = 1 + \frac{\Delta}{k}(\ln \nu - \imath \ln \eta) \tag{4.114}$$

This relationship allows the response functions of the combustion product temperature to be expressed through the response functions of the burning rate

$$\begin{aligned}
\Theta_m &= \frac{\Delta}{k}(U_m - \imath) \\
\Theta_{m,-m} &= \frac{\Delta}{k}(U_{m,-m} - |U_m|^2 + \imath), \quad \Theta_{m,m} = \frac{\Delta}{k}\left[U_{m,m} - \frac{1}{2}(U_m^2 - \imath)\right] \\
\Theta_{n,-m} &= \frac{\Delta}{k}(U_{n,-m} - U_n\overline{U}_m + \imath), \quad \Theta_{n,m} = \frac{\Delta}{k}(U_{n,m} - U_n U_m + \imath)
\end{aligned} \tag{4.115}$$

Note that as the parameter Δ is small the functions Θ are much smaller than the corresponding functions U.

In order to calculate the acoustic admittance conservation of mass and the equation of state of ideal gas

$$\rho u = \rho_g u_g \text{ and } \quad \rho_g = \frac{p\mu_g}{RT_b} \tag{4.116}$$

need be taken into account. These two restrictions lead to the relationship

$$g = \frac{\nu b}{\eta} \tag{4.117}$$

which in turn provides

$$g_e = \nu_e + b_e - \eta_e + \nu_e b_e - (\nu_e + b_e)\eta_e + \eta_e^2 \tag{4.118}$$

This relationship allows the linear and quadratic components of the acoustic admittance to be found

$$\begin{aligned}
G_m &= U_m + \Theta_m - 1 \\
G_{m,-m} &= U_{m,-m} + \Theta_{m,-m} + 2\mathrm{Re}(U_m - 1)(\overline{\Theta}_m - 1) \\
G_{m,m} &= U_{m,m} + \Theta_{m,m} + (U_m - 1)(\Theta_m - 1) \\
G_{n,-m} &= U_{n,-m} + \Theta_{n,-m} + \Theta_n(\overline{U}_m - 1) + \overline{\Theta}_m(U_n - 1) - U_n - \overline{U}_m + 2 \\
G_{n,m} &= U_{n,m} + \Theta_{n,m} + \Theta_n(U_m - 1) + \Theta_m(U_n - 1) - U_n - U_m + 2
\end{aligned} \tag{4.119}$$

Substituting (4.115) we obtain the relationship between the responses of the velocity of combustion products and the propellant burning

$$\begin{aligned}
G_m &= U_m - 1 + \frac{\Delta}{k}(U_m - \imath) \\
G_{m,-m} &= U_{m,-m} - 2\mathrm{Re}U_m + 2 + \frac{\Delta}{k}[U_{m,-m} + |U_m|^2 - 2(1+\imath)\mathrm{Re}U_m + 3\imath] \\
G_{m,m} &= U_{m,m} - U_m + 1 + \frac{\Delta}{k}\left[U_{m,m} - (1+\imath)U_m + \frac{1}{2}(U_m^2 + 3\imath)\right] \\
G_{n,-m} &= U_{n,-m} - U_n - \overline{U}_m + 2 + \frac{\Delta}{k}[U_{n,-m} + U_n\overline{U}_m - (1+\imath)(U_n + \overline{U}_m) + 3\imath] \\
G_{n,m} &= U_{n,m} - U_n - U_m + 2 + \frac{\Delta}{k}[U_{n,m} + U_n U_m - (1+\imath)(U_n + U_m) + 3\imath]
\end{aligned} \tag{4.120}$$

Figures 4.11–4.14 compare the response functions of the burning rate (dashed curves) with the response functions of the velocity of combustion products (solid curves).

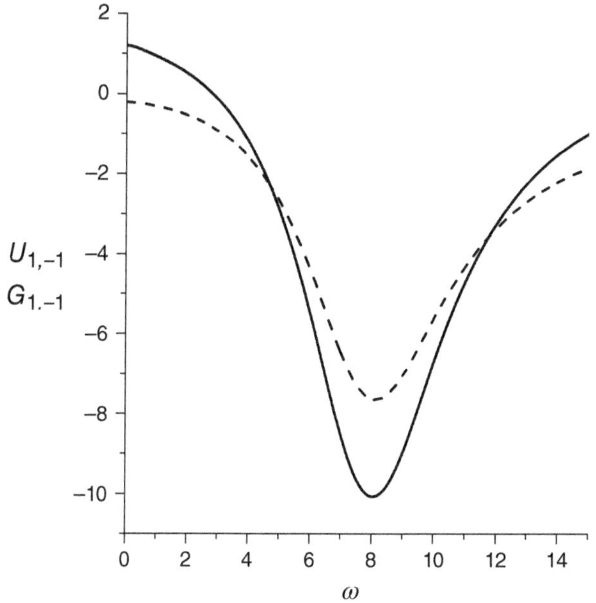

Figure 4.11 Constant components of response functions of burning rate and acoustic admittance.

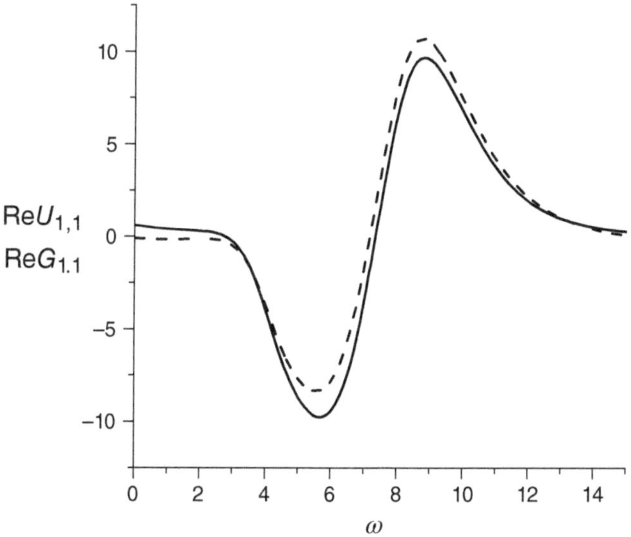

Figure 4.12 Real parts of response functions of burning rate and acoustic admittance for double frequency.

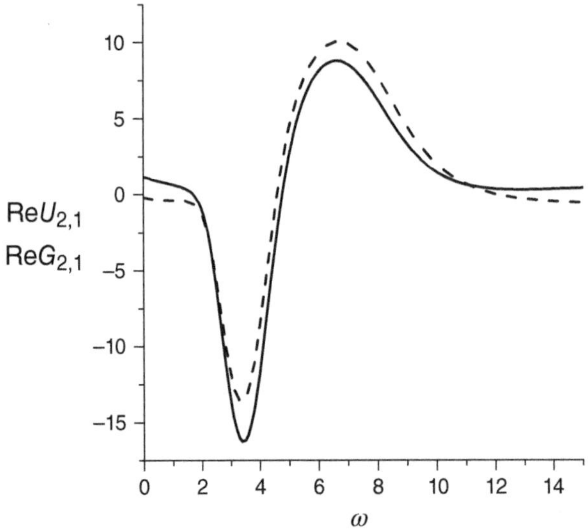

Figure 4.13 Real parts of response functions of burning rate and acoustic admittance for a combinational frequency equal to the tripled frequency of the primary harmonic.

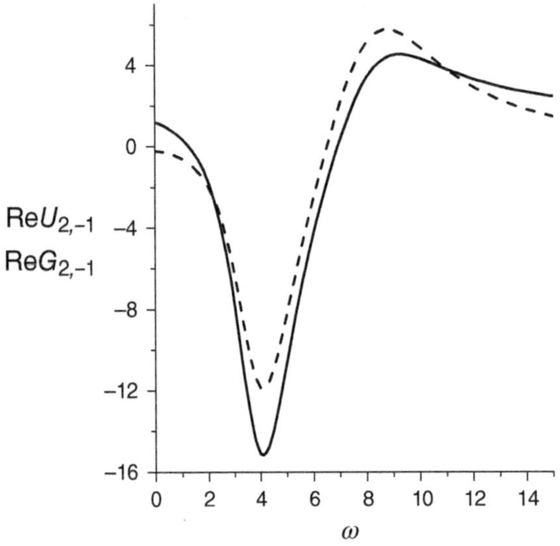

Figure 4.14 Real parts of response functions of burning rate and acoustic admittance for a combinational frequency equal to the frequency of the primary harmonic.

4.5 Nonlinear Resonance

The preceding sections of this chapter considered combustion oscillation regimes in the lower order approximations, namely linear and quadratic. This section investigates some of the most interesting effects of combustion oscillation regimes in the third (with respect to pressure amplitude) approximation. The motivation for such an attempt is driven by the fact that mechanical and electrical oscillating systems demonstrate, once the third degree of nonlinearity is taken into account, qualitatively new properties. The most famous example is dependence of the period of free oscillations of mathematical pendulum on the amplitude of oscillations. It is only in the limit, once the amplitude approaches zero, that the system becomes linear and oscillations are isochronous.

Comprehensive reports on the properties of nonlinear oscillating systems with a finite number of degrees of freedom may be found in Landau and Lifshitz (1982), Stoker (1992), and Andronov et al. (1966). We shall discuss here some of these properties which, as will be demonstrated below, are also attributes of oscillating combustion regimes of solid fuels.

Consider as an example the mechanical system possessing the state of the equilibrium. Deviation from the equilibrium leads to the emergence of the restoring force, the force directed towards the equilibrium state. In the linear approximation, the restoring force is proportional to the magnitude of deviation. Generally, the expression for the restoring force contains additional terms proportional to higher powers of the magnitude of deviation, that is, the second power, third power, and so on. Under moderate amplitudes of free oscillations dynamical equations for such a system may be solved by the method of successive approximations. Deviation from the equilibrium is sought in the form of a series where the first term is a harmonic oscillation with some frequency which itself is a sum of terms of different orders of magnitude. Substitution of this series into dynamical equations leads to the set of differential equations whose solution process starts from the equation for the first approximation containing only linear (with respect to deviation) term in the restoring force. Since the equations for higher approximations contain products of the terms of the first approximation, the dependence of the deviation on time, taking into account nonlinear effects, would contain contributions with frequencies which are multiples of the primary frequency, which are higher harmonics. It turns out to be possible to find corrections to the primary frequency that depend on the oscillation amplitude. Quantitative investigation shows, for example, that the second harmonic as well as an emerging nonzero constant contribution are proportional to the square of the oscillation amplitude. Correction to the primary frequency is also quadratic with respect to the amplitude while the third harmonic varies as the cube of the amplitude.

Significantly new properties emerge on taking into account anharmonic terms in forced oscillations, in particular the resonance region.

In the linear approximation, the amplitude a of forced oscillations

$$a_f \sim \frac{f}{\sqrt{\varepsilon^2 + \lambda^2}} \tag{4.121}$$

where f is amplitude of harmonically varying with time exciting force, λ is damping decrement, and $\varepsilon = \omega - \omega_1$ is the difference between the frequency of the exciting force and the natural frequency of the system.

As discussed above, accounting for nonlinearity results in the emergence of the correction to the natural frequency proportional to the square of the amplitude. Therefore (4.121) is replaced by the relationship

$$a_f^2[(\varepsilon - \alpha a_f^2)^2 + \lambda^2] \sim f^2 \tag{4.122}$$

(where α is a constant coefficient), which is an equation of the third power with respect to the square of the amplitude. Its real roots provide the amplitude of forced oscillations. The dependence of this amplitude on frequency is determined by the magnitude of the amplitude of the exciting force.

In the case of relatively small values, f oscillation amplitude is also not large, therefore only the linear term, with respect to the square of the amplitude, needs be retained in (4.122). This case corresponds to the well-known type of resonance curve that is obtained in the linear approximation and is described by the relationship (4.121). This curve is shown in Figure 4.15a. The shape of the resonance curve changes with an increase in the amplitude of the exciting force. If f is smaller than a certain critical value then the Eq. (4.122) has just one real root at all the frequencies. The corresponding resonance curve is shown in Figure 4.15b. Upon further increase of the amplitude of the exciting

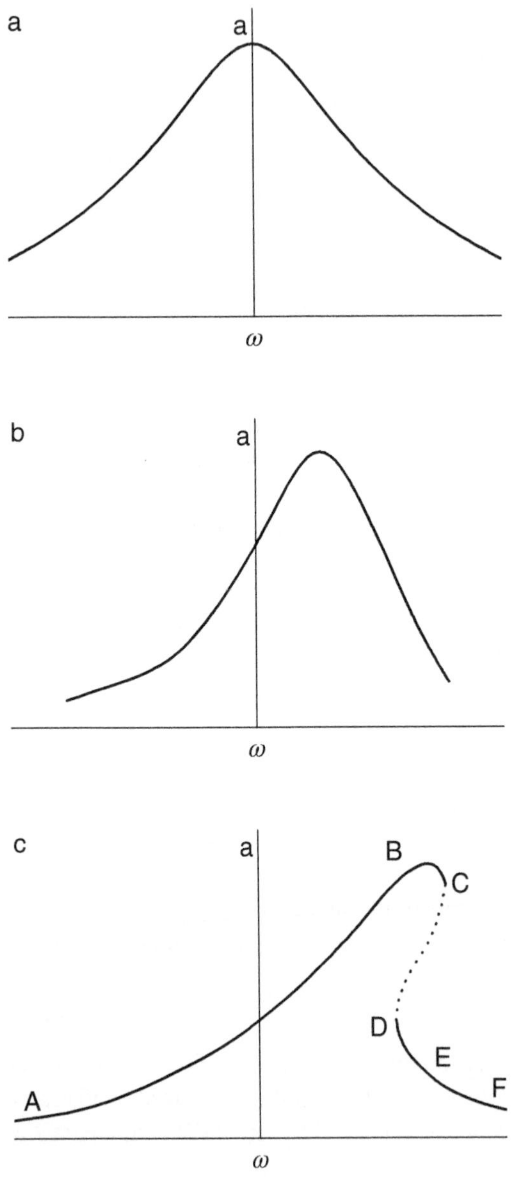

Figure 4.15 Resonance curves for different amplitudes of the exciting force.

force there appears a region with ambiguous dependence of the oscillation amplitude on frequency (Figure 4.15c). At some values of f Eq. (4.122) has three real roots, therefore in this region three different amplitudes of forced oscillations correspond to the same value of frequency.

The theory of nonlinear oscillations (e.g. Bogoliubov and Mitropolsky (1961)) proves that out of the three possible regimes, the stable ones are only those with the maximum and minimum amplitudes. The dashed section of the resonance curve (Figure 4.15c) corresponds to the unstable oscillations of the system, therefore in some region of frequencies stable oscillations with two different amplitudes are possible. The section BC of the resonance curve may be travelled by a gradual increase in the frequency of the exciting force that is on the movement along the branch ABC. Once point C is reached, the amplitude experiences a jump down to the value corresponding to point E. Conversely, on frequency decrease from point F the branch FED is travelled and at point D the amplitude jumps up to the value corresponding to point B.

A peculiar feature of nonlinear oscillating systems with a finite number of degrees of freedom is the possibility of resonances at frequencies different from the natural frequency of the system (so-called ultra- and subharmonic oscillations).

The above examples demonstrate that accounting for nonlinearity allows a number of quite interesting effects in the dynamics of oscillating systems with a finite number of degrees of freedom to be observed.

It is natural to expect that similar phenomena exist in the oscillating combustion regimes of solid fuels under large amplitudes of burning rate. An additional motivation for the study of nonlinear oscillating combustion regimes is the fact that mathematical formulation of the problem is different in this case from the one describing nonlinear oscillations of mechanical and electrical systems. Instead of second-order ordinary differential equations with small nonlinear terms (to which most oscillation problems for systems with a finite number of degrees of freedom can be reduced) one has to investigate nonlinear partial differential equations with the parameter (burning rate) under nonlinear restrictions involving the unknown function (temperature) and its derivative (temperature gradient) at the boundary and control by the exciting force (pressure). This circumstance prevents straightforward application of well-developed methods of the theory of nonlinear oscillations. On the other hand, it may result in the discovery of new significant effects unknown in the theory of conventional mechanical and electrical oscillations.

Section 4.1 considered resonance in the linear approximation. It follows from (4.27) that if the frequency of pressure oscillations is close to the natural propellant burning rate oscillations frequency and the damping decrement is small (i.e. the combustion regime is close to the stability boundary), then the relation between the complex amplitudes of the burning rate and pressure takes the form

$$\nu_1(\varepsilon - i\lambda) \sim \eta_1 \tag{4.123}$$

Resonance investigation in the third approximation is a quite difficult analytical task (Novozhilov 1966). Here we only report the final result. In the cube approximation (4.123) is replaced by a relationship similar to (4.122)

$$\nu_1[\varepsilon - i\lambda + (\alpha_1 - i\alpha_2)|\nu_1|^2] = F\eta_1 \tag{4.124}$$

where α_1 and α_2 are real and F is a complex function of the parameters describing the linear sensitivity of the burning rate and surface temperature to the pressure and initial temperature variations, that is, k, r, ι, and μ.

Taking the modulus of the latter relationship we find the equation for the resonance curve

$$|\nu_1|^2[(\varepsilon+\alpha_1|\nu_1|^2)^2+(\lambda+\alpha_2|\nu_1|^2)^2]=|F||\eta_1| \tag{4.125}$$

Accounting for nonlinear effects therefore leads to dependence of frequency and damping decrement at resonance on the oscillation amplitude. As is the case for conventional nonlinear oscillations of electrical and mechanical systems, the correction to the frequency is proportional to the square of the oscillation amplitude. A significant feature of the ZN theory is existence in (4.124) squared with respect to amplitude correction to the damping decrement.

In the steady-state propellant combustion regime, $\lambda > 0$. In order to investigate properties of the resonance curve it is convenient to introduce the new variables

$$U^2=\frac{|\alpha_2||\nu_1|^2}{\lambda},\quad \Gamma=\frac{\varepsilon}{\lambda},\quad \alpha=\frac{\alpha_1}{|\alpha_2|},\quad P^2=\frac{|F|^2|\alpha_2||\eta_1|^2}{\lambda^2} \tag{4.126}$$

Then the resonance curve equation takes the form

$$(\Gamma+\alpha U^2)^2+(1\pm U^2)^2=\frac{P^2}{U^2} \tag{4.127}$$

where the positive sign corresponds to the positive and the negative sign to the negative values of α_2. The frequency can be expressed from (4.127) as a function of oscillation amplitude

$$\Gamma=-\alpha U^2\pm\sqrt{\frac{P^2}{U^2}-(1+U^2)^2}\ \text{for}\ \alpha_2>0 \tag{4.128}$$

$$\Gamma=-\alpha U^2\pm\sqrt{\frac{P^2}{U^2}-(1-U^2)^2}\ \text{for}\ \alpha_2<0 \tag{4.129}$$

The sign of α does not play a significant role as its reversal (or equivalently the reversal of the sign of α_1) is equivalent to the transformation $\Gamma\to-\Gamma$, that is, to the reflection of the resonance curve with respect to the axis $\Gamma = 0$. In the following we assume that $\alpha > 0$.

Consider first the case $\alpha_2 > 0$, that is, damping increasing with amplitude. For small oscillation amplitudes, neglecting higher powers of U^2 in (4.127) we obtain the expression for the resonance curve in the linear approximation

$$(1+\Gamma^2)U^2=P^2 \tag{4.130}$$

The oscillation amplitude has maximum at $\Gamma = 0$. An increase in the pressure amplitude leads to a shift of the frequency corresponding to the maximum oscillation amplitude. From (4.127)

$$\frac{dU^2}{d\Gamma}=\frac{-2U^2(\Gamma+\alpha U^2)}{3(\alpha^2+1)U^4+4(\alpha\Gamma+1)U^2+1+\Gamma^2} \tag{4.131}$$

Oscillation amplitude reaches a maximum at

$$\Gamma_e=-\alpha U_e^2 \tag{4.132}$$

Substitution of this relationship into (4.127) provides the relation between the maximum amplitude of burning rate oscillations and prescribed pressure amplitude

$$(1 + U_e^2)U_e^2 = P^2 \tag{4.133}$$

The derivative (4.131) is infinite if

$$3(\alpha^2 + 1)U_i^4 + 4(\alpha\Gamma_i + 1)U_i^2 + 1 + \Gamma_i^2 = 0 \tag{4.134}$$

Solving the last equation together with (4.127) we obtain the coordinates of those points on the amplitude curve where the derivative is infinite for given P и α. Let us find conditions for the existence of such points (they play an important role in studying the stability of oscillating regimes). Eq. (4.134) is quadratic with respect to U_i^2 and has two identical positive roots if

$$\Gamma_i^2(\alpha^2 - 3) + 8\alpha\Gamma_i - 3\alpha^2 + 1 = 0 \text{ and } \alpha\Gamma_i + 1 < 0 \tag{4.135}$$

The latter implies

$$(U_i^*)^2 = \frac{2}{\sqrt{3}(\alpha - \sqrt{3})}, \quad \Gamma_i = -\frac{4\alpha + \sqrt{3}(\alpha^2 + 1)}{\alpha^2 - 3} \tag{4.136}$$

Therefore the points with infinite derivatives exist only for $\alpha > \sqrt{3}$, that is, for sufficiently large ratio α_1/α_2. For a given value of α such points only appear at the pressure amplitude exceeding the critical value

$$(P_i^*)^2 = \frac{8(\alpha^2 + 1)}{3\sqrt{3}(\alpha - \sqrt{3})^2} \tag{4.137}$$

which can be obtained by substituting (4.136) into (4.127). If $\alpha < \sqrt{3}$ the resonance curve does not possess points with infinite derivatives for any values of the pressure amplitude.

Figure 4.16 shows resonance curves for $\alpha = 10$. Curves 1–4 are drawn for the values of P^2 equal 0.05, 0.2, 0.4, and 0.8, respectively. The locus of the maxima of resonance curves is curve e while the points with infinite derivatives belong to curve i. The critical value of the square of pressure amplitude is $(P_i^*)^2 = 0.227$.

Resonance curves for $\alpha = 1$ are plotted in Figure 4.17. Curves 1–3 correspond to the values of P^2 equal to 0.25, 1.0, and 4.0.

It is instructive to compare the obtained results with the conventional nonlinear resonance. In nonlinear oscillations of electrical or mechanical systems, the damping decrement does not change with the oscillation amplitude. Therefore $\alpha_2 = 0$ and $\alpha = \infty$, that is, the resonance is of the type shown in Figure 4.16. For sufficiently large amplitude of the exciting force the resonance curves have points with infinite derivatives. The section of the curve between these points correspond to unstable oscillation regimes. In contrast, during propellant combustion, the damping decrement does change with the amplitude. In the presently considered case where $\alpha_2 > 0$, damping increases with the increase in the oscillation amplitude. If $\alpha_1/\alpha_2 < \sqrt{3}$ then widening of the resonance curve with increase in the amplitude compensates its bending resulting from dependence of frequency on the amplitude; consequently there are no points with infinite derivatives. Under the reverse relation between α_1 and α_2 (strong bending and weak widening of the resonance curve)

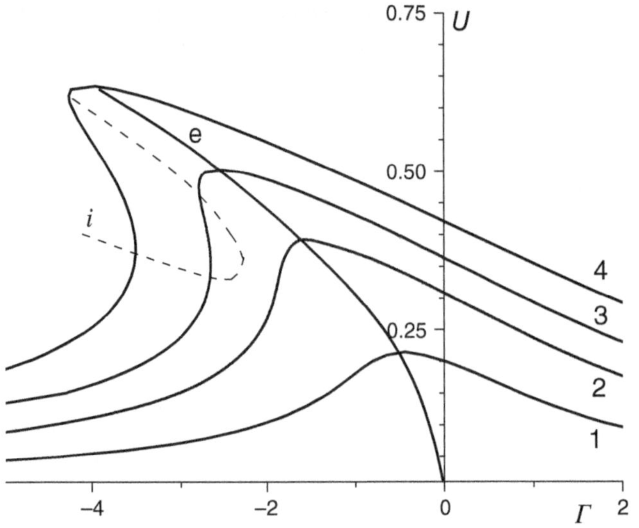

Figure 4.16 First type of resonance curves.

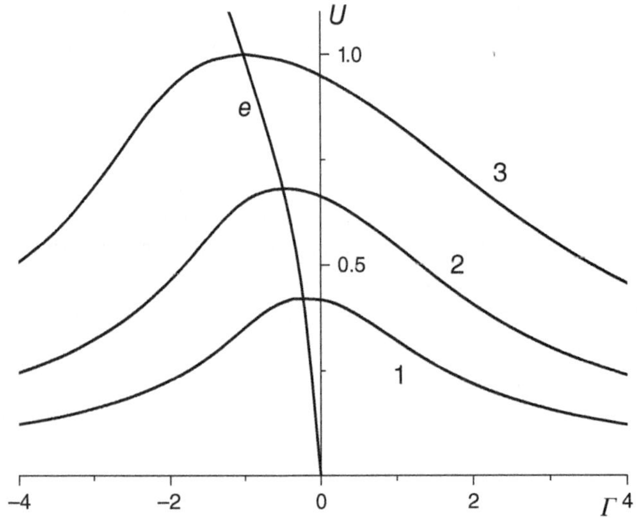

Figure 4.17 Second type of resonance curves.

such points exist. The value $\alpha = \infty$ corresponds to the case where the damping decrement does change with the amplitude and the formulas derived above take the form of the known relationships of the theory of conventional nonlinear oscillations.

Consider now resonance curves for the case $\alpha_2 < 0$ where damping decrement decreases with increase in the oscillation amplitude. Extremum amplitude values are obtained from the relationships

$$\Gamma_e = -\alpha U_e^2, \quad (1 - U_e^2)U_e^2 = P^2 \tag{4.138}$$

The second of these equations may have, depending on the value of P, either one or three positive roots U_e^2. Straightforward analysis shows that for small values of the pressure amplitude

$$P < P_r, \quad P_r^2 = 4/27 \tag{4.139}$$

there are three roots. For $P > P_r$ the root is unique.

Coordinates of the points with infinite derivatives can be found by solving the following set of equations

$$(\Gamma_i + \alpha U_i^2) + (1 - U_i^2)^2 = \frac{P^2}{U_i^2}$$
$$3(\alpha^2 + 1)U_i^4 + 4(\alpha\Gamma_i - 1)U_i^2 + 1 + \Gamma_i^2 = 0 \tag{4.140}$$

For $\alpha^2 > 3$ the second equation describes the curve with branches approaching infinity. The minimum value of Γ_i^* and the corresponding $(U_i^*)^2$ and $(P_i^*)^2$ are equal to

$$\Gamma_i^* = -\frac{4\alpha - \sqrt{3}(\alpha^2 + 1)}{\alpha^2 - 3}, \quad (U_i^*)^2 = \frac{2}{\sqrt{3}(\alpha + \sqrt{3})}$$
$$(P_i^*)^2 = \frac{8(\alpha^2 + 1)}{3\sqrt{3}(\alpha + \sqrt{3})^2} \tag{4.141}$$

For $\alpha^2 < 3$ the curve $U_i(\Gamma_i)$ is closed. Extremum values of Γ_i^* and the corresponding $(U_i^*)^2$ and $(P_i^*)^2$ are given by the relationships

$$\Gamma_i^* = \frac{4\alpha \mp \sqrt{3}(\alpha^2 + 1)}{\alpha^2 - 3}, \quad (U_i^*)^2 = \frac{2}{\sqrt{3}(\sqrt{3} \pm \alpha)}$$
$$(P_i^*)^2 = \frac{8(\alpha^2 + 1)}{3\sqrt{3}(\sqrt{3} \pm \alpha)^2} \tag{4.142}$$

Figures 4.18 and 4.19 show resonance curves for the case $\alpha_2 < 0$. Figure 4.18 is plotted for $\alpha = 3$; curves 1–3 correspond to the values of P^2 equal to 0.05, 4/27, and 0.5, respectively. Curve i consisting of the points with infinite derivatives is unclosed ($\alpha > \sqrt{3}$). Conversely, on Figure 4.19 drawn for $\alpha = 1$ the locus of the points with infinite derivatives is the closed curve. Curves 1–5 correspond to the values of P^2 equal to 0.02, 4.27, 1.00, 4.00, and 8.00. In other words, resonance curves corresponding to large values of the amplitude of the exciting force always have two points with infinite derivatives if $\alpha > \sqrt{3}$. In the opposite case ($\alpha < \sqrt{3}$) there exists a critical value of the exciting force amplitude such that above this value the points with infinite derivatives are absent.

The most interesting property of the resonance curves for $\alpha_2 < 0$ is their separation into two branches (curve 1) if $P < P_r$ ($P_r^2 = 4/27$). The bottom branch with small oscillation amplitudes may be obtained (for sufficiently small P) in the linear approximation. However, the closed loop enclosing the point $(-\alpha, 1)$ is a consequence of the nonlinearity of the theory. In particular, in the absence of the exciting force the resonance curve degenerates into the line $U = 0$ and the point $(-\alpha, 1)$. Indeed for negative values of α_2, that is, for decrease of damping decrement with oscillation amplitude, the condition $\lambda + \alpha_2|v_1|^2 = 0$ may be fulfilled. This condition corresponds to free nonlinear oscillations with amplitude

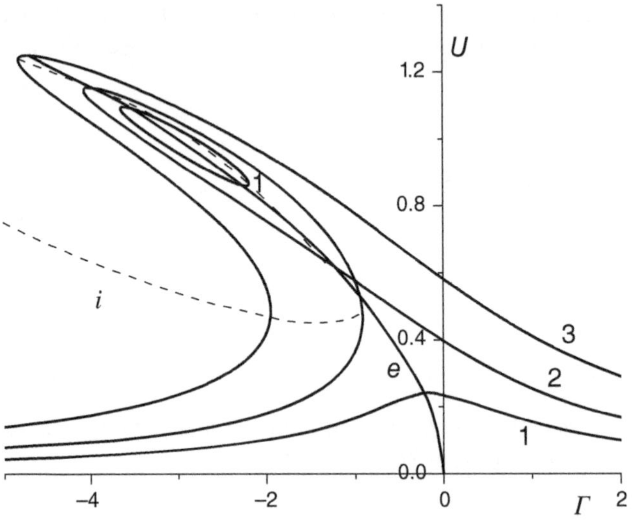

Figure 4.18 Third type of resonance curves.

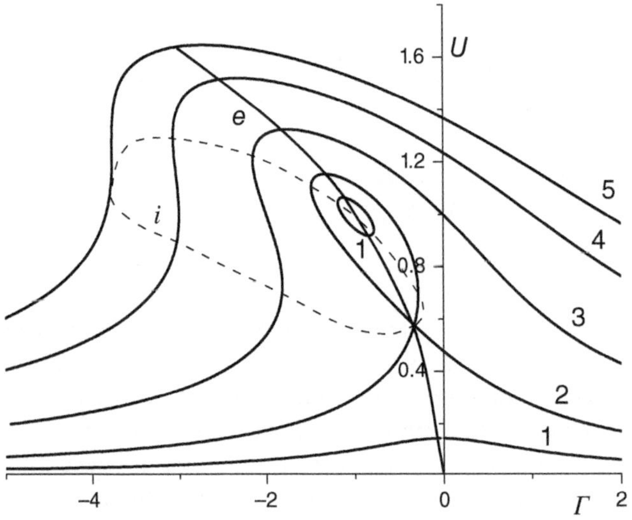

Figure 4.19 Fourth type of resonance curves.

and frequency

$$|v_1| = \sqrt{\lambda/|\alpha_2|}, \quad \varepsilon = -\alpha\lambda \tag{4.143}$$

Such a process may be considered as an auto-oscillation. Resonance curves of the same type are also known in the theory of nonlinear oscillations of electrical systems. For example, Stoker (1992) considers an auto-oscillating system designed as a vacuum tube generator with a feedback loop and an additional source of alternating voltage included in its control grid circuit.

The above analysis refers to the case of the stable steady-state regime $\lambda > 0$. It may turn out that this steady-state regime is not stable ($\lambda < 0$) but the combustion still progresses. In that case, if $\alpha_2 > 0$ then for $\lambda + \alpha_2 |v_1|^2 = 0$ there would exist a nonlinear oscillating regime of propellant combustion.

An important problem of the theory of nonlinear oscillations is the investigation of the stability of the process at different sections of the resonance curve. The solution of this problem requires dedicated analysis. By analogy with conventional nonlinear oscillations (observed in systems with a finite number of degrees of freedom) it may be expected that for the resonance curves shown in Figure 4.16 the middle of the three possible regimes will be unstable.

It is necessary to explain now the occurrence of resonances at frequencies different from the propellant natural frequency. As an example, consider the amplitude of the second harmonic v_2. In the linear approximation the denominator of the expression for v_2 contains the quantity

$$d_2 = 1 + \left(r - \frac{k}{z_2} \right) (z_2 - 1) \tag{4.144}$$

equal to zero at $\omega = \omega_1/2$. Therefore, resonance also occurs at the frequency of the exciting force equal to the half of the natural frequency. The method developed in the present section may be applied to show that resonances must also occur at pressure frequencies equal to $p\omega_1/q$ where p and q are integers. Such ultra- and subharmonic oscillations are well known in the theory of electrical and mechanical oscillations. An example of calculation of the propellant burning rate resonance at the second harmonic is provided by Novozhilov (1992b).

It is instructive to discuss the discovered auto-oscillating propellant combustion regimes from the general viewpoint of qualitative understanding of auto-oscillations.

It is well known (e.g. Kharkevich (1953)) that any auto-oscillation system includes the three elements: the energy source, the valve, i.e. the device regulating the energy supply to the system, and the oscillation system itself. It is easy to identify these elements in our case. Solid fuel, along with gaseous combustion products, serves as an oscillation system. Its parameters, such as burning rate and temperature distribution in condensed and gas phases, change with time. Reaction zones represent energy source. At the same time these zones play the role of a valve since the energy supply to the system is controlled (at constant pressure) by their temperatures.

Furthermore, the three elements of the auto-oscillation system must be arranged in such a way that apart from the direct influence of the valve on the system the reverse influence (feedback) by the system on the valve would occur as well. Feedback is a necessary condition of the existence of auto-oscillation regimes. In propellant combustion feedback appears as dependence of the valve parameters (i.e. temperatures of energy release zones) on the system parameters, namely the surface temperature gradient.

Another necessary property of the auto-oscillation system is its nonlinearity. Established auto-oscillations with finite amplitude may only appear if at least one of the system elements or the system feedback is nonlinear. In propellant combustion both the system itself (due to nonlinearity of the heat transfer equation) and the feedback (due to nonlinearity, in the general case, of nonsteady burning laws) are nonlinear.

Excitation of auto-oscillations may be soft when a regime with finite amplitude is generated due to small accidental perturbation with appropriate frequency, and hard (triggering) when a regime of auto-oscillations may only be achieved by strong disturbance of the system. It seems plausible that hard excitation occurs in a stable (in the linear approximation) combustion regime ($\lambda > 0$) while soft excitation in an unstable one ($\lambda < 0$).

4.6 Response Function Bifurcations

At the beginning of this chapter, the linear and quadratic response functions of the propellant burning rate under harmonically oscillating pressure were investigated. In both cases the period T of burning rate oscillations coincided with the period of pressure oscillations. We shall call such combustion regimes T regimes or T solutions. Naturally, under sufficiently strong nonlinearity T solutions do not only contain a primary harmonic whose frequency coincides with the pressure frequency, but also modes with higher frequencies that are multiples of the frequency of the primary harmonic. What will happen on further penetration into the region of instability, that is if the pressure oscillation amplitude is increased further? It is this question which is considered in the present section.

In Section 3.4, a numerical analysis was applied to investigate nonsteady propellant combustion modes under constant pressure beyond the stability boundary of the steady-state regime. The simplest propellant combustion model, considered within the framework of the ZN theory, contained just two parameters describing the properties of the system, namely k and r. For the fixed value of the parameter r the second one plays the role of a bifurcation parameter. It was demonstrated that changing the bifurcation parameter leads to the system transition from a steady-state combustion regime to chaotic combustion. The transition occurs via the Feigenbaum scenario of cascading period doubling bifurcations that eventually leads to the onset of a chaotic combustion regime.

The present section discusses the results of the numerical analysis of propellant combustion under large amplitudes of harmonically oscillating pressure (Novozhilov 2005). Here, under fixed values of the parameters k and r describing propellant properties, pressure amplitude will be considered as the bifurcation parameter.

The mathematical formulation of the problem is as follows. The heat transfer equation

$$\frac{\partial \theta}{\partial \tau} = \frac{\partial}{\partial \xi}\left(\frac{\partial \theta}{\partial \xi} - v\theta\right), \quad 0 \le \xi < \infty, \quad 0 \le \tau < \infty \tag{4.145}$$

is solved with the boundary conditions

$$\xi = 0, \quad \theta = \vartheta(\tau), \quad \xi \to \infty, \quad \theta = 0 \tag{4.146}$$

Steady-state burning laws are adopted in the form (4.72). The corresponding nonsteady burning laws are

$$v = \eta' \exp\left[k\left(\vartheta - \frac{\varphi}{v}\right)\right], \quad v = \exp\left[\frac{k}{r}(\vartheta - 1)\right], \quad \varphi = -\left.\frac{\partial \theta}{\partial \xi}\right|_{\xi=0} \tag{4.147}$$

Pressure dependence on time is taken as

$$\eta = 1 + h\cos\omega\tau \tag{4.148}$$

The steady-state solution serves as an initial condition

$$\theta^0(\xi, \tau_i) = e^{-\xi}, \quad \text{for} \quad \tau_i = \frac{\pi}{2\omega} \tag{4.149}$$

The other initial conditions are

$$\eta^0(\tau_i) = 1, \quad \varphi^0(\tau_i) = 1, \quad v^0(\tau_i) = 1, \quad \vartheta^0(\tau_i) = 1 \tag{4.150}$$

The adopted formulation represents one of the simplest systems with distributed parameters (i.e. with an infinite number of degrees of freedom). It is described by just one partial differential equation and two, in general case nonlinear, relationships between burning rate, propellant surface temperature, and its gradient at the boundary of the condensed phase. Obviously, full investigation of this nonlinear problem is only possible by means of numerical methods.

The set of Eqs. (4.145)–(4.150) contains five parameters. The values of four of them are fixed

$$k = 1.5, \quad r = 0.15, \quad \iota = 0.3, \quad \omega = 2\pi \tag{4.151}$$

so that $T = 1$. The evolution of the behaviour of this system is investigated by gradually increasing the pressure amplitude h.

The accuracy of numerical modelling has been verified for the case of small amplitudes where an exact analytical solution exists. As an example, for $h = 0.01$ numerical solution gives for the nondimensional amplitude of the burning rate $v_1 = 0.0238$. This value corresponds to the modulus of the response function $|U| = 2.38$ and coincides, with 3% accuracy, with the analytical result $|U| = 2.45$ following from (4.27).

Figures 4.20–4.25 illustrate the response of the system behaviour to variation of the pressure amplitude h.

At small pressure amplitude h, the periods of burning rate and pressure oscillations are the same and equal to T (T regime). This regime is shown in Figure 4.20. Figure 4.20a plots the time history of the burning rate. Figure 4.20b is two-dimensional projection of the phase trajectory of the system on the plane (v, y_0) where

$$y_0(\tau) = \int_0^\infty \theta(\xi, \tau)d\xi \tag{4.152}$$

is the heat content of the condensed phase represented by the first moment of the temperature distribution.

As the limit cycle shown in Figure 4.20b is substantially different from an ellipse, the corresponding regime can be called a nonlinear T solution. Recall that in the case of linear oscillations the limit cycle has the shape of an ellipse which can degenerate into a line or a circle.

On reaching the critical value of $h^{(1)} = 0.16565$, the system undergoes bifurcation from the T to the $2T$ regime. This bifurcation is shown in Figure 4.21.

Under the $2T$ regime (Figure 4.22a) the period of burning rate oscillations becomes double the period of pressure oscillations. The limit cycle starts to look more complicated

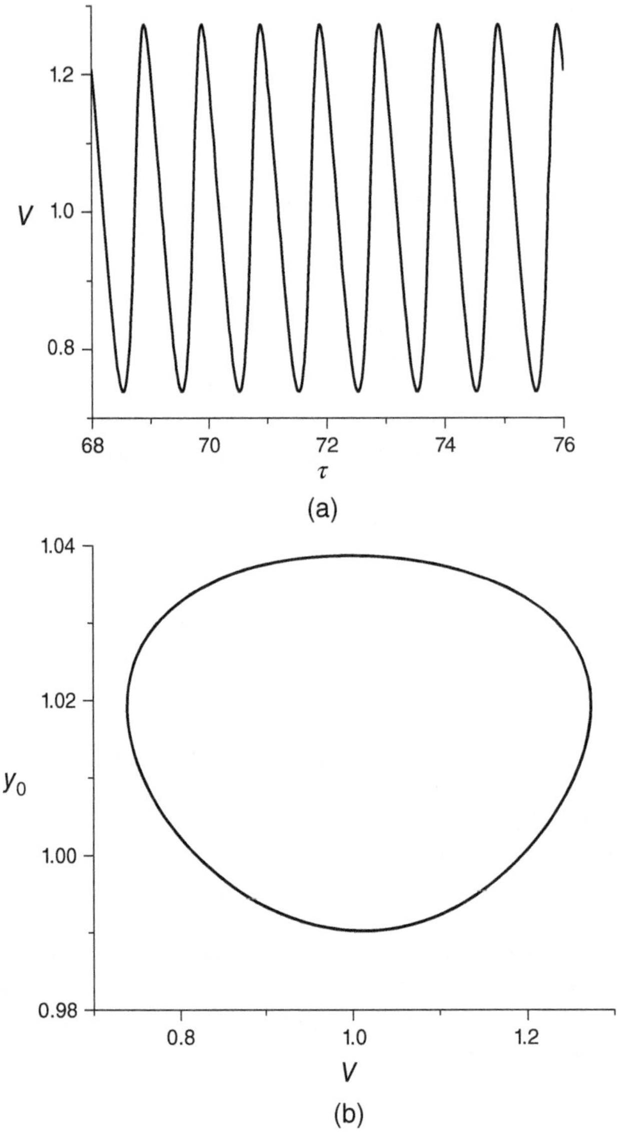

Figure 4.20 T regime. $h < h^{(1)}$, $h = 0.1$.

(Figure 4.22b). A further increase in the bifurcation parameter h leads to a cascade of bifurcations and the period doubles at each of them. The inaccuracies of this numerical analysis do not allow the full infinite sequence of bifurcations to be followed. The first few period doubling bifurcations occur at the following values of h

$$h^{(1)} = 0.16565, \quad h^{(2)} = 0.18095, \quad h^{(3)} = 0.18275, \quad h^{(4)} = 0.18320 \tag{4.153}$$

After the fourth bifurcation, on further slight increase of the bifurcation parameter, a chaotic combustion regime is observed. The phase trajectory fills a certain region of the phase plane nearly uniformly (Figure 4.25b).

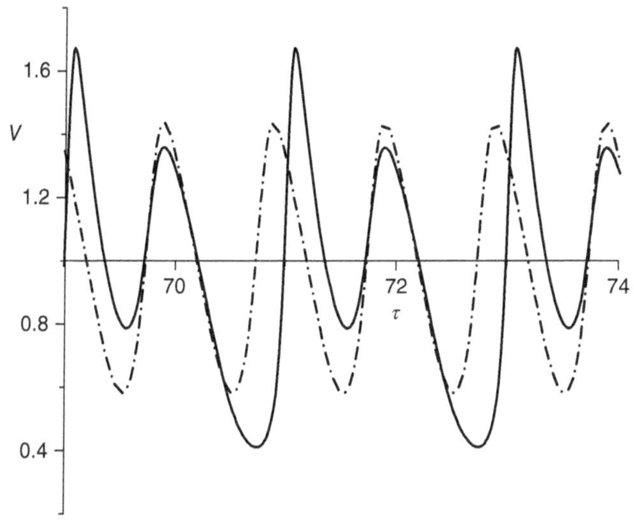

Figure 4.21 Time dependence of burning rate. Dashed curve, $h = 0.16$; solid curve, $h = 0.177$.

The performed numerical simulations confirm that transition from the stable steady-state to a chaotic combustion regime follows the Feigenbaum scenario of cascading bifurcations (Landau and Lifshitz 1987; Arnold 1988). In this scenario the successive bifurcation values $h^{(m)}$ satisfy the simple law

$$\lim_{m \to \infty} \delta_m = \delta, \quad \delta_m = \frac{h^{(m)} - h^{(m-1)}}{h^{(m+1)} - h^{(m)}} \tag{4.154}$$

where $\delta = 4.669\ldots$ is the universal Feigenbaum constant. From the bifurcation values of the parameter $h^{(m)}$ given above (4.153) one can estimate

$$\delta_2 = 8.5 \pm 0.5, \quad \delta_3 = 4 \pm 1 \tag{4.155}$$

Therefore the sequence δ_m, as has been observed previously in many cases, converges very quickly. The significant value of the relative error is due to the necessity to calculate differences between close values involved in (4.154) and also to the intrinsic inaccuracy of the numerical method.

4.7 Frequency–Amplitude Diagram

Section 4.6 considered the behaviour of the system under periodic pressure oscillations with fixed frequency. Nevertheless it is clear that bifurcation properties depend not only on the pressure amplitude but also on its frequency.

Two different ways of representing periodic functions are used in the present section. Numerical calculations involve real numbers and it is convenient therefore to use Fourier series with real coefficients. In this case pressure and burning rate are given by the following formulae

$$\eta = 1 + h \cos \omega\tau, \quad v = 1 - \alpha_0 + \sum_{l=1}^{\infty} \alpha_l \cos(l\omega\tau + \psi_l) \tag{4.156}$$

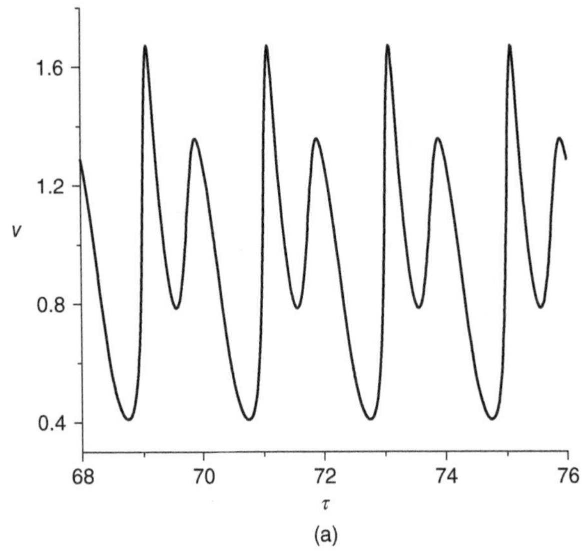

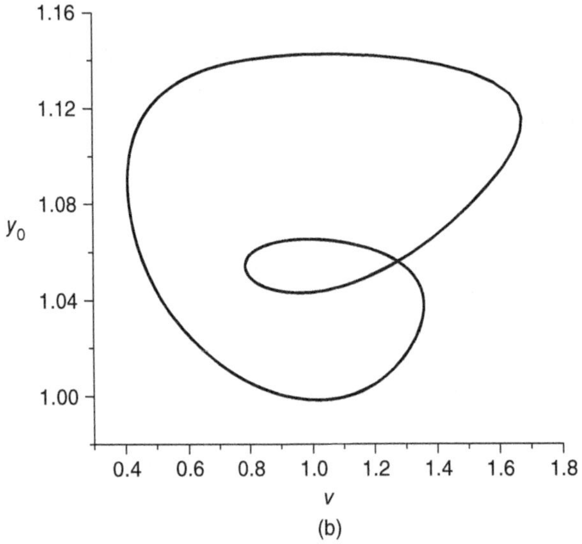

Figure 4.22 $2T$ regime. $h^{(1)} < h < h^{(2)}$, $h = 0.177$.

Here, nonlinear correction to the constant component of the burning rate is written down explicitly. Normally the mean burning rate decreases on accounting for nonlinear effects, therefore $\alpha_0 > 0$. It is assumed that the periods of the pressure and burning rate variations are equal.

For analytical analysis it is more convenient to use the method of complex amplitudes. Pressure and burning rate are written in the following form

$$\eta = 1 + [\eta_1 \exp(i\omega\tau) + c.c], \quad v = 1 - v_0 + \sum_{l=1}^{\infty} [v_l \exp(il\omega\tau) + c.c] \tag{4.157}$$

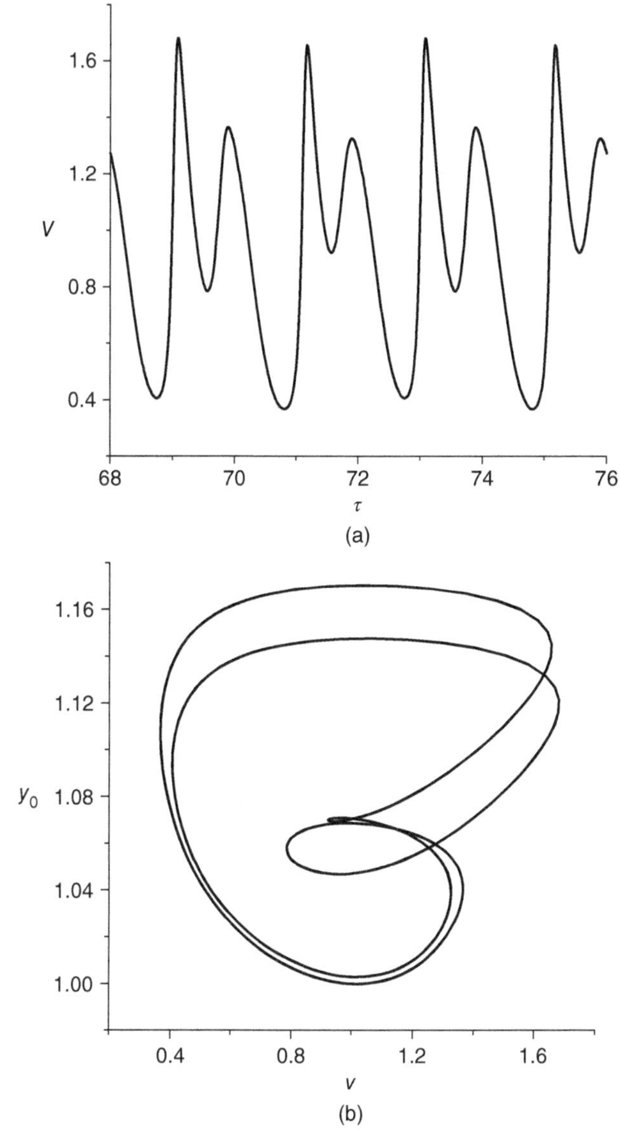

Figure 4.23 4T regime. $h^{(2)} < h < h^{(3)}$, $h = 0.182$.

It is obvious that

$$\eta_1 = \frac{h}{2}, \quad v_0 = \alpha_0, \quad v_l = \frac{1}{2}\alpha_l \exp(i\psi_l) \tag{4.158}$$

If variable pressure contains just one harmonic then the behaviour of the system may be studied on the plane with pressure frequency and amplitude as coordinates (Novozhilov 2005; Novozhilov et al. 2006). We shall call this plane the frequency–amplitude diagram. It is shown in Figure 4.26.

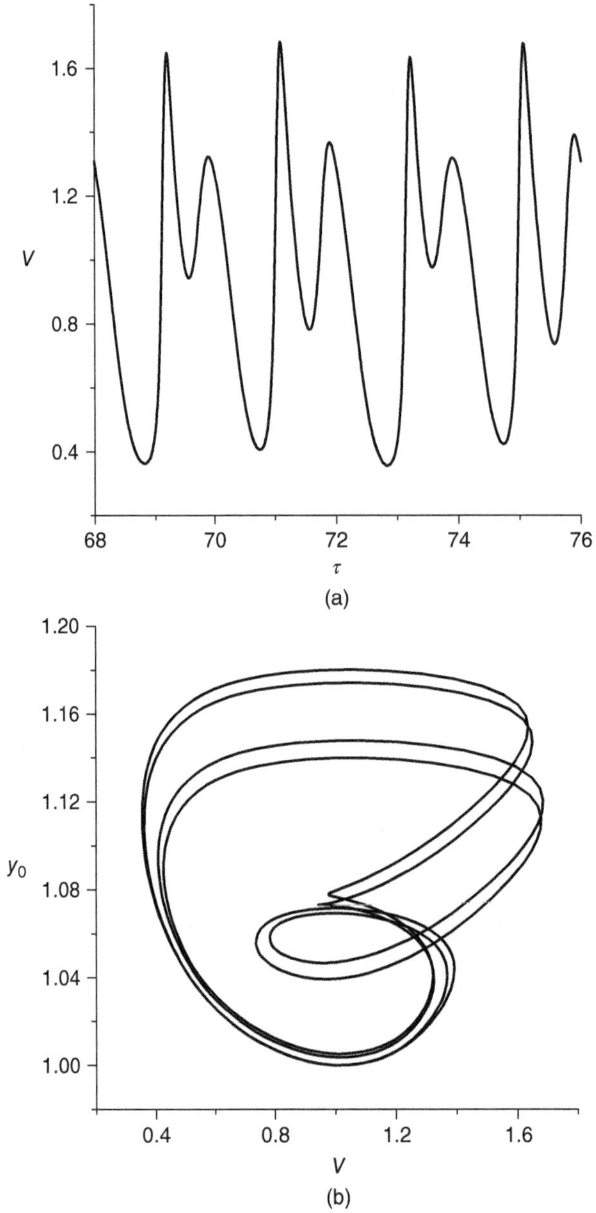

Figure 4.24 8T regime. $h^{(3)} < h < h^{(4)}$, $h = 0.183$.

The solid curve B separates the bifurcation region (above this curve) from the regimes where the burning rate oscillates with the same frequency as pressure. The latter part of the plane may be further divided tentatively into two parts. For small amplitudes of pressure oscillations (below curve L) the linear approximation may be applied and the first-order (i.e. linear) response functions be used. This region is denoted T_l. For high values of the pressure amplitude burning rate contains higher harmonics, the linear approximation does

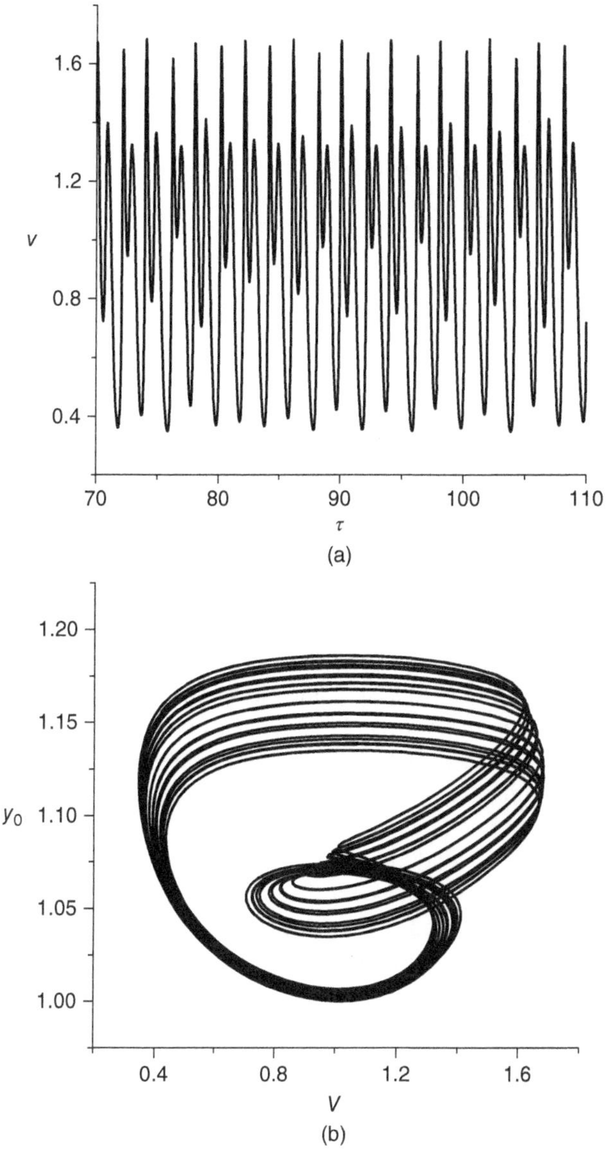

Figure 4.25 Chaotic combustion regime. $h > h^{(4)}$, $h = 0.1835$.

not work and involvement of response functions of higher orders is necessary (T_n region). The exact position of curve L separating linear and nonlinear T regimes depends of course on the criterion of application of the linear approximation that is on an *a priori* accepted error resulting from higher harmonics cut-off.

The curves B and L (as well as all the other curves in Figure 4.26) have a minimum frequency which is close to the natural frequency of the propellant. Resonance is most pronounced in the vicinity of this minimum, therefore as the pressure amplitude increases, the

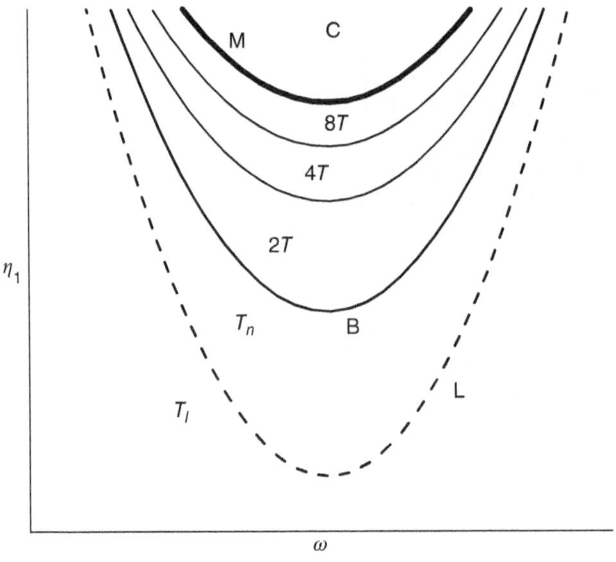

Figure 4.26 Frequency–amplitude diagram.

nonlinear effects in this region manifest themselves early and bifurcations occur at lower values of the amplitude. Boundaries separating various $2^m T$ regimes, that is, the curves of successive bifurcations, are plotted above curve B. On increasing the pressure amplitude the curves become more dense as the differences between successive bifurcation values of the parameter η_1 drop quickly. The thick line M embodies the infinite sequence of period doubling bifurcations which likely ends up with the chaotic combustion regime (region C).

Variation of propellant parameters would not change the diagram qualitatively. It is obvious that if resonance becomes more pronounced (e.g. with a decrease in the difference $r - r^*$) all the curves shown will be shifted down, that is, all the nonlinear and bifurcation effects will become more intensive. This effect is illustrated in Figure 4.27.

The frequency–amplitude diagram discussed above provides general understanding of the interaction between the process of propellant combustion and single-mode variation of pressure. Obviously, the presence of multiple harmonics complicates considerations significantly. In addition, a coupled consideration of the acoustic volume (i.e. engine chamber) and combustion process would lead to new challenges both in the problem formulation and its (even qualitative) investigation.

Let us consider now nonlinear T regimes. Analytical investigation of nonlinear interaction between acoustics and combustion is clearly only achievable in the lower (quadratic or, at best, third order with respect to oscillation amplitude) approximations. The results of numerical simulations and their comparison with analytical results are provided below. The comparison makes use of nonlinear response functions of the burning rate to periodically oscillating pressure. These functions were discussed in detail in Section 4.3. We shall restrict ourselves to quadratic response functions

$$U_{1,-1} = -|U_1|^2 + \imath(2\text{Re}U_1 - 1)$$

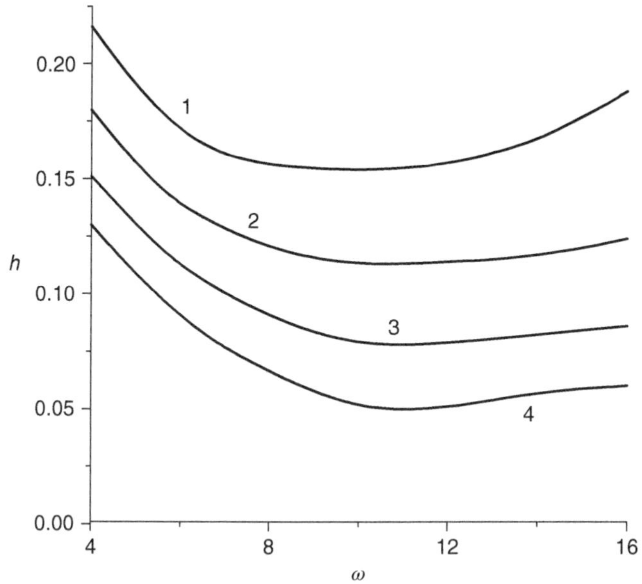

Figure 4.27 Dependence of pressure oscillation amplitude on frequency on bifurcation from the T regime to the $2T$ regime. $k = 1.5$, $\imath = 0.3$, $\delta = 0.1$, $r = 0.15$; 2, $r = 0.14$; 3, $r = 0.13$; 4, $r = 0.12$.

$$U_{1,1} = \frac{U_1^2 U_2}{\imath}\left[-\frac{1}{2} + \frac{k(z_2 - 1)}{2\omega^2}(2z_1 - z_2 - 1) + \frac{r}{2}\left(z_2 - \frac{4z_1}{z_2} + 3\right)\right] + \\ + U_2\left(U_1 - \frac{1}{2}\right) \tag{4.159}$$

where the following notation is introduced

$$U_1 = \frac{\imath + \delta(z_1 - 1)}{1 + (z_1 - 1)(r - k/z_1)}, \quad U_2 = \frac{\imath + \delta(z_2 - 1)}{1 + (z_2 - 1)(r - k/z_2)} \\ z_1 = \frac{1}{2}(1 + \sqrt{1 + 4i\omega}), \quad z_2 = \frac{1}{2}(1 + \sqrt{1 + 8i\omega}) \tag{4.160}$$

Frequency dependencies of the modulus of linear and nonlinear response functions are presented in Figures 4.28–4.30. The plots compare results of analytical and numerical solutions. Curves 1 are drawn using the above formulae (4.159) and (4.160) while curves 2 and 3 are the results of numerical simulation at different values of the pressure amplitude. Quite large values of quadratic response functions, especially in the vicinity of the natural frequency of the propellant, are noticeable. This is a consequence of the resonance behaviour of the linear response functions U_1 and U_2. Quadratic response functions contain products of the response functions of the first order. It is equally evident that the analytical expressions of the response functions provide good accuracy even close to the bifurcation boundary.

Progressive complication of the burning velocity spectrum with growing pressure amplitude can be followed in Figures 4.31–4.34. The first three figures refer to T regimes while Figure 4.34 refers to the $2T$ regime. The frequencies of pressure oscillations $\omega = 8$ are chosen to be close to the natural frequency of the propellant ($\omega_1 = 7.68$). Figures 4.31–4.34 clearly

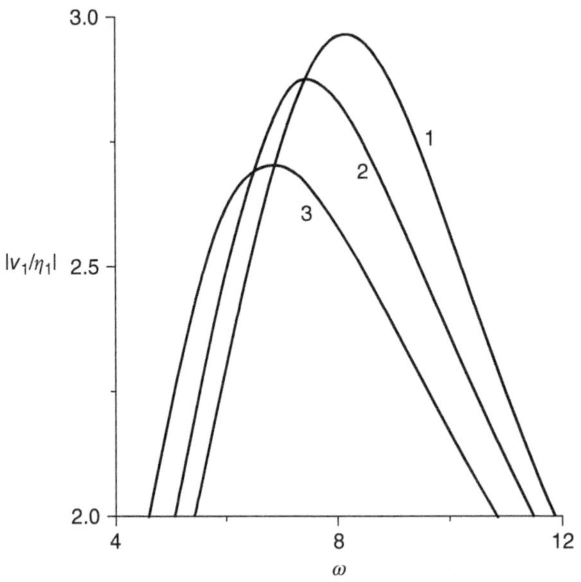

Figure 4.28 Modulus of the relative magnitude of the first harmonic of the burning velocity. 1, linear approximation; 2, $h = 0.1$; 3, $h = 0.15$.

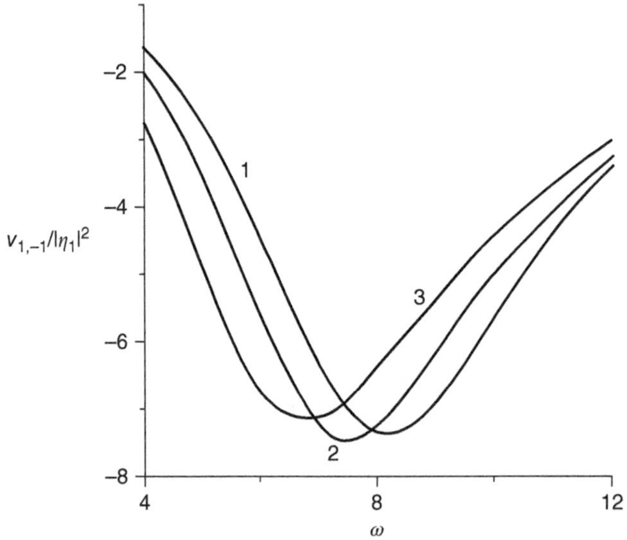

Figure 4.29 Relative magnitude of the zero harmonic of the burning velocity. 1, quadratic approximation; 2, $h = 0.1$; 3, $h = 0.15$.

show that higher harmonics emerge as the pressure amplitude grows and their contribution to the burning velocity spectrum increases. In the $2T$ regime there are also harmonics present with frequencies that equal a multiple of $\omega/2$. Close to the bifurcation point (which is the case for Figure 4.34) the amplitudes of the harmonics with frequencies $m\omega/2$ are small therefore they are expanded in the plot by two orders of magnitude.

The performed calculations allow the accuracy of the linear approximation to be estimated. We shall consider a specific example demonstrating the necessity of accounting for

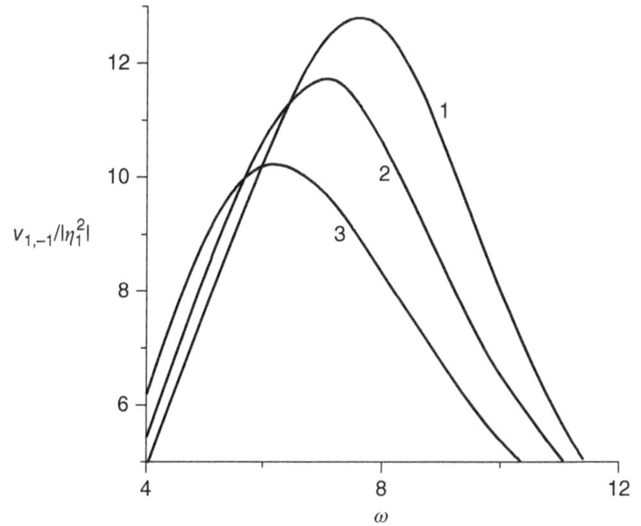

Figure 4.30 Modulus of relative magnitude of the second harmonic of the burning velocity. 1, quadratic approximation; 2, $h = 0.1$; 3, $h = 0.15$.

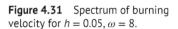

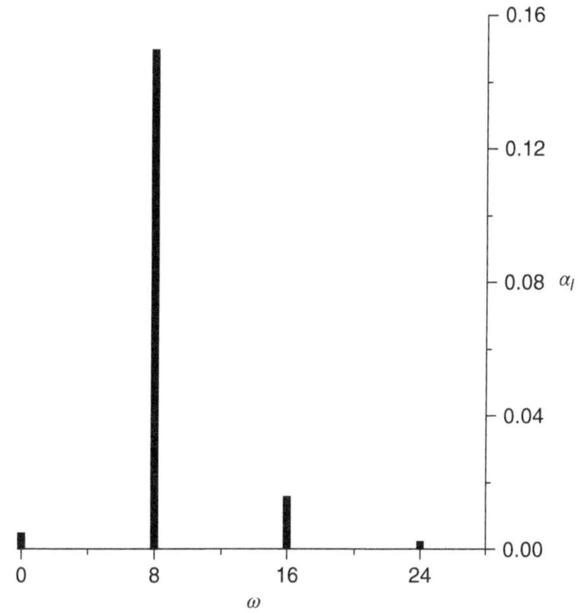

Figure 4.31 Spectrum of burning velocity for $h = 0.05$, $\omega = 8$.

nonlinear interaction in the multimode problem of nonsteady combustion in the chamber of a solid rocket motor.

Consider the results of Culick and Yang (1992) on the response of the burning rate to harmonically oscillating pressure in the linear approximation. The spectrum of pressure oscillations with the values of nondimensional amplitudes $h_1 = 0.151$ and $h_2 = 0.045$ for the first and the second harmonics, respectively, was obtained for a chamber with length

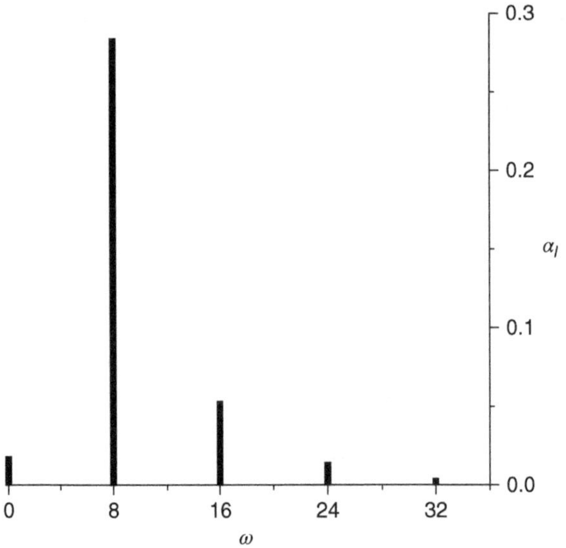

Figure 4.32 Spectrum of burning velocity for $h = 0.1$, $\omega = 8$.

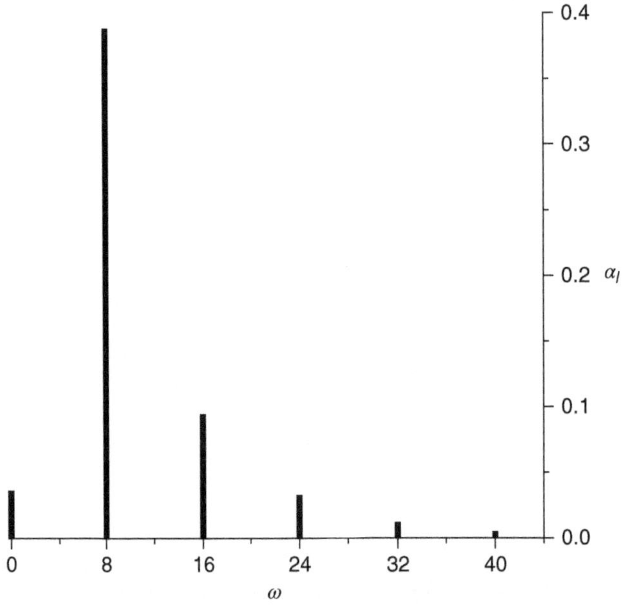

Figure 4.33 Spectrum of burning velocity for $h = 0.15$, $\omega = 8$.

$L = 0.6$ m. With the speed of sound $a = 1075$ m/s the frequency of the primary harmonic with the wavelength $2L$ is equal to $f = a/2L$, or $f = 895$ Hz. The corresponding nondimensional frequency of the primary harmonic $\omega = 2\pi f\kappa/u^2$ is obtained using the values of the linear burning rate $u = 0.0115$ m/s and the thermal diffusivity of the condensed phase $\kappa = 10^{-7}$ m^2/s. As a result, $\omega = 4.25$. The sensitivity of the burning rate and the surface temperature to the changes in initial temperature were described by Culick and Yang (1992)

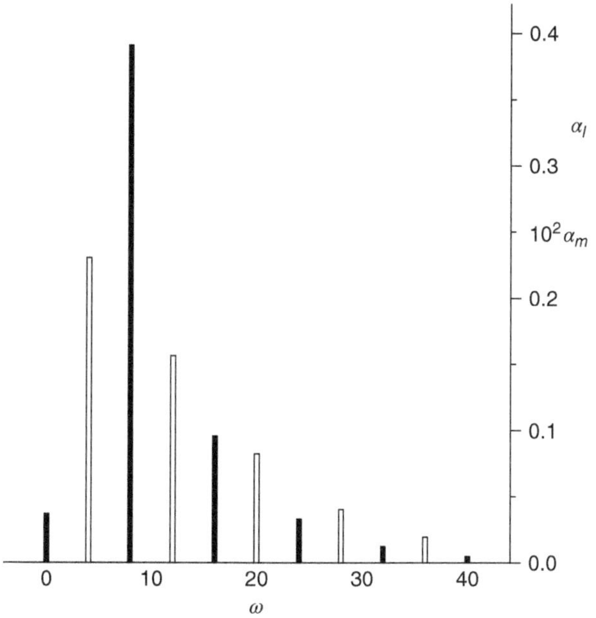

Figure 4.34 Spectrum of burning velocity for $h = 0.152$, $\omega = 8$.

using parameters A and B which relate to parameters k and r quite simply

$$A = \frac{k}{r}, \quad B = \frac{1}{k} \tag{4.161}$$

For the values $A = 6$ and $B = 0.55$ adopted in the paper we find that $k = 1.82$ and $r = 0.303$. Then the natural propellant frequency and the damping decrement are $\omega_1 = 4.34$ and $\lambda = 0.991$, respectively. The parameter describing the sensitivity of the burning rate to the changes in pressure was adopted by Culick and Yang (1992) as $\iota = 0.3$.

Therefore the parameters of the problem considered by Culick and Yang (1992) are chosen in such a way that the response of the burning rate to pressure oscillations is of apparent resonance nature. The problem parameters and the values of the nondimensional amplitudes of the first and the second pressure harmonics allow the importance of the nonlinear effects to be estimated. As an example we shall compare the linear contribution to the second burning rate harmonic v_{2l} and the nonlinear effect of self-interaction of the first harmonic, which also results in the generation of the second burning rate harmonic $v_{1,1}$. These values can be expressed through the corresponding response functions

$$| v_{2l} | = | U_1(2\omega) || \eta_2 |, \quad | v_{1,1} | = | U_{1,1}(\omega) || \eta_1 |^2 \tag{4.162}$$

It follows from the definition of complex amplitudes that their modulus is half the values of nondimensional amplitudes of corresponding harmonics. Therefore

$$| v_{2l} | = | U_1(2\omega) | \frac{h_2}{2}, \quad | v_{1,1} | = | U_{1,1}(\omega) | \frac{h_1^2}{4} \tag{4.163}$$

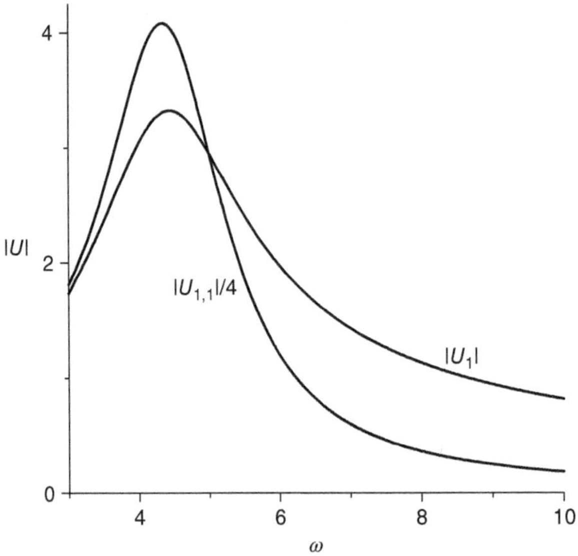

Figure 4.35 Modulus of response functions for the propellant parameters adopted by Culick and Yang (1992).

The response functions involved in the expressions above are plotted in Figure 4.35. From (4.159) and (4.160)

$$| U_1(8.5) | = 1.027, \quad | U_{1,1}(4.25) | = 16.6 \tag{4.164}$$

which gives

$$| v_{2l} | = 0.0231, \quad | v_{1,1} | = 0.0943 \tag{4.165}$$

The linear contribution turns out to be several times less than the nonlinear (quadratic) effect. Obviously, the results of Culick and Yang (1992) cannot even qualitatively describe nonsteady processes in the combustion chamber of a solid rocket engine.

4.8 Comparison with Experimental Data

Let us compare the theoretical results of this chapter with the experiments where nonsteady combustion regimes under harmonically oscillating pressure were observed. However, we will make two remarks before we proceed.

First, an exact quantitative comparison between the experiment and theory is only possible if all the parameters involved in the theoretical formulation are known. Nonsteady propellant combustion theory relies on the steady-state burning laws, which must either be established in sets of independent experiments or derived from steady-state combustion theory. These laws (i.e. the dependencies of the burning rate and surface temperature on pressure and initial temperature) are not sufficiently well studied for real systems. Consequently, the required parameters, in the first place derivatives of the above-mentioned dependencies, cannot be estimated reliably. Therefore, at present rigorous quantitative comparison between the theory and experiment is not possible. The theory may only be tested by establishing whether various effects it predicts are experimentally observable, and if they do whether relevant quantities agree between the theory and experiment by an order of magnitude.

The second remark concerns with the possibility of obtaining certain information about steady-state burning laws by comparing the theory with experimental data (Novozhilov 1973b). If the developed theory is satisfactory then the experimental dependencies may be sufficiently well reproduced by adjusting the model parameters. In the linear case such parameters are the first derivatives of the functions representing steady-state burning laws $u^0(p^0, T_a)$ and $T_s^0(p^0, T_a)$. The values of these parameters obtained by comparison between the theory and experiment may be used further for calculation of other nonsteady phenomena. For example, sufficiently detailed experimental investigation of low-frequency instability in a semi-enclosed volume may provide the necessary information for predicting the value of the acoustic admittance of the surface of the same fuel.

One of the conclusions of the theory of nonsteady solid fuel combustion is the existence of the natural frequency of propellant burning rate oscillations. This effect exists only in the model which takes into account variable surface temperature. Direct measurements of the nonsteady burning rate face substantial difficulties. For this reason the existence of the natural frequency of oscillations may only be confirmed implicitly. Most often, pressure is measured in a chamber whereas solid fuel is burnt.

As discussed previously, the development of acoustic oscillations in a chamber is most probable at natural frequencies of the chamber which are close to the natural frequency of fuel burning rate oscillations. Therefore, the existence of some selective fuel burning rate frequency may be judged by the form of acoustic admittance dependence on frequency. If the natural frequency of propellant burning rate exists then the real part of the acoustic admittance should have a maximum at this frequency. The natural frequency of the propellant is proportional to square of its burning velocity and therefore increases with pressure. Consequently, the maximum value of the real part of the acoustic admittance should shift towards higher frequencies with increasing pressure. Experimental dependencies for ballistites (Horton and Price 1963) have exactly this shape. Moreover, maximum amplification of acoustic waves is observed at frequencies whose order of magnitude $(u^0)^2/\kappa$ is predicted by the theory. Horton and Price (1963) investigated acoustic instability at large values of background pressure (from 14 to 112 atm). The combustion chamber had dimensions of the order of several tens of centimetres.

The existence of selective frequency for various solid propellants at low pressures was demonstrated by Eisel et al. (1964). The combustion chamber was designed as a steel cylindrical tube with a diameter of about 14 cm, assembled from several sections which allowed its length to be varied from 3 m to 18 m. The propellant charge was fitted into either one or both ends of the tube. Ignition was initiated at atmospheric pressure. The pressure subsequently increased due to accumulation of combustion products. In some cases propellant shavings were burnt in the chamber for more rapid increase of pressure. The rate of pressure increase could be controlled by a nozzle. Pressure rise occurred rather slowly (with a large time scale compared to the time scale of propellant thermal layer relaxation) and could be considered as a quasi steady-state process.

Pressure oscillograms confirmed that at fixed chamber length the combustion regime was stable at low pressures. As the pressure rises above a certain critical pressure level oscillations develop and their frequency corresponds to the acoustic frequency of the chamber. On further increasing the mean pressure the oscillation amplitude increases, reaches a maximum, and then the pulsations disappear. Changing the chamber length (and correspondingly its acoustic frequency) led to the development of pulsating combustion at different values of background pressure. In this way the dependence of the natural frequency of burning rate oscillations on pressure was obtained.

While comparing these experimental results with the theory, it should be remembered that the parameters k and r that control the value of the natural frequency are not known and also may vary with pressure themselves. If these parameters are assumed to be constant then the natural frequency must vary with pressure as $[u^0(p)]^2 \sim p^{2\iota}$. The dependence $u^0(p)$ is not explicitly provided by Eisel et al. (1964), but one of their figures demonstrates dependence of a certain parameter, proportional to burning rate, on pressure. From these data the burning rate dependence on pressure may be recovered to be $u^0 = Bp^\iota$ where $B = 0.1$ cm/s, $\iota = 0.5$ (pressure is expressed in atm). Therefore, the natural frequency must grow linearly with pressure, which was exactly what was found by Eisel et al. (1964). The order of magnitude of the natural frequency (several tens of Hertz) conforms with the theoretical prediction. Indeed, with a value of thermal diffusivity of $\kappa \sim 10^{-3}$ cm^2/s, the value $(u^0)^2/\kappa$ belongs to the above indicated interval.

Qualitative consideration of the experimental results obtained on acoustic instability of combustion confirms therefore the conclusion of the theory on the existence of the natural frequency of propellant burning rate oscillations. The order of the natural frequency magnitude (to within a factor of $\sqrt{k/r}$) and its dependence on pressure also conforms with predictions of the theory.

We shall try now to describe experimental dependence of acoustic admittance on the frequency of the sound using the theoretical formulae derived in this chapter. First, it should be noted that currently acoustic admittance of solid fuel surfaces is obtained experimentally. Normally, instead of measuring the velocity of gases leaving the surface (which directly enters the expression for acoustic admittance) measurements are taken of other parameters such as the time history of pressure. Implicit methods of measuring the response of the combustion process to harmonically oscillating pressure is reviewed by Strand and Brown (1992).

One of the most successful designs of resonators which are used to study acoustic admittance turned out to be the T-shaped chamber proposed by Price and Soffers (1958). The chamber is of cylindrical shape, with discs of propellant fitted at both end faces. A nozzle to remove the combustion products is located in the middle of the cylinder. Such chambers are called one-dimensional since the state of the gas almost everywhere in the chamber is determined by only one spatial coordinate. This one-dimensional structure slightly breaks close to the nozzle due to combustion products discharge. However, in the vicinity of fuel discs the acoustic field has the simplest structure, which is longitudinal sound waves propagating orthogonally to the burning surface with the gas and acoustic wave spread velocities being parallel.

On development of the pulsating combustion regime, pressure oscillations are observed in the T chamber. At early stages as deviation from background pressure is small and can be described by the linear approximation, oscillation amplitude grows with time exponentially as $\exp(\alpha_1 t)$, where α_1 is a positive parameter that can be estimated from the pressure oscillogram. After completion of the combustion process, oscillations disappear asymptotically following also the exponential law $\exp(-\alpha_2 t)$. The α_2 embodies the acoustic losses in the chamber and may be estimated using a pressure oscillogram.

It is easy to show that the nondimensional value

$$z = \frac{2(\alpha_1 + \alpha_2)l}{a} \tag{4.166}$$

where l is the chamber length and a is the speed of sound in combustion products is related to the acoustic admittance of the propellant surface by the simple relation

$$z = -\frac{4\rho_g a u^0}{p}\left(\mathrm{Re}\, y - \frac{1}{\gamma}\right) \tag{4.167}$$

Here y is the variable proportional to the acoustic admittance (4.64)

$$y = 1 + \frac{\Delta\iota}{k} - \left(1 + \frac{\Delta}{k}\right)U \tag{4.168}$$

Let us compare experimental and theoretical dependencies of z on frequency (Novozhilov 1973a). The crosses in Figure 4.36 mark experimental measurements (Horton and Price 1963) corresponding to combustion of ballistite JPN at the pressure of 56 atm. The properties of this fuel, available in Wimpress (1950), are pressure dependence exponent in the steady-state burning law $\iota = 0.69$, density $\rho = 1.62\,\mathrm{g/cm^3}$, and thermal diffusivity $\kappa = 10^{-3}\,\mathrm{cm^2/s}$. The burning rate at the above pressure is $u^0 = 1.4\,\mathrm{cm/s}$ and the speed of sound in the combustion chamber is $a = 940$ m/s. The specific heat ratio may be assigned the usually adopted value $\gamma = 1.25$. The value of the parameter Δ is much less than unity and only weakly affects the results. The value adopted below is $\Delta = 0.14$. This value corresponds (for equal specific heats of the gas and the fuel) to a combustion temperature of $T_b^0 = 2900\,\mathrm{K}$ and a surface temperature of 700 K. There are still remain

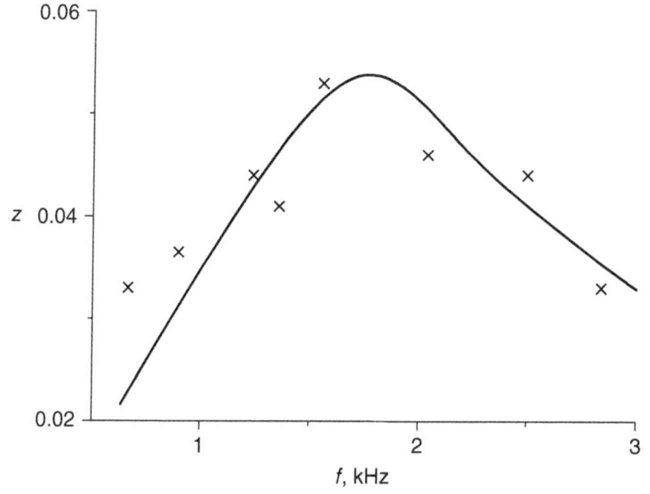

Figure 4.36 Dependence of the acoustic admittance of ballistite JPN on frequency.

three undetermined parameters in the function (4.168), namely k, r, and δ. The last of these parameters may be assumed to be zero since it was demonstrated in Chapter 1 that the burning rate and surface temperature of a ballistite fuel are related uniquely. The parameters k and r may only be estimated approximately as they are related to the surface temperature and its derivative. The latter are measured in experiments with significant errors. At the same time the relationship (4.168) is quite sensitive to the values of k and r due to the resonance behaviour of the response function U. Small variations of these parameters lead to significant distortion of the curve $z(\Omega)$.

As explained by Novozhilov (1973a), the values of k and r may be obtained from nonsteady combustion experiments with much better accuracy than by direct thermocouple measurements. The curve in Figure 4.36 corresponds to $k = 1.45$ and $r = 0.2$. It should be noted that the measurement errors are rather significant; Horton and Price (1963) estimate these to be 40% for low frequencies and 20% in the rest of the spectrum. However, due to the strong sensitivity of the function U to the parameters k and r it may be reasonably assumed that their values are determined with an accuracy that is probably not worse than 10%.

The obtained value of k is equal to the product $\beta(T_s^0 - T_a)$. The temperature coefficient of the burning rate for the JPN propellant is $\beta = 0.0038\ \text{K}^{-1}$ (Wimpress 1950) while $T_s^0 - T_a \sim$ 400 K. The value of $r = 0.2$ clearly indicates the dependence of the surface temperature on the initial temperature. Therefore, the analysis acoustic instability confirms direct thermocouple measurements of the surface temperature and may serve as an implicit method of determining those properties of the propellant which carry certain information on the kinetics of chemical reactions taking place at the propellant surface.

It is instructive to illustrate the nonlinearity of the nonsteady propellant combustion effects under periodically oscillating pressure using mean burning rate as an example.

In the quasi steady-state regime $u \sim p^{\iota}$. Therefore, if pressure follows the law

$$p = p^0(1 + h\cos\omega\tau) \tag{4.169}$$

then the relative change of the mean burning rate would be

$$\frac{\Delta u}{u^0} = \frac{\iota(\iota - 1)}{4}h^2 \tag{4.170}$$

In the general case of nonsteady combustion we obtain from (4.70)

$$\frac{\Delta u}{u^0} = \frac{U_{1,-1}}{4}h^2 \tag{4.171}$$

Both corrections are negative as normally $\iota < 1$, $U_{1,-1} < 0$. In terms of absolute values (4.171) is substantially larger than the quasi steady-state correction (4.170).

Experimental investigation of the influence of pressure oscillations on mean burning rate was reported by Crump and Price (1964). An unstable combustion regime was generated in the cylindrical T-shaped chamber with tubular propellant charge. A standing wave with an amplitude of up to 80% of the background pressure was observed in the chamber. In order to determine the mean burning rate the combustion process was interrupted by a sudden pressure drop (caused by opening an additional nozzle). The partially burnt charge was measured at several locations along its length. The mean burning rate at each location was found by considering the thickness of the burnt layer and the combustion time. This mean rate was compared to the same value measured at the steady-state burning conditions and

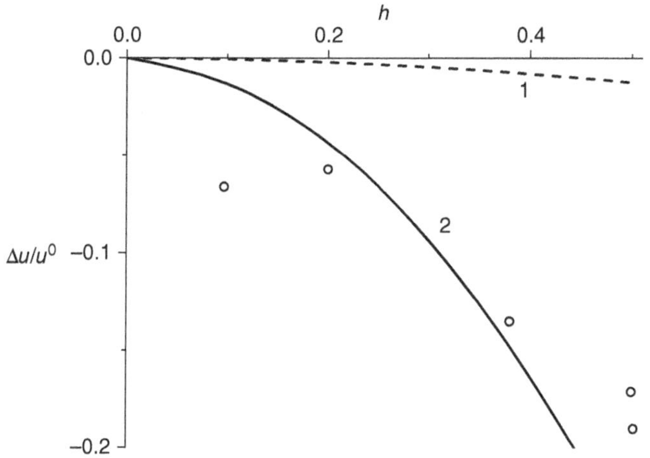

Figure 4.37 Relative variation of the burning rate of ballistite JPN as a function of amplitude of pressure oscillations. 1, quasi steady-state value (4.170) for $\iota = 0.7$; 2, nonsteady theory (4.171) for $U_{1,-1} = 4$.

at the same background pressure. In this way, the relative change of the mean burning rate could be found. As the acoustic field changes along the chamber such a methodology allows influences of pressure oscillations and gas stream velocity to be decoupled. In particular, near the end faces of the chamber where the gas velocity approaches zero (the nozzle is located in the middle of the chamber) the burning rate is influenced by time-dependent pressure only.

Ballistite JPN was investigated under different intensities of pressure oscillations. The experimental data of Crump and Price (1964) shown in Figure 4.37 confirms the conclusion of the theory on the significant change in the mean burning rate calculated at nonsteady conditions, compared to the quasi steady-state case.

Investigation of the nonlinear combustion effects undertaken in this chapter suggests a possibility of existence of auto-oscillating propellant combustion regimes. There have been no systematic experimental investigations of such a phenomenon. The results obtained above can only be compared with experimental data if relevant nonsteady burning laws are known.

Auto-oscillating combustion regimes are frequently observed experimentally. Examples of such processes are the oscillating combustion of ballistites, ammonium perchlorate, and gasless systems mentioned in Chapter 3. Unfortunately, no comparison between the theory and the experiments, apart from the existence of auto-oscillations, can be made at the present time. Moreover, it should be remembered that theoretical predictions relate auto-oscillations to the inertia of the condensed phase, which is not the only possible cause of such regimes. It may turn out that some oscillating regimes occur due to the existence of relaxation processes in the gas phase, or perhaps deviation of the combustion process from the one-dimensional mode.

Section 4.5 considered the region $k > 1$. The possibility of auto-oscillations under the condition of reverse inequality, where free linear oscillations are absent, should be noted. Nonlinear effects may lead to a decrease and nullification of the damping decrement in this case as well. Consider, as an example, a conventional pendulum in the medium with viscosity decreasing from the bottom up. If the viscosity in the vicinity of equilibrium is sufficiently large then a small deviation of the pendulum results in aperiodic motion. Imagine that at sufficiently large amplitudes the pendulum reaches the region of negative viscosity. In that case, free oscillations with large amplitude are possible as the pendulum would receive energy in the region with negative viscosity and spend it in the vicinity of the equilibrium.

5

Nonsteady Erosive Combustion

5.1 Problem Formulation

Interaction of the combustion process with the acoustic field generated inside the combustion chamber plays a decisive role in studying nonsteady operating regimes of solid rocket engines. The two aspects of this problem have been identified since investigations started: the influence of variable pressure on propellant burning rate and nonsteady erosive combustion. The latter process is the burning rate variation under the action of a variable tangential gas stream flowing along the propellant surface. Up to now, only the first aspect has been sufficiently investigated. As an example, we can refer to the studies of Culick and Yang (1992) and Novozhilov (2002), who investigated the stability conditions of the steady-state engine operating regime as well as sustained periodical nonlinear modes generated in the combustion chamber. Pressure oscillations in longitudinal acoustic waves are always accompanied by tangential oscillations of the combustion product velocity in the vicinity of the propellant surface. Therefore the analysis of the above-mentioned studies needs be generalized to take erosive combustion into account.

The present chapter considers, within the framework of the phenomenological theory of nonsteady propellant combustion, that burning rate responds to periodically oscillating pressure and tangential mass flux of combustion products (Novozhilov 2007; Novozhilov et al. 2007).

Within the framework of the Zeldovich–Novozhilov (ZN) theory, the nonsteady process of propellant combustion was investigated by solving the heat transfer equation in the condensed phase

$$\frac{\partial T_\varepsilon}{\partial t} = \kappa \frac{\partial^2 T_\varepsilon}{\partial x^2} - u_\varepsilon \frac{\partial T_\varepsilon}{\partial x}, \quad -\infty < x \leq 0 \tag{5.1}$$

with the boundary conditions

$$x = 0, \quad T_\varepsilon = T_{s\varepsilon}(t); \quad x \to -\infty, \quad T_\varepsilon = T_a \tag{5.2}$$

Here $T_\varepsilon(x, t)$, t, and x are propellant temperature under erosive conditions, and the time and space coordinates respectively, T_a and $T_{s\varepsilon}(t)$ are the initial temperature and temperature at the phase interface, $u_\varepsilon(t)$ is the linear burning rate, and κ is the thermal diffusivity of the condensed phase. The coordinate system origin is fixed at the phase interface so that the condensed phase is moving with speed $u_\varepsilon(t)$ in the positive direction of the x axis.

Theory of Solid-Propellant Nonsteady Combustion, First Edition. Boris V. Novozhilov and Vasily B. Novozhilov.

Companion website: www.wiley.com/go/Novozhilov/solidpropellantnonsteadycombustion

The subscript ε denotes the erosive combustion regime. The following analysis will also use properties without this subscript. They refer to the combustion regime in the absence of erosion. For example, the two notations for the mass burning rate are

$$m = \rho u, \quad m_\varepsilon = \rho u_\varepsilon \tag{5.3}$$

where ρ is density of the condensed phase.

A necessary input into the theory are steady-state burning rate and surface temperature dependencies on initial temperature, pressure p, and mass velocity g of the gas stream parallel to the surface.

The results below were obtained for a specific propellant combustion model. It was chosen to contain minimum number of parameters. The simplest model satisfying this requirement has the form

$$m^0 = A(p^0)^n \exp(\beta T_a), \quad m^0 = B\exp(\beta_s T_s^0) \tag{5.4}$$

where A, B, β, β_s, and n are constants. The upper null index corresponds to steady-state regime.

The parameters of linear sensitivity of the burning rate and surface temperature to variations of initial temperature and pressure

$$k = (T_s^0 - T_a)\left(\frac{\partial \ln m^0}{\partial T_a}\right)_{p^0}, \quad r = \left(\frac{\partial T_s^0}{\partial T_a}\right)_{p^0}$$

$$\iota = \left(\frac{\partial \ln m^0}{\partial \ln p^0}\right)_{T_a}, \quad \mu = \frac{1}{T_s^0 - T_a}\left(\frac{\partial T_s^0}{\partial \ln p^0}\right)_{T_a}$$

$$\delta = \iota r - \mu k \tag{5.5}$$

for the present model take the form

$$k = \beta(T_s^0 - T_a), \quad r = \beta/\beta_s, \quad \iota = n, \quad \mu = \iota r/k, \quad \delta = 0 \tag{5.6}$$

Let us assume that in the presence of erosion the steady-state laws can be written in the form

$$m_\varepsilon^0 = m^0\sqrt{1 + b\left(\frac{g^0}{m^0}\right)^2} \tag{5.7}$$

$$m_\varepsilon^0 = B\exp(\beta_s T_{s\varepsilon}^0) \tag{5.8}$$

where b = constant. It is assumed here that in the presence of the erosion mass the burning rate is still determined by the surface temperature only. On the other hand, the relationship (5.7) should be considered as a one-parameter interpolation of experimental data on the dependence of erosion ratio on the tangential mass flux of the combustion products. The theoretical justification for this relationship was presented in the Section 1.4.

By comparing the steady-state burning laws (5.4), (5.7), and (5.8) it is easy to obtain the relationships between the temperature coefficients of burning rate sensitivity β and β_ε, and the surface temperatures T_s^0 and $T_{s\varepsilon}^0$ without and with erosion

$$\beta_\varepsilon = \frac{\beta}{\varepsilon^2}, \quad \beta_\varepsilon = \left(\frac{\partial \ln m_\varepsilon^0}{\partial T_a}\right)_{p^0, g^0}$$

$$T_{s\varepsilon}^0 = T_s^0 + \frac{1}{\beta_s}\ln\varepsilon \tag{5.9}$$

We note also the helpful relation

$$\frac{g^0}{m^0} = \sqrt{\frac{\varepsilon^2 - 1}{b}} \tag{5.10}$$

The standard procedure of transforming steady-state burning laws into nonsteady ones requires the initial temperature to be expressed as a function of temperature gradient at the surface. Invoking once again the Michelson steady-state temperature distribution in the condensed phase

$$T_\varepsilon^0(x) = T_a + (T_{s\varepsilon}^0 - T_a)\exp\left({}^{u_\varepsilon^0 x}/_{\kappa}\right) \tag{5.11}$$

the following expression for the temperature gradient at the surface (on the condensed phase side) is obtained

$$f_\varepsilon^0 = \left.\frac{\partial T_\varepsilon^0}{\partial x}\right|_{x=0}, \quad f_\varepsilon^0 = \frac{u_\varepsilon^0}{\kappa}(T_{s\varepsilon}^0 - T_a) \tag{5.12}$$

These relationships allows the burning rate in the absence of erosion (5.4) to be expressed in the form

$$m^0 = A(p^0)^n \exp\left[\beta\left(T_{s\varepsilon}^0 - \kappa\frac{f_\varepsilon^0}{u_\varepsilon^0}\right)\right] \tag{5.13}$$

According to the principle idea of the ZN theory the same form of the burning law applies in the nonsteady regime. Dropping the steady-state subscript the following nonsteady burning law is obtained

$$m = A(p)^n \exp\left[\beta\left(T_{s\varepsilon} - \kappa\frac{f_\varepsilon}{u_\varepsilon}\right)\right] \tag{5.14}$$

This relationship allows the steady-state erosion burning laws (5.7) and (5.8) to be transformed into nonsteady counterparts as well

$$m_\varepsilon^2 = m^2 + bg^2, \; m_\varepsilon = B\exp(\beta_s T_{s\varepsilon}) \tag{5.15}$$

In order to close the set of equations of nonsteady combustion theory it is necessary to specify the exact type of acoustic wave. In the following we shall only consider plane monochromatic travelling waves where all the acoustic disturbances (i.e. pressure, density, velocity) are in phase. If h denotes the pressure amplitude then the amplitudes of acoustic disturbances (denoted by primes) are

$$\frac{p'}{p^0} = h, \quad \frac{\rho'}{\rho^0} = \frac{h}{\gamma}, \quad \frac{w'}{a} = \frac{h}{\gamma} \tag{5.16}$$

Here ρ and w are density and tangential gas velocity, a is the speed of sound, and γ is the specific heat ratio. Now the disturbance of the tangential mass flux may be expressed as

$$\frac{g'}{g^0} = \left(1 + \frac{a}{w^0}\right)\frac{h}{\gamma} \tag{5.17}$$

and the time dependencies of pressure and tangential flux as

$$p = p^0(1 + h\cos\Omega t), \quad g = g^0\left[1 + \left(1 + \frac{a}{w^0}\right)\frac{h}{\gamma}\cos\Omega t\right] \tag{5.18}$$

The set of equations (5.1), (5.2), (5.15), and (5.18) allows the process of nonsteady erosion combustion in the field of the travelling acoustic wave to be analysed. In the following we consider only stabilized combustion regimes so that consideration of initial conditions is not necessary. It should be remembered that any acoustic wave may be described as superposition of plane monochromatic travelling waves with different wave vectors and frequencies.

It is convenient to use the nondimensional variables

$$\tau = \frac{(u^0)^2 t}{\kappa}, \quad \omega = \frac{\kappa \Omega}{(u^0)^2}, \quad \xi = \frac{u^0 x}{\kappa}$$
$$v = \frac{u_\varepsilon}{u_\varepsilon^0}, \quad \eta_p = \frac{p}{p^0}, \quad \eta_g = \frac{g}{g^0},$$
$$\theta = \frac{T_\varepsilon - T_a}{T_{\varepsilon s}^0 - T_a}, \quad \vartheta = \frac{T_{\varepsilon s} - T_a}{T_{\varepsilon s}^0 - T_a}, \quad \varphi = \left(\frac{\partial \theta}{\partial \xi}\right)_{\xi=0} \tag{5.19}$$

The heat transfer equation (5.1), boundary conditions (5.2), nonsteady burning laws (5.15), and dependencies (5.18) turn the new variables into

$$\frac{1}{\varepsilon^2}\frac{\partial \theta}{\partial \tau} = \frac{\partial}{\partial \xi}\left(\frac{\partial \theta}{\partial \xi} - v\theta\right), \quad -\infty < \xi \leq 0 \tag{5.20}$$

$$\xi = 0, \quad \theta = \vartheta(\tau); \quad \xi \to -\infty, \quad \theta = 0 \tag{5.21}$$

$$(\varepsilon v)^2 = X^2 + (\varepsilon^2 - 1)\eta_g^2, \quad v = \exp\left[\left(\frac{k}{r} + \ln \varepsilon\right)(\vartheta - 1)\right]$$
$$X = \eta_p^{\iota} \exp\left[(k + r \ln \varepsilon)\left(\vartheta - \frac{\varphi}{v}\right)\right] \tag{5.22}$$

$$\eta_p = 1 + h\cos(\omega\tau)$$
$$\eta_g = 1 + \frac{h}{\gamma}\left(1 + \frac{1}{\mathrm{M}}\sqrt{\frac{b}{\varepsilon^2 - 1}}\right)\cos(\omega\tau) \tag{5.23}$$

where $\mathrm{M} = w/a$ is the Mach number for the normal velocity of combustion products.

Let us make two remarks. First, to derive the above set of equations, the relationships (5.9) and (5.10) must be involved. Second, the steady-state burning rate without erosion is used for time and frequency scales, while all the other variables are non-dimensionalized using the steady-state erosion burning rate. This approach results in variable external conditions (i.e. oscillation frequencies) being independent of the burning regime, and the steady-state combustion regime ($\eta_p^0 = 1$, $\eta_g^0 = 1$) taking the simplest form

$$\theta^0 = e^{\xi}, \quad \varphi^0 = 1, \quad v^0 = 1, \quad \vartheta^0 = 1 \tag{5.24}$$

The above problem formulation applies to a general case where there exists a nonzero steady-state tangential stream, that is, $g^0 \neq 0$ and $\varepsilon \neq 1$. In the opposite case (5.22) must be replaced by

$$v^2 = \eta_p^{2\iota} \exp\left[2k\left(\vartheta - \frac{\varphi}{v}\right)\right] + S(\eta_p - 1)^2$$
$$v = \exp\left[\frac{k}{r}(\vartheta - 1)\right], \quad S = \frac{b}{(\gamma \mathrm{M})^2} \tag{5.25}$$

5.2 Linear Approximation

Under small deviations from the steady-state regime (5.24), the burning rate may be found in the linear approximation analytically. It is convenient to use the method of complex amplitudes.

Let us express pressure and tangential combustion product stream mass velocity as

$$\eta_p = 1 + [\eta_{p1}\exp(i\omega\tau) + c.c.], \quad \eta_g = 1 + [\eta_{g1}\exp(i\omega\tau) + c.c.]$$

$$\eta_{p1} = \frac{h}{2}, \quad \eta_{g1} = \frac{h}{2\gamma}\left(1 + \frac{1}{\mathrm{M}}\sqrt{\frac{b}{\varepsilon^2 - 1}}\right) \tag{5.26}$$

where *c.c.* stands for complex conjugate.

The following analysis is based on the approach by Novozhilov (1965b) where the response function of the burning rate to harmonically oscillating pressure was found in the linear approximation.

All the time-dependent variables are expressed in a form similar to (5.26), that is, as the sum of the steady-state value and a small harmonic perturbation. For example, the burning rate

$$u_\varepsilon = u_\varepsilon^0 + u_{\varepsilon 1}\cos(\Omega t + \psi) \tag{5.27}$$

is written as

$$v = 1 + [v_1\exp(i\omega\tau) + c.c.], \quad v_1 = \frac{u_{\varepsilon 1}}{2u_\varepsilon^0}\exp(i\psi) \tag{5.28}$$

and the temperature distribution within the propellant as

$$\theta(\xi, \tau) = e^{\xi} + [\theta_1(\xi)e^{i\omega\tau} + c.c.] \tag{5.29}$$

Substituting these expressions into the heat transfer Eq. (5.20) and neglecting quadratic contributions the following linear equation

$$\theta_1'' - \theta_1' - \frac{i\omega}{\varepsilon^2}\theta_1 = v_1 e^{\xi} \tag{5.30}$$

is obtained. The solution is

$$\theta_1 = C_1 e^{z_1\xi} - \frac{v_1\varepsilon^2}{i\omega}e^{\xi} \tag{5.31}$$

where

$$z_1 = \frac{1}{2}\left(1 + \sqrt{1 + \frac{4i\omega}{\varepsilon^2}}\right), \quad z_1(z_1 - 1) = \frac{i\omega}{\varepsilon^2} \tag{5.32}$$

This solution satisfies the required boundary condition at $\xi \to -\infty$. From the first of the boundary conditions (5.21) the linear corrections to surface temperature and temperature gradient are found

$$\vartheta_1 = C_1 - \frac{v_1\varepsilon^2}{i\omega}, \quad \varphi_1 = C_1 z_1 - \frac{v_1\varepsilon^2}{i\omega} \tag{5.33}$$

The integration constant C_1 may be eliminated to obtain

$$z_1\vartheta_1 - \varphi_1 + \frac{v_1}{z_1} = 0 \tag{5.34}$$

$$\theta_1 = v_1 \left(a_1 e^{z_1 \xi} - \frac{\varepsilon^2 e^{\xi}}{i\omega} \right), \quad a_1 = \frac{\vartheta_1}{v_1} + \frac{\varepsilon^2}{i\omega} \tag{5.35}$$

Linearization of the nonsteady burning laws (5.22) provides two additional relationships between the above amplitudes and the amplitudes of pressure and tangential mass flux

$$\begin{aligned} &\varepsilon^2 v_1 - (k + r \ln \varepsilon)(\vartheta_1 - \varphi_1 + v_1) = \iota \eta_{p1} + (\varepsilon^2 - 1)\eta_{g1} \\ &v_1 = (k + r \ln \varepsilon)\frac{\vartheta_1}{r} \end{aligned} \tag{5.36}$$

The three equations (5.34), (5.36) allow the response function of burning rate to oscillating pressure, in the presence of erosion, to be determined

$$U_1(\omega) = \frac{v_1}{\eta_{p1}} \tag{5.37}$$

This function may be written in the form

$$U_1(\omega) = \frac{N}{D}, \quad N = \frac{\iota}{\varepsilon^2} + \frac{1}{\gamma \varepsilon^2} \left(\varepsilon^2 - 1 + \frac{\sqrt{b(\varepsilon^2 - 1)}}{M} \right)$$

$$D = 1 + \frac{z_1 - 1}{\varepsilon^2} \left(r - \frac{k + r \ln \varepsilon}{z_1} \right) \tag{5.38}$$

Figures 5.1–5.5 illustrate the dependence of various properties of the response function on frequency ω and erosion ratio ε, and are plotted for the following values of fixed parameters

$$\begin{aligned} &\iota = 0.7, \quad k = 1.8, \quad r = 0.35 \\ &\gamma = 1.25, \quad b = 4.5 \times 10^{-5}, \quad M = 1.3 \times 10^{-3} \end{aligned} \tag{5.39}$$

These values correspond to the combustion of ballistite propellant at a pressure of about 100 atm.

Note the difference in responses to periodically oscillating pressure at large and small values of the erosion ratio. Figure 5.1 shows the dependencies of the modulus of response

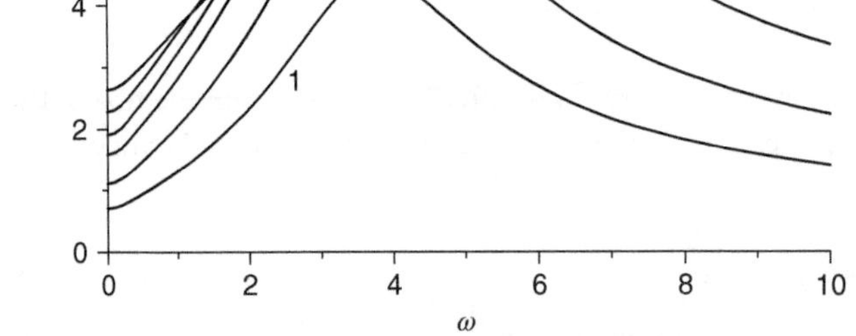

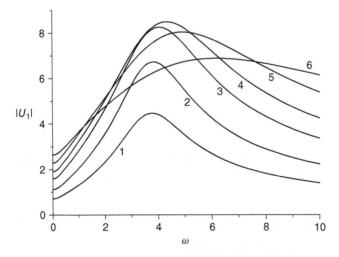

Figure 5.1 Frequency dependence of the modulus of the response function for different values of the erosion ratio ($\varepsilon - 1 << 1$). 1, $\varepsilon = 1$; 2, $\varepsilon = 1.005$; 3, $\varepsilon = 1.025$; 4, $\varepsilon = 1.05$; 5, $\varepsilon = 1.1$; 6, $\varepsilon = 1.2$.

function on frequency under weak steady-state erosion, that is, for $\varepsilon - 1 << 1$. It is evident that the modulus of the response function grows as the erosion ratio increases up to $\varepsilon = 1.05$ (curve 4). At this maximum the value of the response is approximately twice the same value without erosion (curve 1). We note also that under considered values of erosion ratio the resonance shape of the response function is preserved and the resonance frequency stays nearly the same. On a further increase of the erosion ratio ε, the resonance widens and the response value at the resonance drops (curves 5 and 6).

The frequency dependence of the burning rate response under a very pronounced steady-state erosion, that is, for $\varepsilon - 1 \sim 1$, is qualitatively different. This is illustrated in Figure 5.2. In this case the modulus of the response function drops with the growing erosion ratio and the resonance shape of its frequency dependence disappears.

Such behaviour of the response function is easy to understand. Steady-state erosion changes the sensitivity parameters of the burning rate and surface temperature in the steady-state regime to changes in initial temperature k_ε and r_ε. For the adopted combustion model these parameters decrease with the growing erosion ratio. Therefore steady-state erosion substantially changes the denominator of the response function (5.38). As a result, the steady-state regime becomes more stable while the resonance becomes less pronounced. The first term in the numerator of the response function (5.38) is related to pressure change while the second is a contribution from nonsteady erosion. Both also depend strongly on the ratio.

Figure 5.3 shows the numerator as well as the moduli of the denominator and the response function as functions of the erosion ratio for a fixed value of the frequency. The latter is chosen to be equal to the natural frequency of the propellant $\widetilde{\omega}$ at $\varepsilon = 1$

$$\widetilde{\omega} = \sqrt{\frac{k}{r^2} - \lambda^2}, \quad \lambda = \frac{r(k+1) - (k-1)^2}{2r^2} \tag{5.40}$$

For the values of the parameters k and r specified above $\widetilde{\omega} = 3.57$.

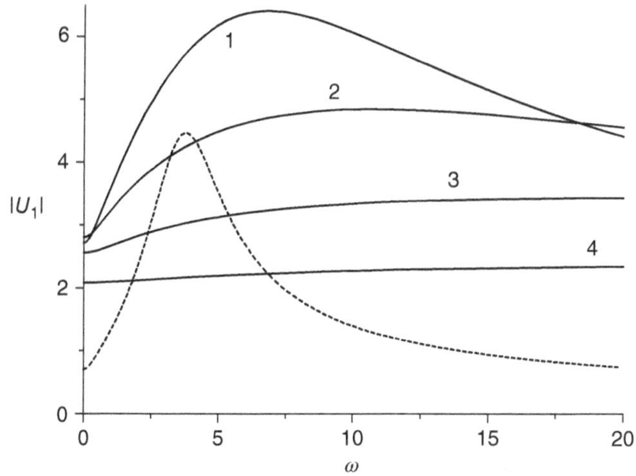

Figure 5.2 Frequency dependence of the modulus of the response function for different values of the erosion ratio ($\varepsilon - 1 \sim 1$). 1, $\varepsilon = 1.25$; 2, $\varepsilon = 1.5$; 3, $\varepsilon = 2$; 4, $\varepsilon = 3$; dotted line, $\varepsilon = 1$.

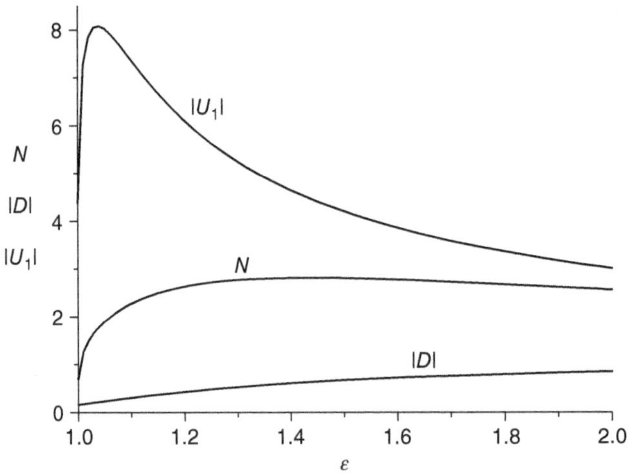

Figure 5.3 Numerator and moduli of the denominator and the response function (5.38) as functions of the erosion ratio. $\tilde{\omega} = 3.57$.

In the region $\varepsilon - 1 \ll 1$, the numerator increases rather rapidly: it has a maximum and then slowly approaches the asymptotic value γ^{-1}. On the other hand, the modulus of the denominator increases monotonically with growing erosion ratio. As a result, the modulus of the response function has a very pronounced maximum close to $\varepsilon = 1$. For other values of frequency, as is evident from Figure 5.4, the modulus of the response function also has a maximum at some value of erosion ratio. Relative contributions of steady-state and nonsteady erosion processes are compared in Figure 5.5. It is clear from the plot that the nonsteady contribution increases the magnitude of burning rate response to oscillating pressure. In contrast, the steady-state contribution may essentially suppress the resonance between pressure and burning rate.

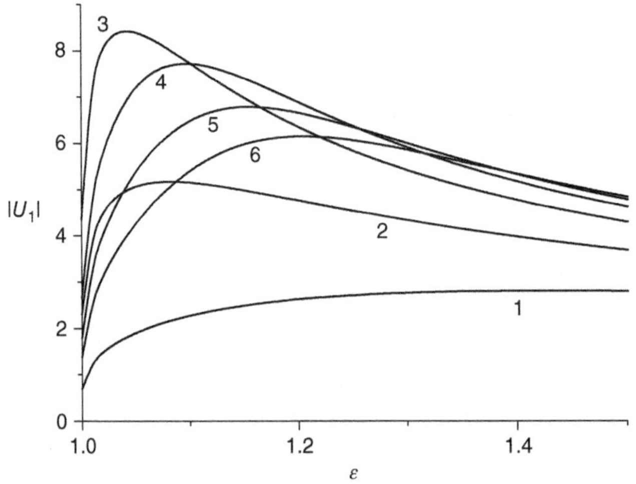

Figure 5.4 Modulus of the response function (5.38) as a function of the erosion ratio. 1, $\omega = 0$; 2, $\omega = 2$; 3, $\omega = 4$; 4, $\omega = 6$; 5, $\omega = 8$; 6, $\omega = 10$.

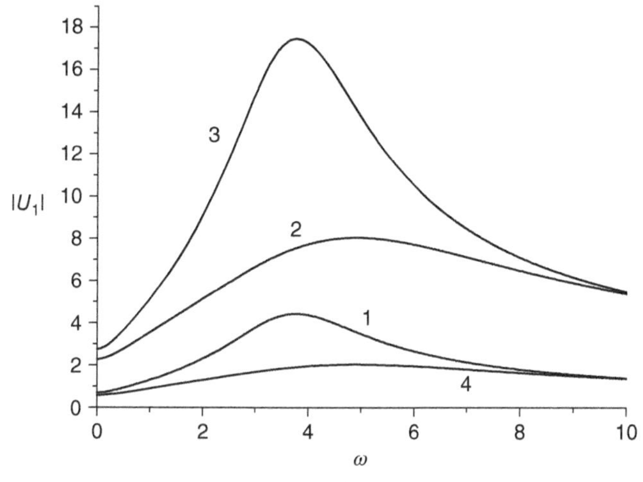

Figure 5.5 Influence of erosion on the response function (5.38). 1, without erosion; 2, with erosion; 3, with nonsteady erosion only; 4, with steady erosion only. $\varepsilon = 1.1$.

As well as the linear response function $U_1(\omega)$ the linear response function at double frequency $U_2(\omega) = U_1(2\omega)$ is used in the following analysis. Since frequency enters the response function expression through characteristic root z only, the expression for $U_2(\omega)$ may be obtained from (5.38) by replacing z_1 with

$$z_2 = \frac{1}{2}\left(1 + \sqrt{1 + \frac{8i\omega}{\varepsilon^2}}\right) \tag{5.41}$$

It should be remembered that the obtained results are valid for specific dependence of the erosion ratio on the tangential flux velocity, and for the simplest acoustic waves only. Different erosion laws and types of acoustic waves may be considered in a similar manner.

5.3 Nonlinear Effects in Nonsteady Erosive Combustion

The linear analysis conducted above may be used to determine the conditions of stability of the steady-state engine operating regime. However, for an investigation of processes with finite pressure amplitudes (such as sustained oscillations and triggering) methods extending beyond the framework of the linear approximation are required.

The periodic acoustic field in the combustion chamber normally contains many harmonics. Their interaction may be described on introduction of nonlinear response functions of burning rate to harmonically oscillating pressure.

This section generalizes the quadratic response functions, obtained in Chapter 4, for the case of erosion combustion. In order to illustrate the notion of quadratic response functions it is sufficient to consider an example where pressure in the vicinity of the burning propellant surface consists of just two harmonics with small amplitudes (assume for certainty $n > m$)

$$\eta = 1 + [\eta_m \exp(im\omega\tau) + \eta_n \exp(in\omega\tau) + c.c.] \tag{5.42}$$

The propellant burning rate in the quadratic with respect to pressure amplitude approximation will contain the same harmonics as well as harmonics with combinational frequencies (including time-independent corrections to the steady-state burning velocity)

$$\begin{aligned} v = 1 + v_{m,-m} + v_{n,-n} + \{ v_m \exp(im\omega\tau) + v_n \exp(in\omega\tau) \\ + v_{m,m} \exp[2im\omega\tau] + v_{n,n} \exp[2in\omega\tau] \\ + v_{n,-m} \exp[i(n-m)\omega\tau] + v_{n,m} \exp[i(n+m)\omega\tau] + c.c.\} \end{aligned} \quad (5.43)$$

Obviously, along with the linear relationships

$$U_m(\omega) = \frac{v_m}{\eta_m}, \quad U_n(\omega) = \frac{v_n}{\eta_n} \quad (5.44)$$

one can also write down the relationships

$$\begin{aligned} v_{m,-m} &= U_{m,-m}(\omega)\eta_m\overline{\eta}_m, \quad v_{n,-n} = U_{n,-n}(\omega)\eta_n\overline{\eta}_n \\ v_{m,m} &= U_{m,m}(\omega)\eta_m^2, \quad v_{n,n} = U_{n,n}(\omega)\eta_n^2 \\ v_{n,-m} &= U_{n,-m}(\omega)\eta_n\overline{\eta}_m, \quad v_{n,m} = U_{n,m}(\omega)\eta_n\eta_m \end{aligned} \quad (5.45)$$

The complex functions

$$\begin{aligned} U_{m,-m}(\omega) &= \frac{v_{m,-m}}{|\eta_m|^2}, \quad U_{m,m}(\omega) = \frac{v_{m,m}}{\eta_m^2} \\ U_{n,-m}(\omega) &= \frac{v_{n,-m}}{\eta_n\overline{\eta}_m}, \quad U_{n,m}(\omega) = \frac{v_{n,m}}{\eta_n\eta_m} \end{aligned} \quad (5.46)$$

are called the quadratic response functions of the burning rate to oscillating pressure. It should be remembered that the two lower indexes show which of the harmonics are interacting while their sum gives the generated burning rate harmonic number.

As an example, consider calculation of the response function $U_{1,1}$ which corresponds to generation of the second harmonics.

Let pressure in the vicinity of the surface contain just one primary harmonic

$$\eta(\tau) = 1 + (\eta_1 e^{i\omega\tau} + c.c.) \quad (5.47)$$

It is obvious that self-interaction of the primary harmonic leads in the quadratic approximation to generation of the harmonic with double frequency. Therefore the burning rate in the quadratic approximation should be written in the form

$$v(\tau) = 1 + v_{1,-1} + (v_1 e^{i\omega\tau} + v_{1,1} e^{2i\omega\tau} + c.c.) \quad (5.48)$$

Here $v_{1,-1}$ is the quadratic correction to the time-independent contribution to the burning rate and $v_{1,1}$ is the complex amplitude of the second harmonic.

By definition

$$U_{1,1} = \frac{v_{1,1}}{\eta_1^2} \quad (5.49)$$

Assume the temperature profile in the condensed phase in the form

$$\theta(\xi,\tau) = e^{\xi} + [\theta_1(\xi)e^{i\omega\tau} + \theta_{1,1}(\xi)e^{2i\omega\tau} + c.c.] \quad (5.50)$$

The heat transfer Eq. (5.20) for the double frequency harmonic written to within second-order accuracy

$$\theta''_{1,1} - \theta'_{1,1} - 2i\omega\theta_{1,1} = \left(v_{1,1} - \frac{v_1^2}{i\omega}\right)e^{\xi} + a_1 z_1 v_1^2 e^{z_1\xi} \tag{5.51}$$

has the solution

$$\theta_{1,1} = C_{1,1}e^{z_2\xi} + A_{1,1}e^{\xi} + B_{1,1}e^{z_1\xi}$$
$$A_{1,1} = -\frac{1}{2i\omega}\left(v_{1,1} - \frac{v_1^2}{i\omega}\right), \quad B_{1,1} = -\frac{a_1 z_1 v_1^2}{i\omega} \tag{5.52}$$

From this solution the quadratic corrections to surface temperature and temperature gradient can be obtained, and eliminating the integration constant $C_{1,1}$ we arrive at

$$z_2\vartheta_{1,1} - \varphi_{1,1} + \frac{v_{1,1}}{z_2} = \frac{v_1^2}{i\omega}\left[\frac{1}{z_2} - a_1 z_1(z_2 - z_1)\right] \tag{5.53}$$

In order to illustrate the role of nonlinearity, consider the case with steady-state erosion absent ($\varepsilon = 1$). Rewrite the nonsteady burning laws (5.25) in the form

$$\varphi = v\left[1 + \frac{1}{2k}\ln\frac{v^{2r}\eta^{2\iota}}{v^2 - S(1-\eta_p)^2}\right], \quad \vartheta = 1 + \frac{r}{k}\ln v \tag{5.54}$$

and substitute (5.47) and (5.48) as well as similar expressions for temperature gradient and surface temperature

$$\varphi(\tau) = 1 + \varphi_{1,-1} + (\varphi_1 e^{i\omega\tau} + \varphi_{1,1}e^{2i\omega\tau} + c.c.)$$
$$\vartheta(\tau) = 1 + \vartheta_{1,-1} + (\vartheta_1 e^{i\omega\tau} + \vartheta_{1,1}e^{2i\omega\tau} + c.c.) \tag{5.55}$$

Straightforward calculations deliver complex amplitudes of the second harmonic for temperature gradient and surface temperature

$$\varphi_{1,1} = \frac{1}{k}\left[sv_{1,1} + \frac{(r-1)v_1^2}{2} + \iota v_1\eta_1 + \frac{(S-\iota)\eta_1^2}{2}\right]$$
$$\vartheta_{1,1} = \frac{r}{k}\left(v_{1,1} - \frac{v_1^2}{2}\right) \tag{5.56}$$

The three relationships (5.53), (5.56) allow the response function for the double frequency harmonic to be found

$$U_{1,1} = \frac{U_1^2 U_2}{2\iota}(-1 + rR_{1,1} + kK_{1,1}) + U_2\left(U_1 - \frac{H-1}{2}\right)$$
$$R_{1,1} = 3 + z_2 - \frac{4z_1}{z_2}, \quad K_{1,1} = \frac{z_2 - 1}{\omega^2}(2z_1 - z_2 - 1), \quad H = \frac{b}{\iota(\gamma\text{M})^2} \tag{5.57}$$

The frequency dependencies of this response function are illustrated in Figure 5.6. It is evident that along with the resonance at the natural frequency, it appears also that there is a resonance at half of that frequency. This phenomenon is related to the dominant role of erosion contribution to the response function, compared to the contribution from oscillating pressure. For comparison, the corresponding response function with erosion absent is

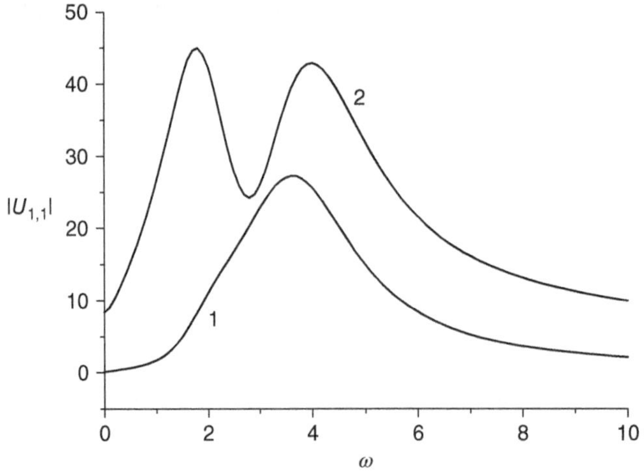

Figure 5.6 Influence of erosion on the response function $|U_{1,1}|$. 1, without erosion; 2, with erosion. $\varepsilon = 1$.

presented in the same figure. In the resonance region the moduli of both functions differ by approximately a factor of two, while at half of that frequency they differ by approximately an order of magnitude. This circumstance considerably limits the range of applicability of the linear approximation, which is only valid if the amplitude of the second harmonic is much smaller than the amplitude of the first. These amplitudes are proportional to $|U_{1,1}|h^2$ and $|U_1|h$, respectively. The ratio of these amplitudes for cases with and without erosion (for relatively small pressure perturbation amplitude $h = 0.05$) is shown in Figure 5.7. The plot demonstrates that it is acceptable to treat the case without erosion in the linear approximation as the second harmonic amplitude is at least an order of magnitude smaller than the amplitude of the first one. In the presence of erosion, even for the specified small pressure

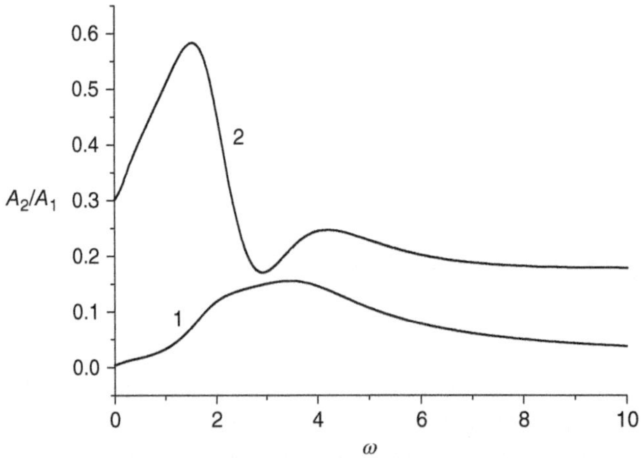

Figure 5.7 Ratio of the amplitudes of the second and first harmonics. 1, without erosion; 2, with erosion. $h = 0.05$, $\varepsilon = 1$.

perturbation amplitude, the ratio of the first and second harmonic amplitudes is comparable by an order of magnitude to one.

Nonlinearity of the burning rate response to harmonically oscillating pressure may be described quantitatively by the distortion factor

$$K = \sqrt{\frac{\sum_{n=2}^{\infty} A_n^2}{A_1^2}} \tag{5.58}$$

where A_n is the amplitude of the nth harmonic. This parameter is adopted in telecommunications where the system is considered to be linear if $K << 1$.

The distortion factor was calculated by numerical integration of the set of original equations (5.20)–(5.23). For several values of the erosion ratio Figure 5.8 shows the frequency dependence of such a pressure amplitude that results in $K = 0.1$. For small values of the erosion ratio the burning rate response may be considered as linear only at very small pressure amplitudes. As the erosion ratio grows, the nonlinearity of the response becomes less pronounced.

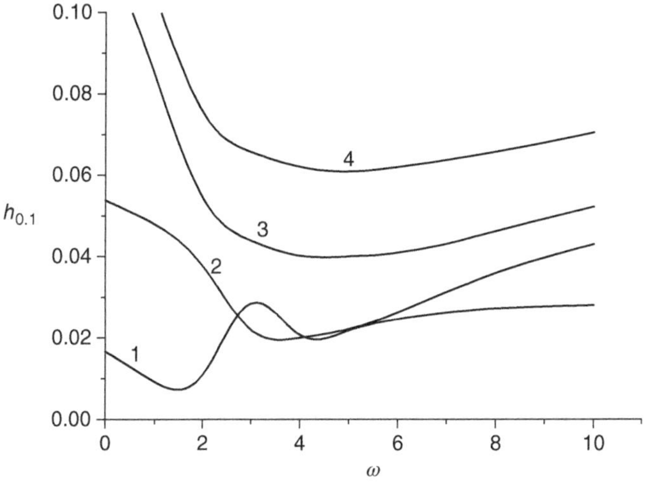

Figure 5.8 Pressure amplitude at which the nonlinearity of the response is 10%. 1, $\varepsilon = 1$; 2, $\varepsilon = 1.05$; 3, $\varepsilon = 1.2$; 4, $\varepsilon = 1.3$.

6

Nonsteady Combustion Under External Radiation

6.1 Steady-state Combustion Regime

This chapter deals with some problems of nonsteady combustion in the presence of radiative energy flux (external or generated by combustion products) received by a propellant surface. In such circumstances, an additional heat source, on spatial scales of the order of the mean free path for radiation absorption, appears in the condensed phase. This additional source significantly affects the nonsteady combustion process. Major outcomes of theoretical investigations concerned with this phenomenon have been reported by Assovskii and Istratov (1971), Ibiruci and Williams (1975), De Luca (1976), Kiskin (1983), Son and Brewster (1992, 1993), De Luca et al. (1995), and Kohno et al. (1995).

We consider first the steady-state regime in the presence of constant radiative heat flux I^0. Let fuel fill the left half-space $x \leq 0$ in the coordinate system with the origin fixed at its surface. For simplicity the energy flux received by the fuel from the right half-space $x \geq 0$ is considered to be normal to the propellant surface. An additional heat source in the condensed phase distorts its temperature profile and consequently changes various parameters of the steady-state combustion regime. In particular, steady-state burning rate and surface temperatures are larger than the corresponding values in the absence of radiation, i.e. $u_I^0 > u^0$, $T_{Is}^0 > T_s^0$.

In the steady-state regime, the heat transfer equation uses Beer's law (Incropera and DeWitt 2002) and the boundary conditions have the form

$$u_I^0 \frac{dT_I^0}{dx} = \frac{d}{dx}\left(\kappa \frac{dT_I^0}{dx} + \frac{I^0}{\rho c}\exp(\alpha x)\right)$$

$$x = -\infty, \quad T_I^0 = T_a; \quad x = 0, \quad T_I^0 = T_{Is}^0 \tag{6.1}$$

Here α is the linear absorption coefficient (it is assumed that the mean free path for absorption α^{-1} is much larger than the thickness of the reaction zone in the condensed phase) and the origin of the coordinate system is attached to the plane of the phase change ($x = 0$). Integration over the total volume of the condensed phase results in the following expression for the temperature gradient at the surface

$$f_I^0 = \frac{u_I^0}{\kappa}\left(T_{Is}^0 - T_a - \frac{I^0}{\rho c u_I^0}\right) \tag{6.2}$$

Theory of Solid-Propellant Nonsteady Combustion, First Edition. Boris V. Novozhilov and Vasily B. Novozhilov.

Companion website: www.wiley.com/go/Novozhilov/solidpropellantnonsteadycombustion

In the absence of external radiative heat flux, Eq. (6.2) reduces to the well-known relation (see Section 2.2)

$$f^0 = \frac{u^0}{\kappa}(T_s^0 - T_a) \tag{6.3}$$

Integration of Eq. (6.1) provides the steady-state temperature profile in the condensed phase

$$T_I^0(x) = T_a + \left(T_{Is}^0 - T_a - \frac{lI^0}{\rho c u_I^0 (l-1)}\right)\exp\left(\frac{u_I^0 x}{\kappa}\right) + \frac{lI^0}{\rho c u_I^0(l-1)}\exp(\alpha x)$$

$$l = \frac{u_I^0}{\alpha\kappa} \tag{6.4}$$

where l is the nondimensional mean free path for radiation absorption in the condensed phase.

Consider the influence of the external radiative flux on the burning rate and the propellant surface temperature (Assovskii and Istratov 1971). Let us assume that the steady-state burning laws in the absence of radiation are known

$$u^0 = F_u(p^0, T_a), T_s^0 = F_s(p^0, T_a) \tag{6.5}$$

The corresponding nonsteady burning laws are

$$u = F_u\left(p, T_s - \kappa\frac{f}{u}\right), T_s = F_s\left(p, T_s - \kappa\frac{f}{u}\right) \tag{6.6}$$

The major postulate of the Zeldovich–Novozhilov (ZN) theory is that in any combustion regime the burning rate and surface temperature are fully determined by the pressure and temperature gradient. Therefore the nonsteady burning laws in the presence of radiation may be written in the form

$$u_I = F_u\left(p, T_{Is} - \kappa\frac{f_I}{u_I}\right), T_{Is} = F_s\left(p, T_{Is} - \kappa\frac{f_I}{u_I}\right) \tag{6.7}$$

Applying these to the steady-state case we obtain

$$u_I^0 = F_u\left(p^0, T_{Is}^0 - \kappa\frac{f_I^0}{u_I^0}\right), T_{Is}^0 = F_s\left(p^0, T_{Is}^0 - \kappa\frac{f_I^0}{u_I^0}\right) \tag{6.8}$$

From here, using (6.2) the steady-state burning laws in the presence of radiation may be found

$$u_I^0 = F_u\left(p^0, T_a + \frac{I^0}{\rho c u_I^0}\right), T_{Is}^0 = F_s\left(p^0, T_a + \frac{I^0}{\rho c u_I^0}\right) \tag{6.9}$$

The difference of these laws from (6.5) is that in the presence of radiation the initial temperature must be assumed to be equal to

$$T_{Ia} = T_a + \frac{I^0}{\rho c u_I^0} \tag{6.10}$$

The steady-state burning laws (6.9) allow burning rate and surface temperature for propellant subjected to external radiative heat flux to be found.

6.2 Heat Transfer Equation in the Linear Approximation

The heat transfer equation in the condensed phase with the heat source resulting from absorption of time-dependent external radiation is

$$\frac{\partial T_I}{\partial t} = \kappa \frac{\partial^2 T_I}{\partial x^2} - u_I \frac{\partial T_I}{\partial x} + \frac{\alpha I(t)}{\rho c} \exp(\alpha x), -\infty < x \leq 0 \tag{6.11}$$

The boundary conditions are

$$x \to -\infty, T_I = T_a; x = 0, T_I = T_{Is}(t) \tag{6.12}$$

This chapter will only consider linear nonsteady phenomena developing under harmonic time variations of pressure and radiative flux

$$p = p^0 + (p_1 \exp i\Omega t + c.c), I = I^0 + (I_1 \exp i\Omega t + c.c) \tag{6.13}$$

Let us introduce nondimensional variables

$$\tau = \frac{(u_I^0)^2 t}{\kappa}, \quad \omega_I = \frac{\Omega\kappa}{(u_I^0)^2}, \quad \xi = \frac{u_I^0 x}{\kappa}$$
$$v = \frac{u_I}{u_I^0}, \quad \eta = \frac{p}{p^0}, \quad \varphi = \frac{f_I}{f_I^0}$$
$$\theta = \frac{T_I - T_a}{T_{Is}^0 - T_{Ia}}, \quad \vartheta = \frac{T_{Is} - T_a}{T_{Is}^0 - T_{Ia}}, \quad s = \frac{I}{\rho c u_I^0 (T_{Is}^0 - T_{Ia})} \tag{6.14}$$

In these variables the heat transfer equation (6.11) and the boundary conditions (6.12) take the form

$$\frac{\partial \theta}{\partial \tau} = \frac{\partial^2 \theta}{\partial \xi^2} - v \frac{\partial \theta}{\partial \xi} + \frac{s(\tau)}{l} \exp\left(\frac{\xi}{l}\right)$$
$$\xi \to -\infty, \quad \theta = 0, \quad \xi = 0, \quad \theta = \vartheta(\tau) \tag{6.15}$$

In the steady state regime Eq. (6.15) reduces to

$$\frac{d^2\theta^0}{d\xi^2} - \frac{d\theta^0}{d\xi} + \frac{s^0}{l} \exp\left(\frac{\xi}{l}\right) = 0$$
$$\xi \to -\infty, \quad \theta^0 = 0, \quad \xi = 0, \quad \theta^0 = 1 + s^0 \tag{6.16}$$

with the solution

$$\theta^0 = \left(1 - \frac{s^0}{(l-1)}\right) \exp \xi + \frac{l s^0}{(l-1)} \exp\left(\frac{\xi}{l}\right), \quad s^0 = \frac{I^0}{\rho c u^0 (T_s^0 - T_a)}$$
$$v^0 = 1, \quad \varphi^0 = 1, \quad \vartheta^0 = 1 + s^0 \tag{6.17}$$

Since the linear approximation $p_1 \ll p^0, I_1 \ll I^0$ is considered, the problem variables can be represented in the form

$$\theta(\xi,\tau) = \theta^0 + [\theta_1(\xi)\exp(i\omega_I\tau) + \text{c.c.}], \quad \vartheta = \vartheta^0 + [\vartheta_1 \exp(i\omega_I\tau) + \text{c.c.}]$$
$$v = 1 + [v_1 \exp(i\omega_I\tau) + \text{c.c.}], \quad \varphi = 1 + [\varphi_1 \exp(i\omega_I\tau) + \text{c.c.}]$$
$$\eta = 1 + [\eta_1 \exp(i\omega_I\tau) + \text{c.c.}], \quad s = s^0 + [s_1 \exp(i\omega_I\tau) + \text{c.c.}] \tag{6.18}$$

Eq. (6.15) can be linearized near the steady-state regime (6.17) to obtain

$$\theta_1'' - \theta_1' - i\omega_I\theta_1 = v_1\left(1 - \frac{s^0}{(l-1)}\right)\exp\xi + \left(\frac{v_1 s^0}{(l-1)} - \frac{s_1}{l}\right)\exp\left(\frac{\xi}{l}\right) \tag{6.19}$$

with boundary conditions

$$\xi \to -\infty,\ \theta_1 = 0;\ \xi = 0,\ \theta_1 = \vartheta_1 \tag{6.20}$$

The solution of this equation is

$$\theta_1 = Ce^{z_I\xi} - \frac{v_1}{z_I(z_I-1)}\left(1 - \frac{s^0}{(1-l)}\right)e^{\xi} + \frac{l^2}{(1-lz_I)[1+l(z_I-1)]}\left(\frac{v_1 s^0}{(l-1)} - \frac{s_1}{l}\right)e^{\frac{\xi}{l}}$$

$$z_I = \frac{1}{2}(1 + \sqrt{1+4i\omega_I}) \tag{6.21}$$

Corrections to surface temperature and the temperature gradient at the surface can easily be found from the above solution

$$\vartheta_1 = C - \frac{v_1}{z_I(z_I-1)}\left(1 - \frac{s^0}{(l-1)}\right) + \frac{l^2}{(1-lz_I)[1+l(z_I-1)]}\left(\frac{v_1 s^0}{(l-1)} - \frac{s_1}{l}\right)$$

$$\varphi_1 = Cz_I - \frac{v_1}{z_I(z_I-1)}\left(1 - \frac{s^0}{(l-1)}\right) + \frac{l}{(1-lz_I)[1+l(z_I-1)]}\left(\frac{v_1 s^0}{(l-1)} - \frac{s_1}{l}\right) \tag{6.22}$$

Eliminating the integration constant we obtain the linear relationship between corrections to burning rate, surface temperature, and temperature gradient at the surface

$$z_I\vartheta_1 - \varphi_1 + \frac{v_1}{z_I}\left[1 + \frac{s^0}{1+l(z_I-1)}\right] = \frac{s_1}{1+l(z_I-1)} \tag{6.23}$$

In the absence of radiative flux Eq. (6.23) reduces to the earlier obtained relation (4.16).

6.3 Linearization of Nonsteady Burning Laws

The procedure of linearization of nonsteady burning laws in the presence of radiation (6.7)

$$u_I = F_u\left(p, T_{IS} - \kappa\frac{f_I}{u_I}\right),\quad T_{IS} = F_s\left(p, T_{IS} - \kappa\frac{f_I}{u_I}\right) \tag{6.24}$$

is somewhat different from the one applicable when external heat flux is absent (Assovskii and Istratov 1971). In the latter case the Taylor expansion at the steady-state regime involve first derivatives and the corresponding Jacobian

$$k = (T_s^0 - T_a)\left(\frac{\partial \ln F_u}{\partial T_a}\right)_{p^0},\quad r = \left(\frac{\partial F_s}{\partial T_a}\right)_{p^0}$$

$$\iota = \left(\frac{\partial \ln F_u}{\partial \ln p^0}\right)_{T_a},\quad \mu = \frac{1}{(T_s^0 - T_a)}\left(\frac{\partial F_s}{\partial \ln p^0}\right)_{T_a}$$

$$\delta = \iota r - \mu k \tag{6.25}$$

Here the role of the symbol T_a is twofold. First, this is the argument in the steady-state burning laws with respect to which differentiation is being carried out. Second, this is the value of initial temperature at which derivatives are taken.

In nonsteady processes where external radiative heat flux is present the argument with respect to which differentiation is carried out and the value of the temperature at which derivatives are taken are different from each other.

For this reason let us denote the initial temperature in the steady-state burning laws as X. This would be the argument with respect to which differentiation occurs. Specific values of the initial temperature are denoted according to their meaning, for example T_a and T_{Ia}.

This generalization leads to some changes in the notation for steady-state burning laws. Instead of the previously used form

$$u^0 = F_u(p^0, T_a), \quad T_s^0 = F_s(p^0, T_a) \tag{6.26}$$

they are now written as

$$u^0 = F_u(p^0, X)|_{X=T_a}, \quad T_s^0 = F_s(p^0, X)|_{X=T_a} \tag{6.27}$$

These are the burning rate and the surface temperature at the initial temperature equal to T_a. For a different initial temperature, for example T_{Ia}, the laws are written in the form

$$u^0 = F_u(p^0, X)|_{X=T_{Ia}}, \quad T_s^0 = F_s(p^0, X)|_{X=T_{Ia}} \tag{6.28}$$

Rewriting (6.25) in the new notation gives

$$\begin{aligned} k &= (T_s^0 - X)\left(\frac{\partial \ln F_u}{\partial X}\right)_{p^0}\Bigg|_{X=T_a}, \quad r = \left(\frac{\partial F_s}{\partial X}\right)_{p^0}\Bigg|_{X=T_a} \\ \iota &= \left(\frac{\partial \ln F_u}{\partial \ln p^0}\right)_X\Bigg|_{X=T_a}, \quad \mu = \frac{1}{(T_s^0 - X)}\left(\frac{\partial F_s}{\partial \ln p^0}\right)_X\Bigg|_{X=T_a} \\ \delta &= \iota r - \mu k \end{aligned} \tag{6.29}$$

The corresponding values in the presence of radiation need to be written as

$$\begin{aligned} k_I &= (T_{Is}^0 - X)\left(\frac{\partial \ln F_u}{\partial X}\right)_{p^0}\Bigg|_{X=T_{Ia}}, \quad r_I = \left(\frac{\partial F_s}{\partial X}\right)_{p^0}\Bigg|_{X=T_{Ia}} \\ \iota_I &= \left(\frac{\partial \ln F_u}{\partial \ln p^0}\right)_X\Bigg|_{X=T_{Ia}}, \quad \mu_I = \frac{1}{(T_{Is}^0 - X)}\left(\frac{\partial F_s}{\partial \ln p^0}\right)_X\Bigg|_{X=T_{Ia}} \\ \delta_I &= \iota_I r_I - \mu_I k_I \end{aligned} \tag{6.30}$$

Making use of this introductory remark, linearization of nonsteady burning laws may be carried out. This linearization determines the relationships between the small corrections v_1, ϑ_1, φ_1, and η_1. Expanding the functions (6.24) into Taylor series the following relationships are easily obtained

$$v_1 = \iota_I \eta_1 + k_I(\vartheta_1 - \varphi_1 + v_1), \quad \vartheta_1 = \mu_I \eta_1 + r_I(\vartheta_1 - \varphi_1 + v_1) \tag{6.31}$$

These relationships obviously allow the analogues of the relationships (4.26) to be obtained.

6.4 Steady-state Combustion Regime Stability

If both pressure and radiative heat flux are constant ($\eta_1 = 0$ and $s_1 = 0$) then (6.23) and (6.31) give the set of homogeneous algebraic equations for the perturbations of burning rate, surface temperature, and temperature gradient

$$z_I\vartheta_1 - \varphi_1 + \frac{v_1}{z_I}\left[1 + \frac{s^0}{1 + l(z_I - 1)}\right] = 0$$
$$v_1 = k_I(\vartheta_1 - \varphi_1 + v_1)$$
$$\vartheta_1 = r_I(\vartheta_1 - \varphi_1 + v_1) \tag{6.32}$$

where

$$z_I = \frac{1}{2}(1 + \sqrt{1 + 4i\omega_I}) \tag{6.33}$$

The condition of solvability for the set of equations in (6.32) leads to the following equation for oscillation frequency ω_I (De Luca 1976)

$$1 + r_I(z_I - 1) - k_I Z = 0 \tag{6.34}$$

where

$$Z = \frac{1}{z_I}\left[z_I - 1 - \frac{s^0}{1 + l(z_I - 1)}\right] \tag{6.35}$$

The complex characteristic equation (6.34) is equivalent to two real equations. Because of the form of representations (6.18), the frequency of oscillations at the stability boundary has to be real. Therefore, it is possible to find a relationship between the parameters k_I and r_I at the stability boundary, as well as dependence of the oscillation frequency on a certain parameter, for example $\omega_I(k_I)$.

In order to describe the stability boundary (6.34), the latter relationships can be multiplied by $(\overline{z}_I - 1)$ (overbar denotes complex conjugate), and then an imaginary part is taken from the product. A similar operation can also be performed taking $\overline{Z}$ as the multiplier, resulting the following relationships

$$k_I = \frac{\mathrm{Im}\overline{z}}{\mathrm{Im}\{(\overline{z} - 1)Z\}}, \quad r_I = \frac{\mathrm{Im}\overline{Z}}{\mathrm{Im}\{(\overline{z} - 1)Z\}} \tag{6.36}$$

from which dependencies $k_I(\omega_I)$ and $r_I(\omega_I)$ are obtained. The union of these dependencies defines (in parametric form) the stability boundary $r_I(k_I)$.

In order to find out how constant radiative flux affects the position of the stability boundary a specific combustion model needs be considered. As an example, consider the model discussed in Section 4.3.

Steady-state burning laws in this model, in the absence of radiative flux, have the form

$$u^0 = A(p^0)^n \exp(\beta T_a), \quad u^0 = B\exp(\beta_s T_s^0) \tag{6.37}$$

where A, B, n, β, and β_s are constants.

The linear sensitivity coefficients for this model are calculated easily

$$\iota = n, k = \beta(T_s^0 - T_a), r = \frac{\beta}{\beta_s}, \mu = \frac{\iota r}{k}, \delta = 0 \tag{6.38}$$

It follows from (6.37) that the steady-state burning laws in the presence of radiative flux are

$$u_I^0 = A(p^0)^n \exp(\beta T_{Ia}), u_I^0 = B\exp(\beta_s T_{Is}^0) \tag{6.39}$$

and the corresponding linear sensitivity coefficients are

$$\iota_I = \iota, k_I = \beta(T_{Is}^0 - T_{Ia}), r_I = r, \mu_I = \frac{\iota r}{k_I}, \delta_I = 0 \tag{6.40}$$

It is convenient to quantify the effect of external heat flux in terms of the coefficient of linear burning rate amplification

$$K = \frac{u_I^0}{u^0} \tag{6.41}$$

instead of its intensity I^0.

Comparing (6.37) and (6.39)

$$T_{Ia} - T_a = \frac{1}{\beta}\ln K, T_{Is}^0 - T_s^0 = \frac{1}{\beta_s}\ln K \tag{6.42}$$

These identities allow all the variables involved in the characteristic equation (6.34) to be expressed via parameters of the combustion regime in the absence of radiation and the coefficient K

$$z_I = \frac{1}{2}\left(1 + \sqrt{1 + 4i\frac{\omega}{K^2}}\right), \quad \omega = \frac{\Omega\kappa}{(u^0)^2}, \quad r_I = r$$
$$k_I = k + (r-1)\ln K, \quad s^0 = \frac{\ln K}{k + (r-1)\ln K} \tag{6.43}$$

Consequently, the characteristic equation (6.34) can be transformed to the form

$$k + rg(\omega) = h(\omega) \tag{6.44}$$

where

$$g(\omega) = \ln K - z_I$$
$$h(\omega) = \frac{z_I}{z_I - 1} + \left[1 + \frac{1}{(z_I - 1)[1 + l(z_I - 1)]}\right]\ln K \tag{6.45}$$

The complex transcendental equation (6.44) is equivalent to two real equations. It is easy to show that

$$r(\omega) = \frac{\operatorname{Im} h(\omega)}{\operatorname{Im} g(\omega)}, \quad k(\omega) = \frac{\operatorname{Im}\{h(\omega)\bar{g}(\omega)\}}{\operatorname{Im} \bar{g}(\omega)} \tag{6.46}$$

The relationships (6.46) allow the stability boundary $r(k)$ and the natural frequency at this boundary $\omega(k)$ (or $\omega(r)$) to be found.

Figures 6.1 and 6.2 show the stability boundaries for different values of the coefficient K. It is clear from the plots that radiation widens the region of stable combustion. This effect depends weakly on the mean free path for radiation absorption in the condensed phase.

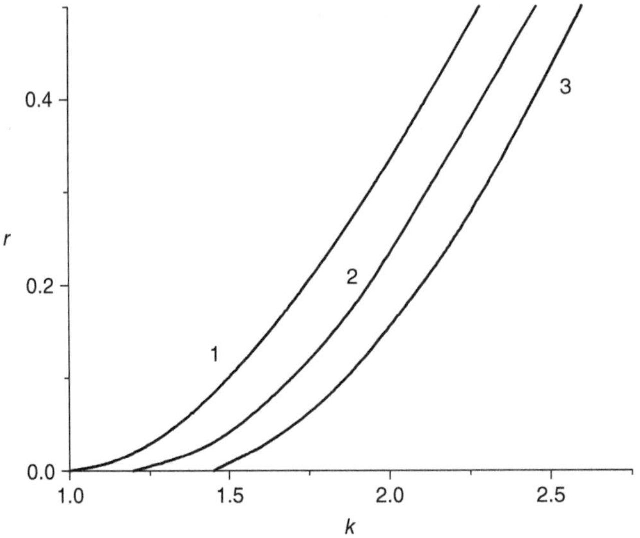

Figure 6.1 Stability boundaries. $l = 0$. 1, $K = 1.0$; 2, $K = 1.2$; 3, $K = 1.4$.

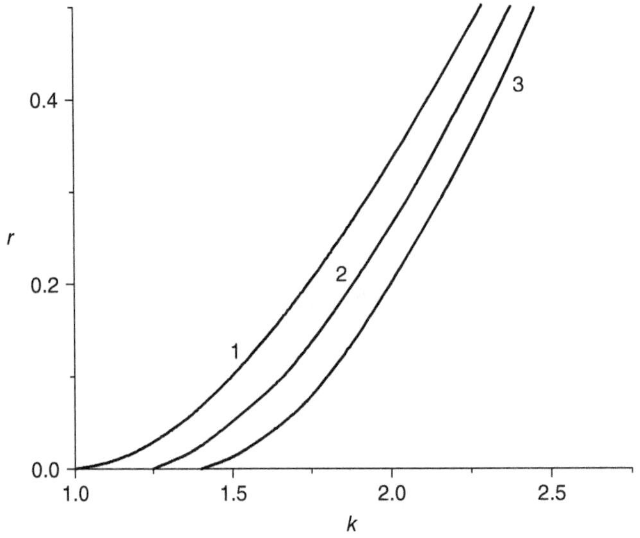

Figure 6.2 Stability boundaries. $l = \infty$. 1, $K = 1.0$; 2, $K = 1.2$; 3, $K = 1.4$.

6.5 Burning Rate Response to Harmonically Oscillating Pressure

If radiative heat flux does not vary with time but pressure does then (6.23) and (6.31) turn into the following set of algebraic equations

$$z_I \vartheta_1 - \varphi_1 + \frac{v_1}{z_I}\left[1 + \frac{s^0}{1 + l(z_I - 1)}\right] = 0$$

$$\begin{aligned} v_1 &= \iota_I \eta_1 + k_I(\vartheta_1 - \varphi_1 + v_1) \\ \vartheta_1 &= \mu_I \eta_1 + r_I(\vartheta_1 - \varphi_1 + v_1) \end{aligned} \tag{6.47}$$

with

$$z_I = \frac{1}{2}(1 + \sqrt{1 + 4i\omega_I}) \tag{6.48}$$

This set of equations (6.47) leads to the following response function of burning rate to oscillating pressure in the presence of constant radiative heat flux (generalization of (4.27)) $U_p = v_1/\eta_1$

$$U_p = \frac{\iota_I + \delta_I(z_I - 1)}{1 + (z_I - 1)\left(r_I - \frac{k_I}{z_I}\right) + \frac{k_I s^0}{z_I[1 + l(z_I - 1)]}} \tag{6.49}$$

For the combustion model considered in Section 6.4, the function (6.49) may be written in the form

$$\begin{aligned} U_p &= \frac{\iota}{1 + (z_I - 1)\left(r - \frac{k}{z_I}\right) + \varphi(\omega) \ln K} \\ z_I &= \frac{1}{2}\left(1 + \sqrt{1 + 4i\frac{\omega}{K^2}}\right), \quad \varphi = \frac{1}{z_I}\left[(z_I - 1)(1 - r) + \frac{1}{1 + l(z_I - 1)}\right] \end{aligned} \tag{6.50}$$

In the absence of radiation ($K = 1$), the response function of the burning rate to harmonically oscillating pressure was found by Novozhilov (1965b). For the considered combustion model it has the form

$$U = \frac{\iota}{1 + (z - 1)\left(r - \frac{k}{z}\right)}, \quad z = \frac{1}{2}(1 + \sqrt{1 + 4i\omega}) \tag{6.51}$$

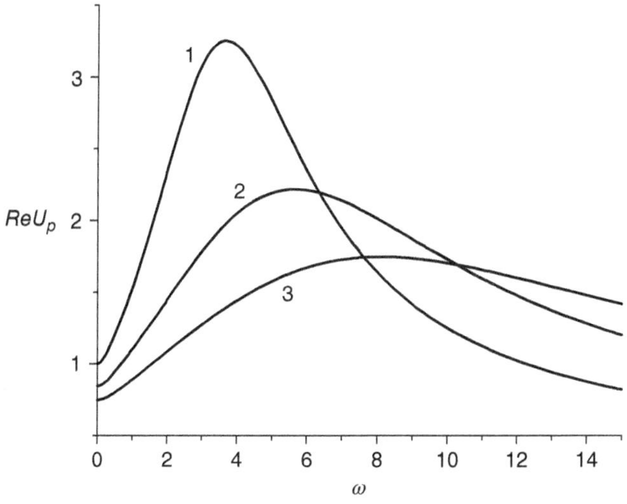

Figure 6.3 Real parts of the response function. $l = 0.1$, $K = 1.0$; 2, $K = 1.2$; 3, $K = 1.4$.

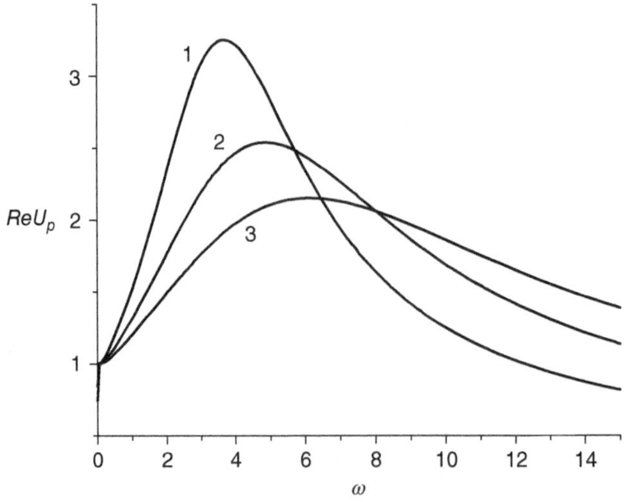

Figure 6.4 Real parts of the response function. $l = \infty$. 1, $K = 1.0$; 2, $K = 1.2$; 3, $K = 1.4$.

Figures 6.3 and 6.4 show the frequency dependencies of real parts of the response function for different values of the amplification coefficient. The values of the other parameters are fixed at $\iota = 1$, $k = 1.5$ and $r = 0.3$.

Curve 1 corresponds to $K = 1$. The plots demonstrate that small changes in the amplification coefficient result in changes in the response function of the order of tens of percent. Accounting for radiation leads to a decrease in the real part of the response function. This behaviour conforms with the improved stability of the steady-state regime in the presence of radiation which was noted at the end of Section 6.4. Similar to the effect on stability boundaries (Figures 6.1 and 6.2), the mean free path for radiation absorption in the condensed phase weakly affects the values of the response function.

6.6 Burning Rate Response to Harmonically Oscillating Radiative Flux

Let radiative flux vary with time in harmonic fashion. In the linear approximation, the burning rate would change with time similarly

$$s = s^0 + [s_1 \exp(i\omega_I \tau) + c.c.], \quad v = 1 + [v_1 \exp(i\omega_I \tau) + c.c.] \tag{6.52}$$

By analogy with the linear response of the burning rate to harmonically oscillating pressure, the value $U_r = v_1/(s_1/s_0)$ is called the response function of burning rate to oscillating radiative flux.

Under constant pressure, (6.23) and (6.31) turn into the following set of algebraic equations for the amplitudes of perturbations

$$z_I \vartheta_1 - \varphi_1 + \frac{v_1}{z_I}\left[1 + \frac{s^0}{1 + l(z_I - 1)}\right] = \frac{s_1}{1 + l(z_I - 1)}$$

$$v_1 = k_I(\vartheta_1 - \varphi_1 + v_1)$$

$$\vartheta_1 = r_I(\vartheta_1 - \varphi_1 + v_1) \tag{6.53}$$

where

$$z_I = \frac{1}{2}(1 + \sqrt{1 + 4i\omega_I}) \tag{6.54}$$

The following expression for the response function is obtained from the set of equations (6.53)

$$U_r = \frac{k_I s^0}{[1 + l(z_I - 1)]\left[1 + (z_I - 1)\left(r_I - \frac{k_I}{z_I}\right)\right] + \frac{k_I s^0}{z_I}} \tag{6.55}$$

Within the framework of the ZN theory, the response function of the propellant burning rate to harmonically oscillating radiative heat flux was first obtained by Kiskin (1983). The final result of this approach is very complicated due to the unfortunate way the problem variables are scaled. De Luca et al. (1995) demonstrated that the response function may be expressed in a simpler form (6.55).

Consider now the response function (6.55) for the specific combustion model discussed in Section 6.4. Using formulae in (6.43)

$$z_I = \frac{1}{2}\left(1 + \sqrt{1 + 4i\frac{\omega}{K^2}}\right), \quad r_I = r$$

$$k_I = k + (r - 1)\ln K, \quad k_I s^0 = \ln K \tag{6.56}$$

Therefore (6.55) takes the form

$$U_r = \frac{\ln K}{[1 + l(z_I - 1)]\left[1 + (z_I - 1)\left(r - \frac{k_I}{z_I}\right)\right] + \frac{\ln K}{z_I}} \tag{6.57}$$

The response function of propellant burning rate to harmonically varying radiative heat flux (6.55) and its specific form (6.57) depend on many parameters. Let us consider qualitatively the dependencies of the response function on those parameters. First of all consider the list of parameters.

Nondimensional frequency ω is determined (6.43) by its dimensional counterpart and the two variables that are known from experiments. These are the linear burning rate in the absence of radiation u^0 and the thermal diffusivity of the condensed phase κ.

The root z_I of the characteristic equation contains, besides nondimensional frequency, the amplification coefficient K. Quantification of this coefficient requires measuring, along with u^0, the linear burning rate in the presence of radiative flux u_I^0. The amplification coefficient also enters the expression for the response function explicitly.

Fuel properties are described in the linear approximation by the sensitivity parameters k and r. Finally, the expression for the response function involves the nondimensional mean free path for radiation absorption in the condensed phase l, which is determined by the linear absorption coefficient α and the parameters u^0 and κ.

Consider first how the frequency dependence $U_r(\omega)$ is affected by the sensitivity parameters. Note that nullification of the denominator (6.55) corresponds to the stability boundary of the steady-state combustion regime under constant radiative flux (6.34). Similar correspondence exists for expressions (6.57) and (6.44).

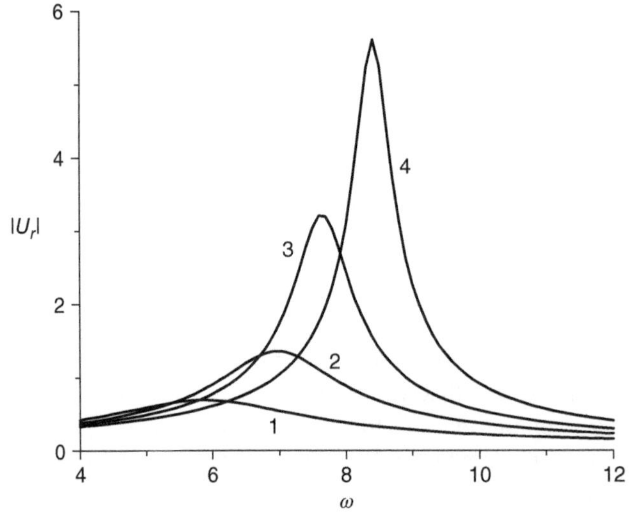

Figure 6.5 Real parts of the response function. $l = 1$. $k = 2$, $K = 1.2$. 1, $r = 0.35$; 2, $r = 0.3$; 3, $r = 0.275$; 4, $r = 0.25$.

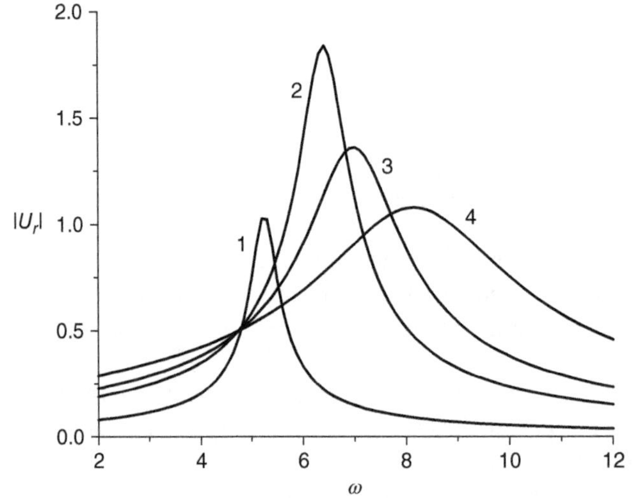

Figure 6.6 Real parts of the response function. $l = 1$. $k = 2$, $r = 0.3$. 1, $K = 1.05$; 2, $K = 1.15$; 3, $K = 1.2$; 4, $K = 1.3$.

As an example, Figure 6.5 demonstrates the real part of the response function for different values of the sensitivity parameter r. Figure 6.6 shows the frequency dependencies of the real part of the response function for various values of amplification coefficient.

It is evident that small changes in the amplification coefficient cause variations of the response function of the order of tens of percent.

The value of the mean free path does not affect the results significantly.

The role of radiation in the combustion of highly metalized propellants was investigated by Kiskin (1993) using a similar method.

6.7 Relation Between Burning Rate Responses to Harmonically Oscillating Pressure and Radiative Flux

Significant efforts have been made in developing experimental methodologies that could determine the response function of the burning rate to harmonically oscillating pressure. Since measurements following developed techniques still encounter significant relative errors, the search for alternative methods to quantify the response function experimentally is underway. One these implicit methods is an attempt to use the response function of the burning rate to oscillating radiative heat flux in order to obtain the desired estimate of response to oscillating pressure. This idea was first formulated over 40 years ago by Mihlfeith et al. (1972). Since then, several attempts to relate the two functions analytically have been made (De Luca 1976; Finlinson et al. 1991; Son and Brewster 1992; Kiskin 1993), but without much success. Expressions of one of the functions through another have been obtained both for the trivial case, in the limit of very small radiation intensity and null mean free path for radiation absorption (De Luca 1976; Finlinson et al. 1991; Son and Brewster 1992), and for a significantly restricted class of fuels (Kiskin 1993).

The reason for the lack of success is rather obvious. The constant component of radiative heat flux alters the steady-state temperature profile in the condensed phase. As a result, the steady-state values of the burning rate and surface temperature are larger than they are in the absence of radiative heat flux. Therefore, the burning rate oscillations caused by the variable component of radiative flux occur around a different steady-state background level compared to oscillations caused by variable pressure.

In the formal mathematical treatment, this results in the fact that the two response functions of the burning rate (to oscillating pressure and to oscillating radiative flux), calculated at the same background pressure and initial temperature, become nevertheless dependent on different parameters. As shown in Chapter 4, within the linear approximation of the ZN theory, any nonsteady process is characterized by the four first partial derivatives of the steady-state burning rate and the propellant surface temperature with respect to pressure and initial temperature. In the absence of radiative flux, derivatives with respect to initial temperature are taken at the true actual value of this temperature. In the presence of radiative flux, these derivatives, as shown by Assovskii and Istratov (1971), have to be found at a higher initial temperature which is determined by the constant component of radiative heat flux as well as by the new value of the steady-state burning rate. For these reasons, the linear theory of any nonsteady phenomenon occurring in the presence of radiative heating would involve propellant parameters different from those which control combustion in the absence of radiation. Obviously, in any general case, the analytic relationship between two functions depending on different arguments and parameters does not exist.

Novozhilov et al. (2002b, 2003) and Novozhilov (2005) demonstrated that it is possible to find a response function to oscillating pressure at given values of initial temperature T_a and pressure using the response function to oscillating radiative flux, calculated at the same pressure but at some lower initial temperature T_e. The value of the temperature T_e should be chosen in such a way that the steady-state burning rate at this temperature, in the presence of constant radiative heat flux, is equal to the steady-state burning rate in the absence of radiation at the initial temperature T_a. In this case, the response function to variable radiative heat flux, calculated at temperature T_e, would depend on the same parameters as the response function to variable pressure calculated at the higher initial temperature T_a. Naturally, such two functions may be analytically related to each other (Novozhilov et al. 2002b, 2003) as follows.

The linear response function of burning rate to oscillating pressure was obtained in Section 4.1

$$U = \frac{\iota + \delta(z-1)}{1 + (z-1)(r - k/z)} \tag{6.58}$$

where

$$k = (T_s^0 - T_a)\left(\frac{\partial \ln F_u}{\partial T_a}\right)_{p^0}, \quad r = \left(\frac{\partial F_s}{\partial T_a}\right)_{p^0}$$

$$\iota = \left(\frac{\partial \ln F_u}{\partial \ln p^0}\right)_{T_a}, \mu = \frac{1}{(T_s^0 - T_a)}\left(\frac{\partial F_s}{\partial \ln p^0}\right)_{T_a}$$

$$\delta = \iota r - \mu k, \quad z = \frac{1}{2}(1 + \sqrt{1 + 4i\omega}), \quad \omega = \frac{\kappa\Omega}{(u^{(0)})^2} \tag{6.59}$$

For the purposes of this section it is convenient to change the symbol for initial temperature. Previously the symbol T_a played a twofold role. It was used as an argument in the steady-state burning laws with respect to which derivatives can be taken, and also as the specific value of initial temperature at which the considered process occurred. This section compares the combustion regimes occurring at different initial temperatures. For this reason, let us denote the steady-state burning laws argument corresponding to the initial temperature as X. The relevant derivatives will be taken with respect to this argument. On the other hand, specific values of initial temperature will be denoted using the conventional symbol and subscripts, for example T_a and T_I.

The new variable steady-state burning laws, instead of the previous form

$$u^0 = F_u(p^0, T_a), \ T_s^0 = F_s(p^0, T_a) \tag{6.60}$$

are written as

$$u^0 = F_u(p^0, X)|_{X=T_a}, \ T_s^0 = F_s(p^0, X)|_{X=T_a} \tag{6.61}$$

These are the burning rate and surface temperature at the initial temperature T_a. For a different initial temperature, for example T_I, one would need to write

$$u^0 = F_u(p^0, X)|_{X=T_I}, \quad T_s^0 = F_s(p^0, X)|_{X=T_I} \tag{6.62}$$

The response functions (6.58) and (6.59) may be rewritten in the new notation. Let us introduce also the subscript p indicating response to variable pressure, and a superscript indicating the value of the initial temperature. The results are

$$U_p^{(a)} = \frac{\iota^{(a)} + \delta^{(a)}(z^{(a)} - 1)}{1 + (z^{(a)} - 1)(r^{(a)} - k^{(a)}/z^{(a)})} \tag{6.63}$$

where

$$k^{(a)} = ((T_s^0)^{(a)} - X)\left(\frac{\partial \ln F_u}{\partial X}\right)_{p^0}\Bigg|_{X=T_a}, \quad r^{(a)} = \left(\frac{\partial F_s}{\partial X}\right)_{p^0}\Bigg|_{X=T_a}$$

$$\iota^{(a)} = \left(\frac{\partial \ln F_u}{\partial \ln p^0}\right)_X\Bigg|_{X=T_a}, \quad \mu^{(a)} = \frac{1}{((T_s^0)^{(a)} - X)}\left(\frac{\partial F_s}{\partial \ln p^0}\right)_X\Bigg|_{X=T_a}$$

$$\delta^{(a)} = \iota^{(a)} r^{(a)} - \mu^{(a)} k^{(a)}, \quad z^{(a)} = \frac{1}{2}(1 + \sqrt{1 + 4i\omega^{(a)}})$$
$$\omega^{(a)} = \frac{\kappa\Omega}{[(u^0)^{(a)}]^2} \tag{6.64}$$

The linear response function of the burning rate to the harmonically oscillating radiative heat flux is given in Section 6.6.

In the new notation, the response function (6.55) assumes the form

$$U_r^{(a)} = \frac{k^{(I)} s^{(I)}}{[1 + l^{(I)}(z^{(I)} - 1)]D^{(I)} + \frac{k^{(I)} s^{(I)}}{z^{(I)}}} \tag{6.65}$$

where

$$D^{(I)} = 1 + (z^{(I)} - 1)\left(r^{(I)} - \frac{k^{(I)}}{z^{(I)}}\right), \quad z^{(I)} = \frac{1}{2}(1 + \sqrt{1 + 4i\omega^{(I)}})$$
$$k^{(I)} = ((T_s^{(0)})^{(I)} - X)\left(\frac{\partial \ln F_u}{\partial X}\right)_{p^0}\bigg|_{X=T_I}, \quad r^{(I)} = \left(\frac{\partial F_s}{\partial X}\right)_{p^0}\bigg|_{X=T_I}$$
$$s^{(I)} = \frac{I^0}{\rho c (u^{(I)})^0 ((T_s^{(0)})^{(I)} - T_I)}, \quad l^{(I)} = \frac{(u^{(I)})^0}{\alpha\kappa}, \quad \omega^{(I)} = \frac{\kappa\Omega}{[(u^{(I)})^0]^2} \tag{6.66}$$

Here α is the linear absorption coefficient of radiation in the propellant body, the superscript I refers to values calculated at radiation conditions, and $(u^{(I)})^0$ and $(T_s^{(0)})^{(I)}$ are the steady-state burning rate and the propellant surface temperature, respectively. The latter two values can be obtained from equations (6.2) and (6.8).

It is evident by comparing (6.63) and (6.65) that these two response functions depend on different parameters characterizing burning propellant. Specifically,

the function $U^{(a)}$ is controlled by the parameters $k^{(a)}$ and $r^{(a)}$, and depends on nondimensional frequency $\omega^{(a)}$ as an argument
the function $U_r^{(a)}$ is controlled by the parameters $k^{(I)}$ and $r^{(I)}$, and depends on nondimensional frequency $\omega^{(I)}$ as an argument.

Obviously, in any general case it is impossible to obtain an analytical relation between two functions controlled by different parameters and dependant on different arguments.

Nevertheless such a relation may be established between the response function to oscillating pressure $U^{(a)}$, measured at the initial temperature T_a, and the response function to oscillating radiative flux $U_r^{(e)}$, measured at some lower temperature $T_I = T_e$.

Let us choose the initial temperature T_e in such a way that the steady-state burning rate at this initial temperature, under constant heat flux I^0, is equal to the steady-state burning rate at temperature T_a with external heat flux being absent

$$u_I^0(p, T_e, I^0) = u^0(p, T_a) \tag{6.67}$$

This requirement would be satisfied if

$$T_e = T_a - {}^{I^0}/_{\rho c u^0} \tag{6.68}$$

At the same time the equality between surface temperatures in the considered regimes would also hold

$$(T_s^0)^{(I)}(p, T_e, I^0) = T_s^0(p, T_a) \tag{6.69}$$

It also follows from comparison of (6.64) and (6.66) that

$$k^{(e)} = k^{(a)}, r^{(e)} = r^{(a)}, \omega^{(e)} = \omega^{(a)} \tag{6.70}$$

Therefore the response of burning rate to radiative heat flux measured at the temperature in (6.68) is expressed through exactly the same parameters as the response of burning rate to oscillating pressure measured at temperature T_a, that is

$$U_r^{(e)} = \frac{\beta\Delta}{[1 + l^{(a)}(z^{(a)} - 1)]D^{(a)} + \frac{\beta\Delta}{z^{(a)}}} \tag{6.71}$$

Here the product $k^{(a)}s^{(a)}$ is expressed through the temperature coefficient of the burning rate at temperature T_a and the initial temperature difference $\Delta = T_a - T_e$.

By comparing (6.63) and (6.71) the response to varying pressure may be expressed through the response to varying radiative heat flux

$$U_p^{(a)} = \frac{[\iota + \delta^{(a)}(z^{(a)} - 1)][1 + l^{(a)}(z^{(a)} - 1)]}{\beta\Delta\left(\frac{1}{U_r^{(e)}} - \frac{1}{z^{(a)}}\right)} \tag{6.72}$$

In a more symmetric form, the relation between the two functions may be written as

$$\frac{\iota + \delta(z - 1)}{U_p^{(a)}} = \frac{\beta\Delta}{[1 + l(z - 1)]}\left(\frac{1}{U_r^{(e)}} - \frac{1}{z}\right) \tag{6.73}$$

Application of this relationship requires prior measurements of the mean free path for radiation absorption in the condensed phase, the steady-state regime parameters ι and δ, and the temperature coefficient of burning rate β.

The initial temperature difference should be set taking into account the capabilities of the experimental rig. The simplest way of choosing this value follows these steps:

1. Find the steady-state burning rate u^0 at given pressure p^0 and initial temperature T_a, in the absence of radiation.
2. Lower the sample temperature at the same pressure down to some chosen new value of initial temperature T_e.
3. Measure the burning rate u_I^0 at the initial temperature T_e in the presence of constant external radiative flux. Then choose the intensity of the radiation source according to (6.67) so that the equality $u_r^0 = u^0$ is fulfilled.

The radiation sources currently used have the power of the order of a megawatt per square metre. Depending on pressure, at normal values of propellant density and heat capacity, the required initial temperature difference ranges from tens of degrees to 200 K.

7

Nonacoustic Combustion Regimes

7.1 Acoustic and Nonacoustic Combustion Regimes

Considerable attention in the preceding chapters was devoted to problems associated with nonsteady propellant burning under harmonically oscillating pressure (Sections 4.5 and 4.7). The importance of such considerations stems from the profound role that acoustic waves may play in determining the regimes of operation of rocket motors. The ability of the burning surface to amplify pressure waves upon their reflection dictates the need to consider a wide range of issues associated with the acoustic properties of the surface, such as acoustic admittance. Consequently, burning regimes occurring under harmonically oscillating pressure may be called acoustic regimes.

Along with such acoustic regimes, other modes on nonsteady combustion behaviour play an important role in theoretical and practical engineering design. These are nonsteady combustion regimes occurring under pressure variation laws different from harmonic. Such regimes may be called nonacoustic. A very important class of such problems are combustion processes occurring in the course of rocket motors transitioning from one steady-state regime of operation to another, at either a higher or lower pressure compared to the initial one.

Furthermore, specific nonacoustic transitional combustion regimes are associated with propellant extinction which occurs on sufficiently rapid and deep pressure drop in a combustion chamber.

Another class of relevant problems is formed by propellant ignition phenomena. Such processes include, for example, ignition of virgin or extinguished propellant by impulse of thermal radiation, a stream of hot combustion products, and others. The ignition of propellants is not discussed in this book. Generally, predicting ignition does not require an application of the methods of nonsteady combustion theory, such as Zeldovich–Novozhilov (ZN) theory. In essence, ignition is associated with heating and inception of a chemical reaction in the thin layer of propellant adjacent to the surface. This is in contrast to the extinction process which normally starts from the steady-state regime then progresses according to the laws of nonsteady combustion until its end, and therefore requires the application of the nonsteady combustion theory. With regard to propellant ignition, the reader is encouraged to consult the book by Vilyunov and Zarko (1989). A review on theoretical and experimental studies on ignition has also been carried out by Hermance (1984).

Theory of Solid-Propellant Nonsteady Combustion, First Edition. Boris V. Novozhilov and Vasily B. Novozhilov.

Companion website: www.wiley.com/go/Novozhilov/solidpropellantnonsteadycombustion

As the heat transfer equation accounting for the thermal inertia of the condensed phase is nonlinear, mathematical investigation of nonacoustic regimes under arbitrary nonsteady burning laws and pressure dependence on time faces considerable difficulties. For this reason, small variations of pressure, that is, the linear approximation, is considered first.

7.2 Linear Approximation

Consider the time history of the burning rate induced by small pressure variation from some initial value p^0 to the final $p_\infty = p^0(1+h)$, where $h \ll 1$. It is assumed that the process starts from the steady-state regime existing at pressure p^0.

Temperature, surface temperature, temperature gradient, burning rate, and pressure may be represented in the nondimensional form as

$$\theta = e^\xi + \theta_1, \quad \vartheta = 1 + \vartheta_1, \quad v = 1 + v_1$$
$$\varphi = 1 + \varphi_1, \quad \eta = 1 + \eta_1 \tag{7.1}$$

The heat transfer equation

$$\frac{\partial\theta}{\partial\tau} = \frac{\partial}{\partial\xi}\left(\frac{\partial\theta}{\partial\xi} - v\theta\right), \quad -\infty < \xi \le 0, \tag{7.2}$$

may be linearized to give

$$\frac{\partial\theta_1}{\partial\tau} = \frac{\partial^2\theta_1}{\partial\xi^2} - \frac{\partial\theta_1}{\partial\xi} - v_1 e^\xi \tag{7.3}$$

Boundary and initial conditions in these variables take the form

$$\xi = 0, \quad \theta_1 = \vartheta_1(\tau); \quad \xi \to -\infty, \quad \theta_1 = 0$$
$$\tau = 0, \quad \theta_1(\xi, 0) = 0 \tag{7.4}$$

Nonsteady burning laws in the linear approximation can be written in the form (4.26)

$$v_1 = \frac{k}{k+r-1}\varphi_1 + \frac{(\delta - \iota)}{k+r-1}\eta_1, \quad \vartheta_1 = \frac{r}{k+r-1}\varphi_1 - \frac{(\delta+\mu)}{k+r-1}\eta_1 \tag{7.5}$$

Pressure correction $\eta_1(\tau)$ changes from zero at $\tau = 0$ to h at $\tau \to \infty$.

Within the framework of the constant surface temperature assumption, the problem was solved by Zeldovich (1964), and burning rate evolution under small periodic pressure variations was found (arbitrary dependence $\eta_1(\tau)$ may be decomposed into superposition of various harmonics).

In a similar manner, the problem is considered within the ZN theory with variable surface temperature (Novozhilov 1967a).

Using the Laplace–Carson transform

$$\bar{f}(p) = p\int_0^\infty e^{-p\tau} f(\tau)\mathrm{d}\tau \tag{7.6}$$

the heat transfer Eq. (7.3) can be written in the transform space as

$$\bar{\theta}_1'' - \bar{\theta}_1' - p\bar{\theta}_1 = \bar{v}_1 e^\xi \tag{7.7}$$

It possesses the solution (finite at $\xi \to -\infty$)

$$\overline{\theta}_1(p,\xi) = Ce^{z\xi} - \frac{\overline{v}_1}{p}e^{\xi} \tag{7.8}$$

where C is the integration constant and

$$z = \frac{1}{2} + \sqrt{p + \frac{1}{4}} \tag{7.9}$$

Boundary conditions at $\xi = 0$ give

$$\overline{\vartheta}_1 = C - \frac{\overline{v}_1}{p}, \quad \overline{\varphi}_1 = Cz - \frac{\overline{v}_1}{p} \tag{7.10}$$

and (7.5) is transformed to

$$\overline{v}_1 = \frac{k}{k+r-1}\overline{\varphi}_1 + \frac{\delta - \iota}{k+r-1}\overline{\eta}_1, \quad \overline{\vartheta}_1 = \frac{r}{k+r-1}\overline{\varphi}_1 - \frac{\delta + \mu}{k+r-1}\overline{\eta}_1 \tag{7.11}$$

The last four relationships allow the integration constant and Laplace–Carson transforms of the linear corrections to the steady-state values to be found

$$\overline{v}_1 = \frac{\iota + \delta(z-1)}{D}\overline{\eta}_1, \quad \overline{\vartheta}_1 = \frac{\mu z + \delta(z-1)}{Dz}\overline{\eta}_1$$
$$\overline{\varphi}_1 = \frac{\iota + \mu z^2 + \delta(z^2-1)}{Dz}\overline{\eta}_1, \quad D = 1 + \left(r - \frac{k}{z}\right)(z-1) \tag{7.12}$$

This result may also be obtained from integral formulation of the theory (Novozhilov 1970). Indeed, on substitution of decompositions (7.1) into the integral equation (2.49) the latter takes the form

$$1 + \vartheta_1(\tau) = \frac{1}{\sqrt{\pi}}\int_0^{\tau}\left[\frac{1}{2} + \varphi_1 - v_1 - \frac{\vartheta_1}{2} + \left(\frac{1}{\tau - \tau'} - \frac{1}{2}\right)\frac{I_1}{2}\right]$$
$$\exp\left(\frac{-(\tau-\tau')}{4}\right)\frac{d\tau'}{\sqrt{\tau - \tau'}} +$$
$$+ \frac{1}{\sqrt{\pi\tau}}\int_{-\infty}^{0}\left\{\left[1 - \frac{(\tau+\xi)}{2\tau}J_1\right]\exp\frac{-(\tau-\xi)^2}{4\tau}\right\}d\xi$$
$$I_1 = \int_{\tau'}^{\tau} v_1(\tau'')d\tau'' \quad J_1 = \int_0^{\tau} v_1(\tau'')d\tau'' \tag{7.13}$$

On integration (the order of integration is changed in the term containing I_1)

$$\vartheta_1(\tau) = \frac{1}{\sqrt{\pi}}\int_0^{\tau}\left(\varphi_1 - \frac{\vartheta_1}{2}\right)\exp\left(\frac{-(\tau-\tau')}{4}\right)\frac{d\tau'}{\sqrt{\tau-\tau'}} - \int_0^{\tau}\operatorname{erfc}\left(\frac{\sqrt{\tau-\tau'}}{2}\right)v_1 d\tau' \tag{7.14}$$

Substituting the constraints

$$v_1 = \frac{k}{k+r-1}\varphi_1 + \frac{(\delta - \iota)}{k+r-1}\eta_1, \quad \vartheta_1 = \frac{r}{k+r-1}\varphi_1 - \frac{(\delta+\mu)}{k+r-1}\eta_1 \tag{7.15}$$

the following Volterra equation of the second kind is obtained

$$v_1(\tau) = \frac{1}{\sqrt{\pi}}\int_0^\tau \frac{e^{-u/4}}{\sqrt{u}}\left[\frac{(2(k-1)+r)}{2r}v_1(\tau-u) - \frac{(\delta-2\imath)}{2r}\eta_1(\tau-u)\right]du+$$
$$+\int_0^\tau \mathrm{erfc}\left(\frac{\sqrt{u}}{2}\right)v_1(\tau-u)\mathrm{d}\tau u + \frac{\delta}{r}\eta_1(\tau) \tag{7.16}$$

This equation can be solved in a conventional manner using the Laplace–Carson transform to deliver the first of the relationships (7.12).

Consider, first of all, the burning rate, surface temperature, and temperature gradient behaviour under very slow pressure variation (quasi steady-state regime).

The case of slow pressure variation is described by the relationships (7.12) for $p \ll 1$. Retaining just the linear term of expansion with respect to p

$$z = 1 + p,\quad D = 1 + (r-k)p$$
$$\overline{v}_1 = [\imath + k(\imath - \mu)p]\overline{\eta}_1,\qquad \overline{\vartheta}_1 = [\mu + r(\imath - \mu)p]\overline{\eta}_1$$
$$\overline{\varphi}_1 = [\imath + \mu + (k + r - 1)(\imath - \mu)p]\overline{\eta}_1 \tag{7.17}$$

The conditions (7.4) imply that at the initial time instant perturbations are absent, therefore the inverse transforms are

$$v_1(\tau) = \imath\eta_1(\tau) + k(\imath - \mu)\frac{\mathrm{d}\eta_1(\tau)}{\mathrm{d}\tau}$$
$$\vartheta_1(\tau) = \mu\eta_1(\tau) + r(\imath - \mu)\frac{\mathrm{d}\eta_1(\tau)}{\mathrm{d}\tau}$$
$$\varphi_1(\tau) = (\imath + \mu)\eta_1(\tau) + (k + r - 1)(\imath - \mu)\frac{\mathrm{d}\eta_1(\tau)}{\mathrm{d}\tau} \tag{7.18}$$

Therefore, under a slow pressure variation burning rate, the surface temperature and temperature gradient at the surface differ from their steady-state values at the instantaneous pressure value $\eta_1(\tau)$ by quantities proportional to the change rate of pressure. For the case $r = \mu = 0$ the expressions (7.18) transforms into the ones obtained by Zeldovich (1964).

Consider now the process in the case of sharp pressure change. Let pressure change from $\eta_1 = 0$ to $\eta_1 = h$ at the time instant $\tau = 0$ and remain constant thereafter. Such a process is idealized. In reality, sudden jumps of pressure cannot be achieved. Moreover, in the present consideration, relaxation of the processes occurring in the gas phase and in the reaction zone of the condensed phase is assumed to occur instantaneously. In reality, relevant relaxation time scales are different from zero. Nevertheless, the consideration of such simple idealized pressure dependence on time reveals important features in the behaviour of burning rate, surface temperature, and temperature gradient at the surface. It will help then to consider a real case where pressure changes with finite rate.

In the model with a constant surface temperature, the profile in the condensed phase, on sharp change of pressure, does not change due to inertia of the preheat layer. The temperature gradient at the surface remains constant as well whilst the burning rate experiences a jump from its initial value to the value determined by the *old* value of the gradient and the *new* value of pressure.

Since the initial value of the gradient correction $\varphi_1 = 0$ does not change, the burning rate correction according to (7.5) jumps from $v_1 = 0$ to $v_1 = \imath h/(1-k)$. Then, the system evolves in time to the new steady-state regime where $v_1 = \varphi_1 = \imath h$.

Consider now the behaviour of the propellant with variable surface temperature. If pressure rise occurs sufficiently quickly then the temperature profile in the condensed phase, similar to the case with constant surface temperature, does not change considerably over the time interval of pressure variation. This does not mean, however, that under a sharp change of pressure both the temperature gradient and the surface temperature remain the same. In the model with variable surface temperature there exists the relationship $\vartheta(\varphi, \eta)$ which indicates that under a change of pressure at least one of the variables ϑ and φ has to experience a jump. It is easy to understand that it is the temperature gradient that experiences the jump. Indeed, a small increase of surface temperature (while the temperature in the body of the propellant remains the same) may cause a sharp change in the gradient. In the limit of pressure, the jump surface temperature must remain the same (the whole temperature profile does not have enough time to change) whilst the temperature gradient would change at a single point that is at the surface. The relationships (7.5) provide exact jumps of the burning rate and temperature gradient

$$\Delta v = \frac{\delta}{r}h, \quad \Delta\varphi_1(\tau) = \frac{\delta + \mu}{r}h \tag{7.19}$$

Evolution of the disturbed process to the final different steady-state regime may be exactly described in the following manner. Using the Laplace–Carson transform rules it is easy to show that the inverse transform of the Laplace–Carson-transformed burning rate (7.12) can be written in the form

$$v_1(\tau) = h\frac{\delta}{r}\frac{e^{-\tau/4}}{\sqrt{\pi\tau}}\int_0^{\infty} e^{-u^2/4\tau}\phi(u)\mathrm{d}u \tag{7.20}$$

where $\phi(u)$ is inverse transform of the function

$$\overline{\phi}(p) = \frac{p^2\,(p + {}^{v}/_{\delta} - {}^{1}/_{2})}{p^3 + {}^{(2-2k-r)p^2}/_{2r} + {}^{(4k-r)p}/_{4r} - {}^{(2+2k-r)}/_{8r}} \tag{7.21}$$

The exact form of the function $\phi(u)$ is determined by the roots of the denominator of $\overline{\phi}(u)$, which are

$$p_1 = \frac{1}{2}, \quad p_{2,3} = \frac{k-1}{2r} \pm \frac{ir}{k-1}\omega \tag{7.22}$$

where $\omega = \sqrt{\omega_1^2 - \lambda^2}$, ω_1 and λ being the natural frequency and the damping decrement of the burning rate oscillations (3.28), respectively. Complex roots lead to the existence of harmonic functions in the expression for $\phi(u)$, and consequently in the expression for $v_1(\tau)$.

Calculations give the following expression for the burning rate

$$\begin{aligned}
\frac{v_1}{h} &= \frac{\imath}{2}\operatorname{erfc}\left(-\frac{\sqrt{\tau}}{2}\right) + \left(\frac{\delta}{r} - \frac{\imath}{2}\right) \times \\
&\times \left[2e^{-\lambda\tau}\cos\omega\tau - e^{-\tau/4}U\left(\frac{\omega r\sqrt{\tau}}{(k-1)}, \frac{(k-1)\sqrt{\tau}}{2r}\right)\right] + \\
&+ \frac{(k-1)}{2\omega r^2}\left[\frac{k(\delta+\mu)}{r} + \frac{\imath(1-r+k)}{2}\right] \times \\
&\times \left[2e^{-\lambda\tau}\sin\omega\tau + e^{-\tau/4}V\left(\frac{\omega r\sqrt{\tau}}{k-1}, \frac{(k-1)\sqrt{\tau}}{2r}\right)\right]
\end{aligned} \tag{7.23}$$

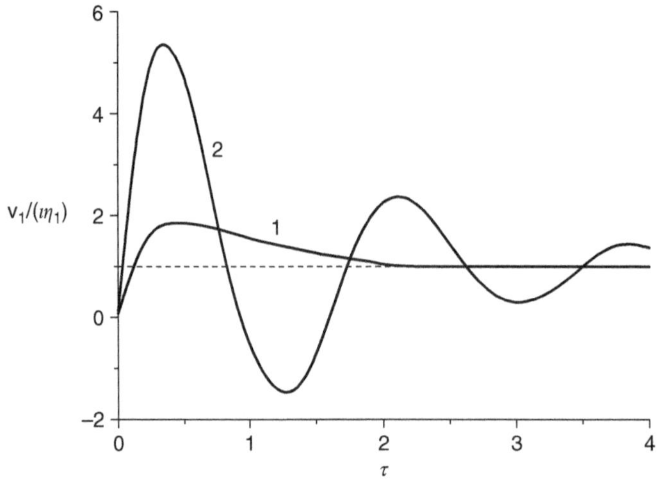

Figure 7.1 Burning rate evolution in a transient process.

where the functions $U(x, y)$ and $V(x, y)$ are related to the error function of the imaginary argument as

$$U(x,y) + iV(x,y) = W(z), \quad z = x + iy \tag{7.24}$$

and

$$W(z) = e^{-z^2}\left(1 + \frac{2i}{\sqrt{\pi}}\int_0^z e^{-t^2}\mathrm{d}t\right) \tag{7.25}$$

The shape of the function $v_1(\tau)$ is determined by the magnitude of the damping decrement λ. For large values of λ, the oscillating term rapidly diminishes. Conversely, at small values of λ, burning rate oscillations occur for a long period of time. Variations of the surface temperature and the temperature gradient in the latter case are also of oscillatory nature.

Figure 7.1 shows typical burning rate dependencies on time. Curve 1 describes the case with the following values of parameters: $k = 1.5$, $r = 0.5$, $\iota = 2/3$, and $\mu = 1/6$. Here $\omega_1 = \sqrt{6}$, $\lambda = 2$, and $\omega = \sqrt{2}$. Parameters for the case described by curve 2 are chosen in such a way that the damping decrement is small compared to the frequency of oscillations. In this case, $k = 2$, $r = 0.4$, $\iota = 2/3$, $\mu = 0$, $\omega_1 = 2.5\sqrt{2}$, $\lambda = 0.625$, and $\omega \approx 3.5$.

Both cases demonstrate significant overshooting of the burning rate before it reaches the final value $v_1(\infty) = \iota h$ (although the burning rate is below the final value initially). Curve 1 exhibits just one maximum of the burning rate. In contrast, in the oscillatory case (curve 2), the burning rate passes the value $v_1(\infty)$ multiple times. The minimum value of the burning rate during this process may actually be below its initial value.

7.3 Approximate Approach in the Theory of Nonsteady Combustion

Linear approximation fails in handling transitional regimes with significant pressure variation. In such cases, the nonlinear heat transfer equation, in conjunction with nonlinear burning laws restrictions, needs to be solved.

The qualitative features of the burning rate behaviour in time may be understood relatively easy. In fact, these will be similar to the cases of small pressure jumps considered in Section 7.2. Consider, for example, instant pressure rise. In the Zeldovich model (constant surface temperature), the temperature gradient at the surface would initially remain constant whilst the burning rate would rise instantly. In the ZN theory (variable surface temperature) the conclusion that surface temperature does not change at $\tau = 0$ would still hold.

It is obvious also that the burning rate asymptotic at $\tau >> 1$ would have a form of decaying exponent in the case of constant surface temperature. In the case of variable surface temperature, there will also be a decaying component oscillating with the frequency ω.

However, nondecaying or even growing in time nonlinear oscillations may develop in the case of large burning rate oscillation amplitudes. This is because the damping decrement, as discussed in Chapter 4, depends on the oscillation amplitude if the nonlinear properties of the system are taken into account.

The exact solution of a nonlinear problem describing the burning rate evolution under pressure variation according to prescribed law may be obtained numerically. It is useful, however, to consider in this section an approximate approach based on the heat balance integral method.

The following analysis applies the method to the case of constant surface temperature (Istratov et al. 1964). The essence of the method is replacing the partial differential equation (PDE) with the ordinary differential equation (ODE) obtainable on integration of the former over the entire half-space occupied by the propellant. In other words, local energy balance is replaced approximately by a global one.

Integration of (2.38) gives

$$\frac{d}{d\tau}\int_{-\infty}^{0}\theta(\xi,\tau)d\xi = \varphi - v \tag{7.26}$$

This equation may be satisfied by the profile $\theta(\xi, \tau)$ of assumed shape containing an unknown function of time which has to be determined. This approximate solution must also satisfy boundary conditions and qualitatively correctly describe the temperature profile evolution in time.

Consider the transition from the steady-state regime

$$\tau = 0, \quad \eta = v = 1, \quad \theta(\xi,0) = \exp(\xi) \tag{7.27}$$

to another steady-state regime corresponding to the final value of pressure

$$\tau = 0, \quad \eta = \eta_1, \quad v = v_1, \quad \theta(\xi,\infty) = \exp(v_1\xi) \tag{7.28}$$

It is sensible to assume the following shape of the profile

$$\theta(\xi,\tau) = [1-\psi(\tau)]\exp(\xi) + \psi(\tau)\exp(v_1\xi) \tag{7.29}$$

where the function $\psi(\tau)$ varying from zero at $\tau = 0$ to one at $\tau = \infty$ needs be determined.

Substitution of (7.29) into (7.26) provides the following ODE which serves to determine the function $\psi(\tau)$

$$\frac{(v_1-1)}{v_1}\frac{d\psi}{d\tau} = v - 1 - (v_1-1)\psi \tag{7.30}$$

Temperature gradient at the surface, according to (7.29), is

$$\varphi = 1 + (v_1 - 1)\psi \tag{7.31}$$

In order for further analysis to be carried out the specific steady-state burning law $u^0(T_0, p)$ and the specific law of pressure variation with time must be imposed.

The two cases are considered below, namely linear and exponential dependence of the burning rate on initial temperature.

If the steady-state burning rate depends on initial temperature linearly

$$u^0(T_a, p) = Bp^{\iota}(1 + \alpha T_a) \tag{7.32}$$

then the corresponding nonsteady burning law $v = v(\varphi, \eta)$ has the form

$$v = P\frac{(1+\beta)}{2}\left[1 + \sqrt{1 - \frac{4\beta\varphi}{P(1+\beta)^2}}\right] \tag{7.33}$$

where

$$P = \eta^{\iota}, \quad \beta = \frac{\alpha(T_s - T_a)}{(1 + \alpha T_a)} \tag{7.34}$$

The relations (7.30) and (7.33) allow the following differential equation for the unsteady burning rate to be derived

$$\frac{1}{v}\left(1 + \beta - 2\frac{v}{P}\right)\frac{dv}{d\tau} + \frac{v}{P^2}\frac{dP}{d\tau} = v_1\left(\frac{v}{P} - 1\right) \tag{7.35}$$

It is convenient to introduce the new function $V = v_1/P$. In the steady-state regime $V = 1$. Deviation of the value V from unity describes the degree of unsteadiness of the regime. The equation which this function satisfies follows from (7.35)

$$\frac{1}{V}(1 + \beta - 2V)\frac{dV}{d\tau} + (1 + \beta - V)\frac{1}{P}\frac{dP}{d\tau} = v_1(V - 1) \tag{7.36}$$

Consider transition from the steady-state regime $v = 1$ to the steady-state regime $v_1 = P_1$ for the following pressure dependence on time

$$\begin{aligned} &P = \exp(-s\tau), \quad 0 < \tau < \tau_0 \\ &P = P_1 = v_1 = \exp(-s\tau_0), \quad \tau_0 < \tau < \infty \end{aligned} \tag{7.37}$$

The choice (7.37) is motivated by the fact that equation (7.36) can in this case be integrated explicitly, and also by the observation that transition processes in the combustion chamber quite often follow exponential dependence of pressure on time. By varying the parameter s both the rapid pressure change ($|s| >> 1$) and the quasi- steady-state regime ($|s| \ll 1$) may be investigated. The cases where $s > 0$ correspond to pressure decrease, while the cases where $s < 0$ correspond to pressure rise.

The solution of equation (7.36) with the initial condition $V(0) = 1$ has the form

$$\tau = \frac{1}{(s(1+\beta) - v_1)}\left[\begin{array}{r}(1+\beta)\ln V - \frac{(s(1+\beta) - v_1(1-\beta))}{(v_1 - s)}\cdot \\ \cdot\ln\left(1 + \frac{(s - v_1)(1-V)}{s\beta}\right)\end{array}\right], \quad \tau < \tau_0 \tag{7.38}$$

$$\tau = \tau_0 + \frac{1}{v_1}\left[(1-\beta)\ln\left(\frac{V_0 - 1}{V - 1}\right) + (1+\beta)\ln\left(\frac{V_0}{V}\right)\right], \quad \tau > \tau_0 \tag{7.39}$$

Here V_0 is the value $V(\tau = \tau_0)$, which is determined from (7.38).

For the instant pressure change ($s = \pm\infty$) the solution is given by the expression (7.39) with $\tau_0 = 0$. In order for V_0 to be determined in this particular case the solution (7.38) needs be considered at $s = \pm\infty$ with the condition $s\tau_0 = -\ln v_1$. Alternatively, $P = P_1 = v_1$ can be substituted into (7.33) with the condition $\varphi = 1$. The resulting expression for V_0 is

$$V_0 = \frac{(1+\beta)}{2}\left[1 + \sqrt{1 - \frac{4\beta}{P_1(1+\beta)^2}}\right] \tag{7.40}$$

The required function $\psi(\tau)$ can be expressed in terms of V and P using (7.31) and (7.33)

$$\psi = \frac{PV(1+\beta-V)-\beta}{\beta(P_1-1)} \tag{7.41}$$

This solution is valid for both decreasing ($s > 0$, $P_1 < 1$) and rising pressure ($s < 0$, $P_1 > 1$).

The obtained results are illustrated in Figure 7.2 for the instant (curve 1) and exponential (curve 2) pressure drop. The dashed line shows the steady-state burning rate (which is proportional to pressure raised to power ι). The curves correspond to the following set of parameters: $\beta = 0.6$, $v_1 = 0.95$, $s = \infty$ (curve 1), and $s = 0.25$ (curve 2). The burning rate deviation from the steady-state value grows with the rate of pressure drop. The steady-state burning rate is larger than the nonsteady one since the latter is controlled by a less heated fuel layer near the surface.

In Figure 7.3 dashed lines show the time dependencies of the burning rate $V = v/v_1$ under rapid pressure increase for $\eta_1 = 10$ (curve 1) and $\eta_1 = 200$ (curve 2) ($\iota = 2/3$, $\beta = 0.74$). Solid lines show the same dependencies obtained by numerical solution of the full nonlinear problem (Istratov et al. 1964). It is evident that the approximate method delivers good agreement with the accurate numerical results.

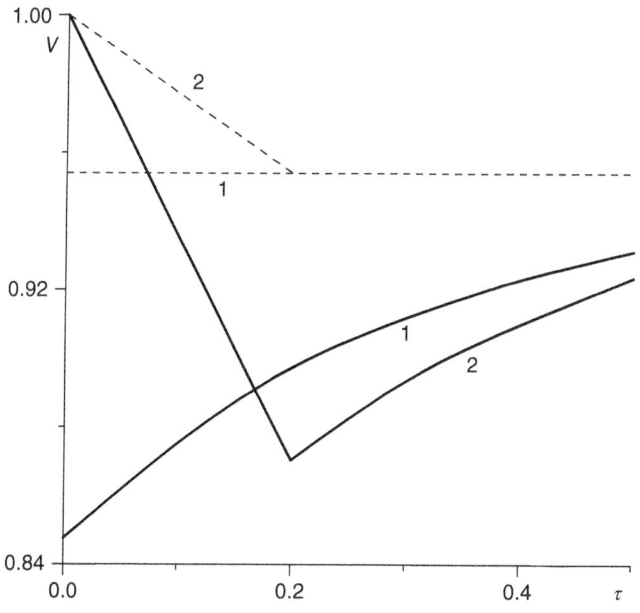

Figure 7.2 Burning rate under pressure drop.

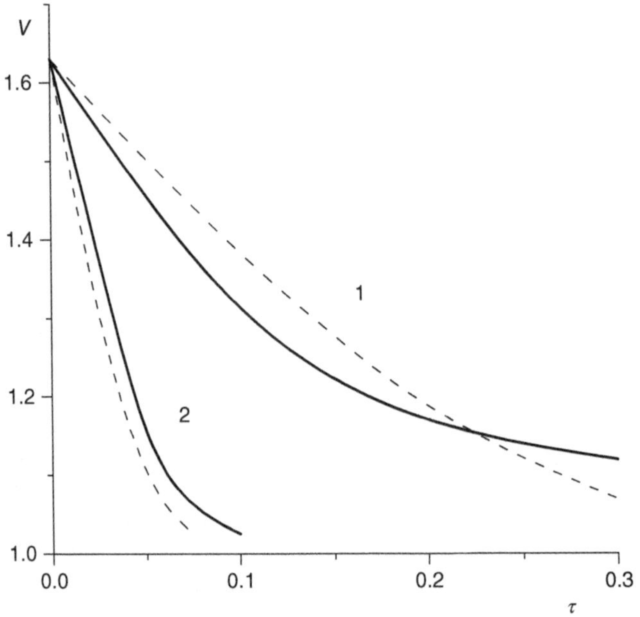

Figure 7.3 Burning rate under rapid pressure increase.

The dashed curves in Figure 7.4 show propellant temperature profiles, calculated using the formulas (7.29), (7.39), and (7.41), for two different instants of time. The corresponding numerical results are shown by solid lines.

In a similar manner, the problem of unsteady propellant combustion in the case of the exponential steady-state burning rate dependence on initial temperature

$$u^0(T_a, p) = Bp^{\prime} \exp(\beta T_a), \quad v = P \exp\left(k\left(1 - \frac{\varphi}{v}\right)\right) \tag{7.42}$$

may be studied.

The burning rate is determined from the equation

$$(k - 1 - \Lambda)\frac{d\Lambda}{d\tau} + (k - \Lambda)\frac{1}{P}\frac{dP}{d\tau} = v_1 \Lambda, \quad \Lambda = \ln\left(\frac{v}{P}\right) \tag{7.43}$$

It is easy to show that for $k \ll 1$ the exponential law (7.42) turns into the linear one with $\beta \ll 1$ as it should be. Indeed the exponential law must degenerate into the linear law under exactly these inequalities. Correspondingly, Eq. (7.43) turns into (7.36). If pressure time dependence follows (7.37), Eq. (7.43) can be easily integrated to give the relationship between the burning rate and time

$$\tau = \frac{(s - v_1(1 - k))}{(v_1 - s)^2} \ln\left[1 + \frac{(v_1 - s)\Lambda}{sk}\right] - \frac{\Lambda}{(v_1 - s)}, \quad \tau < \tau_0 \tag{7.44}$$

$$\tau = \tau_0 + \frac{1}{v_1}\left[(1 - k)\ln\left(\frac{\Lambda_0}{\Lambda}\right) + (\Lambda_0 - \Lambda)\right], \quad \tau > \tau_0 \tag{7.45}$$

Here the value Λ_0 must be determined from (7.44) at $\tau = \tau_0$.

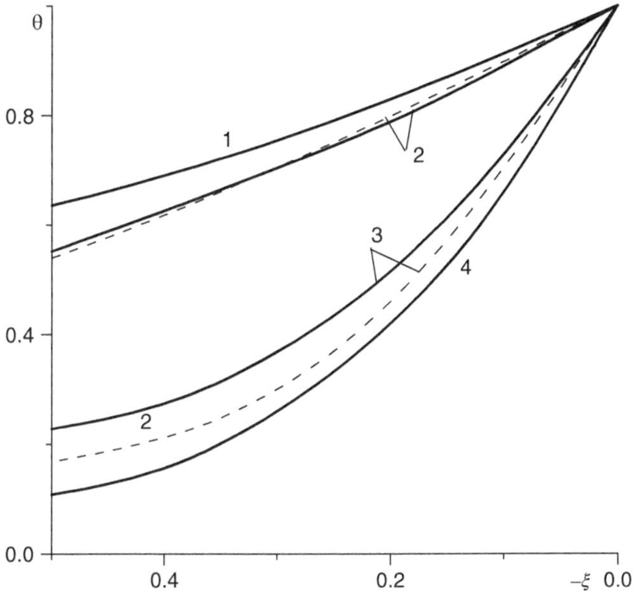

Figure 7.4 Temperature profile evolution under rapid pressure increase. $\beta = 0.74$, $\eta_1 = 10$. 1, $\tau = 0.0$; 2, $\tau = 0.016$; 3, $\tau = 0.237$; 4, $\tau = \infty$.

Under the instant pressure change Λ_0 must be determined from the equation

$$\Lambda_0 = -\ln\left[v_1\left(1 - \frac{\Lambda_0}{k}\right)\right] \tag{7.46}$$

As a final remark, it should be noted that while good accuracy of the method has been demonstrated by comparison with numerical results in Figures 7.3 and 7.4, it may still be improved.

For this purpose integration of the heat transfer equation with the weights ξ^n reduces the problem to the set of ODEs for higher moments of the temperature distribution

$$J_n(\tau) = \int_{-\infty}^{0} \xi^n \theta(\xi, \tau) d\xi \tag{7.47}$$

The heat balance integral method utilizing higher moments of distribution provides more accurate results (Gostintsev 1967).

7.4 Self-similar Solution

Zeldovich (1964) obtained the solution of the nonlinear problem of the nonsteady propellant burning rate in the special case where the steady-state burning rate is related to pressure and initial propellant temperature as

$$u^0 = Bp^{\iota} \exp(\beta T_a)\ (B, \iota, \beta = \text{constant}, \iota < 1, B > 0) \tag{7.48}$$

and pressure, in turn, decreases with time according to

$$p \sim t^{-1/2\iota}\ (A = \text{constant} > 0) \tag{7.49}$$

Consider now the case of arbitrary dependence $u^0(p, T_a)$ assuming, following Zeldovich (1964), that the surface temperature remains constant (Librovich and Novozhilov 1971).

It is well known (Tikhonov and Samarskii 1963) that the heat transfer equation

$$\frac{\partial\theta}{\partial\tau} = \frac{\partial^2\theta}{\partial\xi^2} - v\frac{\partial\theta}{\partial\xi} \tag{7.50}$$

with boundary conditions

$$\xi \to -\infty, \theta = 0; \qquad \xi = 0, \theta = 1 \tag{7.51}$$

admits a self-similar solution if velocity v varies with time as

$$v = C\tau^{-1/2} \; (C = \text{constant} > 0) \tag{7.52}$$

Indeed, by introducing the variable

$$y = \frac{\xi}{2\sqrt{\tau}} \tag{7.53}$$

the equation is reduced to the following ODE

$$\frac{d^2\theta}{dy^2} - 2(y - C)\frac{d\theta}{dy} = 0 \tag{7.54}$$

The latter has the following solution satisfying the boundary conditions (7.51)

$$\theta = \frac{\text{erfc}(C - y)}{\text{erfc}(C)}, \qquad \text{erfc}(z) = 1 - \frac{2}{\sqrt{\pi}}\int_0^z e^{-x^2}\text{d}x \tag{7.55}$$

This self-similar solution should be considered as an intermediate asymptotic (Barenblatt 2005) only of the exact solution describing transition of propellant combustion from one steady-state regime to another, provided pressure forces the burning rate dependence (7.52) over some time interval. The nature of such an asymptotic reflects the fact that the system ceases to feel the influence of initial conditions, being at the same time far from the final steady-state regime where pressure variation deviates from that corresponding to the self-similar solution, that is to the dependence (7.52).

Pressure dependence on time providing the burning rate law (7.52) may be found in the following way. The function $u^0(p, T_a)$ is considered to be known. Substituting initial temperature as a function of surface temperature and temperature gradient at the surface in this expression leads to

$$u = u\left[p(t), T_s - \frac{\kappa}{u}f\right] \tag{7.56}$$

For the self-similar solution (7.55)

$$v = \frac{C}{\sqrt{\tau}} = v[\eta(\tau), T^*], \quad T^* = T_s - \frac{(T_s - T_a)e^{-C^2}}{\sqrt{\pi}C\,erfc(C)} = \text{constant} \tag{7.57}$$

Normally the steady-state burning rate increases with pressure, and a self-similar solution may be realized at decreasing pressure.

If the burning rate is taken in the form (7.48) then the dependence (7.57) turns into

$$\frac{C}{\sqrt{\tau}} = \frac{B}{u^0}p^{\iota}\exp(\beta T^*) \tag{7.58}$$

Taking $U = Bp_0^{\iota} \exp(\beta T_a)$ (p_0 is the unit of pressure measurement in (7.48)) as a characteristic, the burning rate, determined by the relation (7.58) (with the help of (7.57)) may be written in the form

$$\frac{C}{\sqrt{\tau}} = \eta^{\iota} \exp k\left(1 - \frac{\exp(-C^2)}{\sqrt{\pi} C\, erfc(C)}\right) \tag{7.59}$$

This dictates the following law of pressure change in nondimensional form

$$\eta = \frac{D^{1/\iota}}{\tau^{1/2\iota}} \tag{7.60}$$

as well as the relation between the constants C and D in the dependencies (7.52) and (7.60), respectively

$$\frac{C}{D} = \exp\left\{k\left(1 - \frac{\exp(-C^2)}{\sqrt{\pi} C\, erfc(C)}\right)\right\} \tag{7.61}$$

If the constant D in the pressure variation law (7.60) is fixed then the relationship (7.61) determines the burning rate constant C as a function $C = C(D)$. Figure 7.5 illustrates the solution of this equation by means of plotting the left-hand side A (straight lines) and the right-hand side B as functions of C at the fixed value of the parameter k and different values of the constant D. At large values of D Eq. (7.61) has two solutions for the unknown C, one of which corresponds to weak unsteadiness of the burning rate (large value of C). The other solution corresponds to strong unsteadiness (small C). A solution is absent for $D < D_*$ where the critical value D_* (and the corresponding unique solution C_*) is found from the requirement of tangency of the curves in Figure 7.5

$$C_* = \frac{k \exp(-C_*^2)}{\sqrt{\pi}\mathrm{erfc}(C_*)}\left(1 + 2C_*^2 - \frac{2C_* \exp(-C_*^2)}{\sqrt{\pi}\mathrm{erfc}(C_*)}\right) \tag{7.62}$$

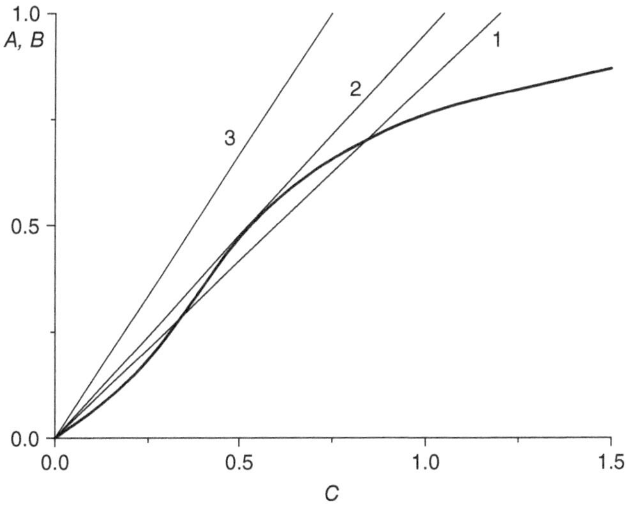

Figure 7.5 Possibility of existence of a different number of self-similar solutions. $k = 0.8$; 1, $D = 1.33$; 2, $D = 0.91$; 3, $D = 0.75$.

The value of D_* is found then from (7.61).

Therefore, for the propellant combustion model with constant surface temperature there generally exist two self-similar solutions. An important problem is therefore stability analysis of these solutions. The relevant analysis (Librovich and Novozhilov 1971) shows that the upper intersection of the curves in Figure 7.5 corresponds to the stable solution, while the lower one corresponds to the unstable solution. This analysis is considered in detail in Section 7.5.

Consider now the self-similar solution in the case of variable surface temperature (Librovich and Novozhilov 1971). Specific pressure dependence on time as well as particular dependencies relating to surface temperature, pressure, and initial temperature need be found in order for a self-similar solution to exist.

It is convenient to use the integral formulation of the theory for this purpose (Section 2.4). Influence of initial conditions is taken care of by the last term on the right-hand side of Eq. (2.49). These initial conditions are irrelevant for a self-similar solution, which approximates real solutions at large times. Taking into account that the integral in (2.49) containing initial temperature distribution approaches zero at $\tau \to \infty$, and omitting this term accordingly, the integral relation which is satisfied by the self-similar solution is obtained. This relation involves burning rate, surface temperature, and temperature gradient at the surface

$$\vartheta(\tau) = \frac{1}{\sqrt{\pi}} \int_0^\tau \left(\varphi - v\vartheta + \frac{\vartheta I}{2(\tau - \tau')} \right) \exp\left(\frac{-I^2}{4(\tau - \tau')} \right) \frac{d\tau'}{\sqrt{\tau - \tau'}} \tag{7.63}$$

The self-similar solution may be sought in such a form that the burning rate is described by the dependence (7.52) while surface temperature and its relation with temperature gradient at the surface are

$$\vartheta = F\tau^n \tag{7.64}$$

$$\frac{\varphi}{\vartheta} = \frac{G}{\sqrt{\tau}} \tag{7.65}$$

where C, F, G, and n are arbitrary real numbers.

Substituting these relationships into (7.63) and introducing the new variable $\sigma = \tau'/\tau$ gives

$$\sqrt{\pi} = \int_0^1 \sigma^n \left(\frac{(G-C)}{\sqrt{\sigma}} + \frac{C}{(1+\sqrt{\sigma})} \right) \exp\left[-C^2 \frac{(1-\sqrt{\sigma})}{(1+\sqrt{\sigma})} \right] \frac{d\sigma}{\sqrt{1-\sigma}} \tag{7.66}$$

Therefore, time completely disappears from the integral equation which confirms the correct choice of the time dependencies (7.52), (7.64), and (7.65). The obtained constraint (7.66) serves to determine the relation between the constants C and G.

In the particular case $n = 0$ corresponding to a combustion model with constant surface temperature the right-hand side of (7.66) can be calculated explicitly. This results in the relationship between the constants C and G that conforms with the results obtained above for that case. In a general case, a similar relationship may be obtained by numerical integration.

Furthermore, the pressure variation law that warrants the existence of the self-similar solution for the case of variable surface temperature may be found. The relationship (7.57) is replaced by

$$\frac{C}{\sqrt{\tau}} = \nu\left[\eta(\tau), T_a + F\left(1-\frac{G}{C}\right)(T_s^0 - T_a)\tau^n\right] \tag{7.67}$$

where $T_s^0 - T_a$ is the characteristic temperature difference which was used to make the problem nondimensional.

In particular, under the constraint (7.48) pressure must change as

$$\eta = \frac{C^{1/\nu}}{\tau^{1/2\nu}}\exp\left[-\frac{kF}{\iota}\left(1-\frac{G}{C}\right)\tau^n\right] \tag{7.68}$$

Consider now the conditions that must be satisfied by the steady-state dependence of surface temperature on initial temperature and pressure in order for the nonsteady surface temperature variation to be of the form in (7.64). The transformation from steady-state to nonsteady relation between surface temperature, pressure, and temperature gradient is similar to the corresponding transformation for the burning rate (7.56)

$$T_s = T_s\left[p(t), T_s - \frac{\kappa}{u}f\right] \tag{7.69}$$

The use of (7.64), (7.52), and (7.65) shows that the identity

$$F\tau^n = \vartheta\left[\eta(\tau), T_a + F\left(1-\frac{G}{C}\right)(T_s^0 - T_a)\right] \tag{7.70}$$

must hold, where $\eta(\tau)$ is given by (7.68). Therefore, for a self-similar solution to exist the function $T_s^0(p, T_a)$ must have a specific form.

Consider the more specific relation between the burning rate and surface temperature

$$u = H\exp\left(-E/RT_s\right) \tag{7.71}$$

This approximation is often used in experimental data processing. Using the steady-state value

$$u^0 = H\exp\left(-E/RT_s^0\right) \tag{7.72}$$

as a scale for the burning rate, the nondimensional problem variables may be written as

$$\nu = \exp\left(\varepsilon\Delta\frac{\vartheta - 1}{1+\Delta\vartheta}\right), \quad \varepsilon = \frac{E}{RT_s^0}, \quad \Delta = \frac{T_s^0 - T_a}{T_s^0} \tag{7.73}$$

Assuming that $|\vartheta - 1| \ll 1$, this formula may be rewritten in the approximate form

$$\nu \simeq \vartheta^m, \quad m = \frac{\varepsilon\Delta}{1+\Delta} \tag{7.74}$$

In a self-similar solution the burning rate and surface temperature must follow time dependencies (7.52) and (7.64), respectively. Substitution of these functions into (7.74) shows that this relation gives a self-similar solution if the constants F and n are chosen as

$$F = C^{1/m}, \quad n = -\frac{m}{2} < 0 \tag{7.75}$$

that is, surface temperature decreases with time.

Since activation energies are usually large, then normally $|n| \ll 1$. A self-similar solution with constant surface temperature ($n = 0$) may be considered therefore as corresponding to an infinitely large value of E.

Temperature distribution within the body of the propellant may be obtained by solving the heat transfer equation (7.50). It reduces to ODE by adopting the variable y and introducing the new function Y

$$\theta = \tau^n Y(y), \quad y = {}^{\xi}/_{2\sqrt{\tau}} \tag{7.76}$$

This procedure leads to

$$\frac{d^2Y}{dy^2} + 2(y - C)\frac{dY}{dy} - 4nY = 0 \tag{7.77}$$

with the boundary conditions

$$y = -\infty, \quad Y = 0; \quad y = 0, \quad Y = F = \text{constant} \tag{7.78}$$

Eq. (7.77) reduces to the particular case of the Whittaker equation

$$Y = \frac{\exp\left[\frac{-(y-C)^2}{2}\right]}{\sqrt{y-C}} \times \\ \times \left\{ C_1 W\left[\left(n+\frac{1}{4}\right), \frac{1}{4}, -(y-C)^2\right] + C_2 W\left[-\left(n+\frac{1}{4}\right), \frac{1}{4}, (y-C)^2\right]\right\} \tag{7.79}$$

where W is the Whittaker function (Whittaker and Watson 2002), and C_1 and C_2 are integration constants which must be determined from the conditions (7.78). For $n < 0$ the first of the functions W on the right-hand side grows infinitely at $y \to -\infty$. Therefore, the first of the conditions (7.78) requires $C_1 = 0$. The second integration constant is expressed through the constant F

$$C_2 = \frac{i\sqrt{C}}{\exp(-C^2/2)} W^{-1}\left[-\left(n+\frac{1}{4}\right), \frac{1}{4}, C^2\right], \quad F^m = C \tag{7.80}$$

The considered class of self-similar solutions with variable surface temperature may be widened by the use of the following construction. So far the heat transfer equation has been considered on the half-space $(-\infty, 0)$. The obtained solutions (7.55) and (7.79) may be extended on the interval $-\infty < \xi < \infty$, and the propellant surface may be considered to be located at $\xi = \xi_0$. For $\xi_0 = 0$ the solution (7.55) corresponds to the combustion model with constant surface temperature, while the solution (7.79) to the surface temperature varies with time according to the power law (7.64). For $\xi_0 \neq 0$ surface temperature is a function of time, which is different from both constant and (7.64). In particular, if $n = 0$ this function has the form

$$T_s(\xi_0, \tau) = T_a + \frac{(T_s^0 - T_a)}{\text{erfc}(C)}\left(1 - \frac{2}{\sqrt{\pi}}\int_0^y e^{-z^2}dz\right), \quad y = C - \frac{\xi_0}{2\sqrt{\tau}} \tag{7.81}$$

If $\xi_0 < 0$ then surface temperature rises with time and approaches at the limit of large times the constant value T_s^0. On the other hand, if $\xi_0 > 0$ then surface temperature decreases with time but also approaches T_s^0.

Calculating the temperature gradient from (7.81) and substituting, as before, into (7.57) the following expression dictating the necessary law of pressure change is obtained

$$\frac{C}{\sqrt{\tau}} = v\left\{\eta(\tau), T_s^0 - \frac{(T_s^0 - T_a)}{\sqrt{\pi}C\,\text{erfc}(C)}\exp\left[-\left(C - \frac{\xi_0}{2\sqrt{\tau}}\right)^2\right]\right\} \tag{7.82}$$

The relation (7.73) provides the following constraint on the surface temperature variation, which must be satisfied in order for self-similar solutions to exist

$$\vartheta = \vartheta \left\{ \eta(\tau), T_s^0 - \frac{(T_s^0 - T_a)}{\sqrt{\pi} C \operatorname{erfc}(C)} \exp\left[-\left(C - \frac{\xi_0}{2\sqrt{\tau}} \right)^2 \right] \right\} \tag{7.83}$$

The relation (7.83) contains the free parameter ξ_0 which can, along with T_s^0, be used for approximation of experimental data. A similar procedure is also applicable in the case where $n \neq 0$.

7.5 Self-similar Solution Stability

It was shown in the previous section that there exist two self-similar solutions for the model with constant propellant surface temperature. The question that arises is which of the two solutions is stable and therefore may occur in reality.

Among steady-state solutions those which amplify over time the modulus of small perturbations imposed on them are considered unstable. Studying the stability of self-similar solutions requires slightly different approach. In the latter case one needs to follow the relative magnitude of perturbations, that is, the ratio of the amplitudes of perturbed and unperturbed (which is also a function of time) solutions. Such an approach was used, for example, by Istratov and Librovich (1966) in the study of the stability of a laminar spherical flame. Solutions whose trajectories remain in the vicinity of the unperturbed solution are considered to be stable.

The heat transfer equation (7.50) may be transformed from the independent variables (τ, ξ) to the new independent variables $(\varsigma = \ln(\tau), y)$, and from the unknown function θ to the new unknown function Y, according to (7.76). On subsequent linearization the following equation for perturbations, which does not contain the variable ς explicitly, is obtained

$$4\frac{\partial Y_1}{\partial \zeta} = \frac{\partial^2 Y_1}{\partial y^2} + 2(y - C)\frac{\partial Y_1}{\partial y} - 2Cv^1\frac{\partial Y_0}{\partial y} - 4nY_1 \tag{7.84}$$

Here Y_1 are small perturbations, $v^1 = v_1/v$ is the relative value of the burning rate perturbation, and Y^0 is the unperturbed solution (7.79).

The solution of Eq. (7.84) may be sought, as usual, in the exponential form with respect to ς. However, it is also necessary that the additional relationships for perturbations of burning rate, surface temperature, and temperature gradient that follow from algebraic constraints of burning laws do not contain time explicitly.

It turns out that this is only possible for a specific relation between the steady-state burning rate, pressure, and initial temperature. Indeed, the function (7.67) may be written in the form

$$v = \frac{C}{\sqrt{\tau}} = \Psi(\eta, j), \quad j = \vartheta - \frac{\varphi}{v} \tag{7.85}$$

Linearizing (7.85) and making use of (7.64) and (7.65)

$$v^1 = \frac{\Psi_j}{\Psi}\frac{G}{C}F\tau^n\left[1 - \frac{\Psi_j}{\Psi}F\tau^n\left(\frac{1}{m} + \frac{G}{C}\right)\right]^{-1}, \quad \Psi_j = \frac{\partial \Psi}{\partial j} \tag{7.86}$$

In order for the solution method with exponential perturbation dependencies on ς to be applied it is necessary that

$$\frac{\Psi_j}{\Psi} F\tau^n = \frac{\Psi_j}{\Psi} F\left(\frac{C}{\Psi}\right)^{2n} = \text{constant} \tag{7.87}$$

Integrating this expression it is found that

$$\Psi = \left[Z(\eta) - \text{constant}\left(\vartheta - \frac{\varphi}{\nu}\right)\right]^{-1/2n}; \quad n \neq 0$$
$$\Psi = Z(\eta)\exp\left[\text{constant}\left(\vartheta - \frac{\varphi}{\nu}\right)\right]; \quad n = 0 \tag{7.88}$$

where $Z(\eta)$ is an arbitrary function of the nondimensional pressure η. Note that in particular the combustion law (7.48) leads to the second of the functional dependencies (7.88).

If solutions for perturbations are expressed as exponents of logarithmic time

$$Y_1 = f_1(y)\exp(\omega\varsigma), \qquad \nu^1 = \text{constant}\cdot \exp(\omega\varsigma) \tag{7.89}$$

then the solution is considered to be unstable if $\mathrm{Re}(\omega) > 0$. In the original time variable τ, the relations (7.89) correspond to the power laws

$$Y_1 \sim \tau^{\omega}, \quad \nu_1 = \nu \cdot \nu^1 = \text{constant} \cdot \tau^{\omega - 1/2}, \tag{7.90}$$

and with transformation to the perturbation θ_1 to

$$\theta_1(\tau, y) = f^{(2)}(y)\tau^{n+\omega} \tag{7.91}$$

The function $f^{(2)}(y)$ needs be found from solution of the heat transfer equation. The surface temperature perturbation can be obtained from (7.91) as

$$\varphi_1 = \left.\frac{\partial\theta_1}{\partial\xi}\right|_{\xi=0} = \frac{1}{2}\tau^{n+\omega-1/2}\left.\left(\frac{\mathrm{d}f^{(2)}}{\mathrm{d}y}\right)\right|_{y=0} \tag{7.92}$$

The stability boundary for the model with constant surface temperature ($n = 0$) may be found using the integral equation (7.63). The latter may be written for the perturbed self-similar solution as

$$\int_0^1 \left\{\varphi_1(x\tau) - \nu_1(x\tau) + \frac{1}{2(1-x)}\int_x^1 \nu_1(\tau s)\mathrm{d}s \left[1 + 2C(C-G)\left(\frac{1}{\sqrt{x}} - 1\right) - \right.\right.$$
$$\left.\left. -2C^2\frac{(1-\sqrt{x})}{(1+\sqrt{x})}\right] \exp\left(-C^2\frac{(1-\sqrt{x})}{(1+\sqrt{x})}\right)\right\} \frac{\mathrm{d}x}{\sqrt{1-x}} = 0 \tag{7.93}$$

Here the variable $x = \tau'/\tau$ is introduced and the exact form of the unperturbed solution (7.52) is used.

The relation (7.93) shows that according to (7.90) and (7.92) the integral would not depend on time τ if the burning rate and the temperature gradient perturbations are expressed in the form

$$\varphi_1(\tau) = (\varphi_1)_0\tau^l, \quad \nu_1(\tau) = (\nu_1)_0\tau^l \tag{7.94}$$

where l is an arbitrary real number. Comparison with (7.90) gives $l = \omega - 1/2$.

Substitution of (7.94) into (7.93) allows the relation between the amplitudes $(\varphi_1)_0$ and $(v_1)_0$ to be obtained. At the stability boundary $\mathrm{Re}(\omega) = 0$. It may be further assumed that $\mathrm{Im}(\omega) = 0$ at the stability boundary as well. Calculations show that the obtained relationships may be satisfied under this assumption, which proves its correctness. Calculation of the integrals in (7.93) leads to the following relationship between the temperature gradient and burning rate perturbations

$$\frac{(\varphi_1)_0}{(v_1)_0}\pi \exp(-C^2)(1 - erf(C)) = \frac{2\exp(-C^2)}{erf(C)} - 2\sqrt{\pi}C \tag{7.95}$$

The second relationship between $(\varphi_1)_0$ and $(v_1)_0$ may be obtained from the steady-state combustion model (7.48)

$$\frac{(\varphi_1)_0}{(v_1)_0} = \frac{G}{C} - \frac{1}{k} \tag{7.96}$$

(it should be assumed under variation that η = constant).

The conditions (7.95) and (7.96) provide the relation between C and k at the stability boundary. This relation is identical to condition (7.62) of tangency of the curves describing the spectrum of self-similar solutions (Figure 7.5). The tangency point is the critical point separating stable and unstable solutions. Both the real and the imaginary parts of the characteristic problem frequency ω become zero at this point. The upper intersection point on Figure 7.5 describes in the limit of large C a weakly unsteady combustion regime. By the continuity argument, it may be concluded that this corresponds to the stable solution. The lower intersection point corresponds to the unstable solution.

The obtained results may be presented using a plot of the burning rate versus temperature gradient at the surface (Figure 7.6). The dependence of the steady-state burning rate on pressure and initial temperature is considered here as given by (7.48), which under transformation to unsteady conditions gives

$$v = \eta' \exp\left(k\left(1 - \frac{\varphi}{v}\right)\right) \tag{7.97}$$

The dependencies (7.97) are shown in Figure 7.6 for the three values of pressure ($\eta' = 0.5$, 1.0, 2.0). The curves have infinite derivatives in the points which lie, as can easily be checked, on the line $v = k\varphi$ (the line 1). Steady-state propellant combustion under different pressures and fixed initial temperature (which is chosen as the characteristic temperature) is described by line 2. The tangent of this line changes with the change in the initial temperature. In particular, the tangent of line 1 corresponds to the temperature T_0^* such that $\beta(T_s - T_0^*) = 1$, that is, to the stability boundary of steady-state solutions. Therefore the bottom (unstable) branches of the curves $v(\varphi)$ in Figure 7.6 are shown by dashed lines.

Self-similar solutions are also presented in Figure 7.6 by straight lines. Indeed, results of Section 7.4 show that

$$v = \sqrt{\pi}C\exp(C^2)erfc(C)\varphi \tag{7.98}$$

For large C, the line (7.98) approaches line 2 (Figure 7.6) corresponding to steady-state combustion (weak unsteadiness). For $C \to 0$ (strong unsteadiness) the line (7.98) approaches the φ axis and ends up in the region of unstable steady-state solutions.

The stability boundary of self-similar regimes is obtained by substituting the critical value C_* defined by (7.62) into the expression (7.98). This is shown in Figure 7.6 by a dashed

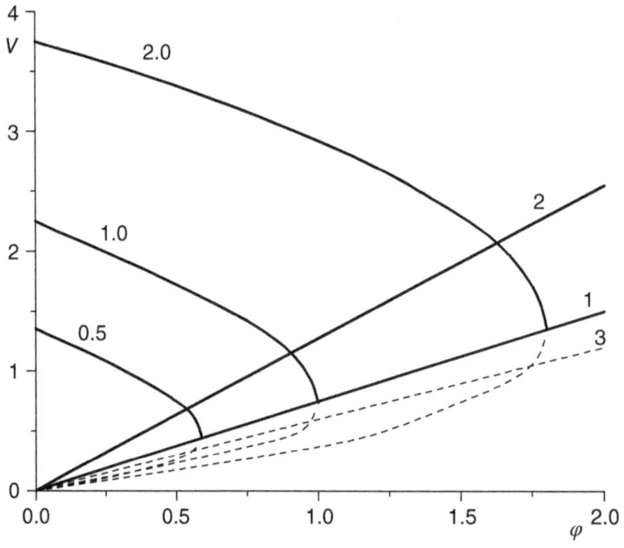

Figure 7.6 Stability boundary of self-similar solutions.

straight line. It is evident that the stability region of self-similar solutions is wider than the stability region of steady-state solutions. This is due to a different definition of unstable perturbation.

It should be noted that Zeldovich (1964) defined the stability boundary for self-similar solutions by the condition of achieving a critical temperature gradient at the propellant surface. Such an approach does not take into account how exactly the perturbation varies in time, relative to the time variation of unperturbed value.

7.6 Propellant Combustion and Extinction Under Depressurization: Constant Surface Temperature

Before considering outcomes of the complete ZN theory pertaining to transient combustion and extinction under depressurization, it is instructive to consider the case of constant surface temperature. This consideration is simpler than the general case, but at the same time clearly illustrates the essential features of the extinction process.

Zeldovich (1964) was the first to consider the problem of propellant extinction under a rapid decrease of pressure. Propellant surface temperature was assumed to be constant in this consideration. In general, extinction is related to the fact that at steady state the temperature gradient at the surface grows with increasing pressure. On the other hand, if the burning rate depends on initial temperature exponentially $u^0 \sim \exp(\beta T_a)$, then at any pressure the curve $f(u)$ has a maximum. If on pressure drop the temperature gradient exceeds its maximum value at final pressure then combustion becomes impossible and propellant extinction occurs.

This process is illustrated in Figure 7.7. Curves 1 and 2 correspond to nonsteady dependencies $u(f)$ for the pressure values p_1 and $p_2 < p_1$. Lines leaving the origin of the coordinate

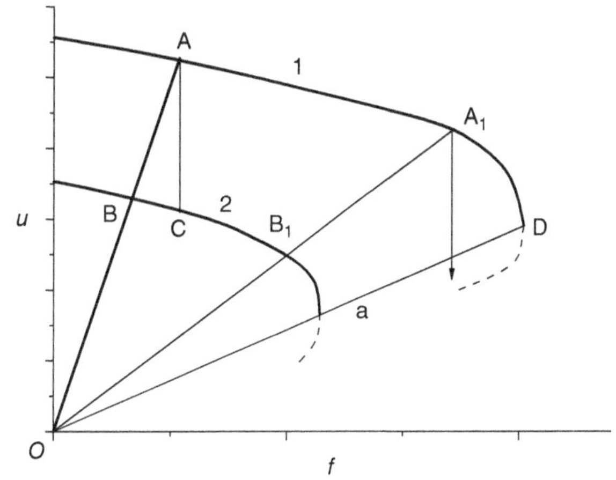

Figure 7.7 Extinction in the model with constant surface temperature.

system describe steady-state combustion regimes at given initial temperature and various pressures.

The limit of the steady-state combustion is described by curve a, where $k = 1$ and the initial temperature is $T_a^* = T_s - 1/\beta$.

If the pressure changes slowly from p_1 to p_2, that is, the process is quasi steady state, then the transition from the initial state (point A) to the final (point B) occurs along the line AB. In the other limit, under sudden pressure change the burning rate first drops to the value $u(C)$ so that the gradient does not change, and then relaxes to the value $u(B)$.

Under large pressure difference and low initial temperature the initial and final points may be in such a position (the points A_1 and B_1) that the initial temperature gradient is larger than the maximum gradient at final pressure. In such a case a sudden pressure drop sends the system to the region of controlling parameters where combustion is not possible.

Zeldovich (1964) found the criterion for propellant extinction under sudden pressure decrease, that is, the minimum magnitude of pressure drop sufficient for extinction. For the initial temperature of the propellant T_a the temperature gradient at the original steady-state regime is

$$f_1 = u^0(T_a, p_1)(T_s - T_a)/\kappa \tag{7.99}$$

On the other hand, the maximum value of the gradient at pressure p_2 is

$$f_2^* = u^0(T_a^*, p_2)(T_s - T_a^*)/\kappa \tag{7.100}$$

For the burning law $u^0 = Bp^{\iota} \exp(\beta T_a)$ the extinction criterion $f_1 \geq f_2^*$ is written in the form

$$\left(\frac{p_1}{p_2}\right)^{\iota} \geq \frac{\exp[\beta(T_s - T_a) - 1]}{\beta(T_s - T_a)} \tag{7.101}$$

This criterion is applicable in the case of sudden pressure drop. If pressure varies gradually then it is necessary to solve the nonlinear heat transfer equation in the condensed phase.

Approximate investigation of extinction may be carried out by the heat balance integral method. The two cases are considered again below, that is, the linear and exponential dependence of the burning rate on initial temperature. Pressure dependence on time is taken as following Eq. (7.37).

In the case of linear dependence (7.32) it follows from the nonsteady burning law (7.33) that under sudden ($s = \infty$) pressure change extinction occurs if

$$P_1 < \frac{4\beta}{(1+\beta)^2} \tag{7.102}$$

Indeed, during a sharp pressure drop the gradient φ (having initial value unity) does not have time to change, and the restriction (7.102) forces the expression under the root in (7.33) to be negative. Under gradual pressure change extinction will occur at $V = V^* = (1+\beta)/2$. This follows directly from (7.33). At the moment of extinction the temperature gradient at the surface reaches its maximum. It also follows from (7.36) that the derivative $dV/d\tau$ at this moment is infinite.

If the magnitude of the pressure drop, that is, the final value of the pressure P_1, is fixed then there are two possibilities. For small values of s (slow decrease of pressure) there will always be $V > V^*$ and extinction will not occur. In the opposite case (s is large and pressure decreases rapidly) the critical value V^* implying extinction will be achieved at a certain moment in time.

The critical value s^* can be defined as a minimum value of s such that propellant extinction occurs at a given magnitude of pressure drop P_1. It may be found on taking $V = V^*$ at the most dangerous extinction time moment $\tau = \tau_0$ when $P_1 = \exp(-s\tau_0)$ and $V = (1+\beta)/2$. Substitution of these expressions into the relationship (7.38) between τ and V provides the following equation to determine the dependence of the critical rate of pressure drop s^* on the magnitude of pressure drop P_1

$$P_1 = \exp\left\{\frac{s^*}{(P_1-(1+\beta)s^*)}\left[(1+\beta)\ln\left(\frac{(1+\beta)}{2}\right) - \right.\right.$$
$$\left.\left.-\frac{((1+\beta)s^* - P_1(1-\beta))}{(P_1-s^*)}\ln\left(1+\frac{(s^*-P_1)(1-\beta)}{2\beta}\right)\right]\right\} \tag{7.103}$$

From here the earlier established criterion (7.102) follows for $s^* = \infty$.

Asymptotically at $s^* \gg 1$

$$P_1 = \frac{4\beta}{(1+\beta)^2}[1 - \frac{4\beta}{(1+\beta)^2 s^*}\left(\ln\left(\frac{(1+\beta)}{2}\right)+\right.$$
$$\left.+(1+2\beta)\ln\left(\frac{(1+\beta)}{2\beta}\right) - 1 + \beta\right)] \tag{7.104}$$

The essential feature of the dependence (7.103) is existence of the minimum value s^*_{min}. For $s < s^*_{min}$ propellant always burns irrespectively of any arbitrary large variations of pressure. Under this condition the derivative of V turns into zero before the value V^* is reached. The value s^*_{min} and the corresponding value P_1^* may be found by assigning $\tau = \tau_0$, $V = V^* = (1+\beta)/2$, and $dV/d\tau = 0$. Then from (7.36)

$$P_1^*/s^*_{min} = (1+\beta)/(1-\beta) \tag{7.105}$$

and (7.103) takes the form

$$P_1^* = \exp\left(\frac{s^*_{min}}{(P_1^*-(1+\beta)s^*_{min})}(1+\beta)\ln\left(\frac{(1+\beta)}{2}\right)\right) \tag{7.106}$$

The last two relationships allow the coordinates of the point where the extinction curve ends to be determined as

$$P_1^* = \left(\frac{1+\beta}{2}\right)^{\frac{(1-\beta)}{\beta}}, \quad s^*_{\min} = \left(\frac{1-\beta}{1+\beta}\right)\left(\frac{1+\beta}{2}\right)^{\frac{(1-\beta)}{\beta}} \tag{7.107}$$

Figure 7.8 shows the dependencies $P_1(s^*)$ for different values of the parameter β. The dashed curve is the locus of the end points of extinction curves.

The case of exponential dependence of the steady-state burning rate on initial temperature (7.42) may be considered in a similar manner.

Under sudden pressure drop ($s = \infty$) extinction occurs under the condition $P_1 < k\exp(1-k)$. This result is obtained from (7.42) by setting $\varphi = 1$ and looking for the minimum value of P_1 such that a solution for v still exists. Pressure drop from $P = 1$ to $P = P_1 = k\exp(1-k)$ leads to a decrease in the burning rate down to the value $v_1 = k$. At this point, the maximum value $\Lambda^* = \ln\left(\frac{v_1}{P_1}\right) = k - 1$ is achieved. A larger magnitude of pressure drop leads to extinction.

Under gradual pressure decrease extinction will also occur at the same value $\Lambda = \Lambda^* = k - 1$, as for smaller Λ (7.42) cannot be satisfied with any value of burning rate. The substitution of Λ^* into (7.44) at $\tau = \tau_0$ gives the following equation, which determines critical pressure drop P_1 as a function of s^*

$$P_1 = \exp\left\{\frac{s^*}{(s^* - P_1)}\left[1 - k - \frac{(s^* - P_1(1-k))}{(s^* - P_1)}\ln\left(1 + \frac{(1-k)(s^* - P_1)}{ks^*}\right)\right]\right\} \tag{7.108}$$

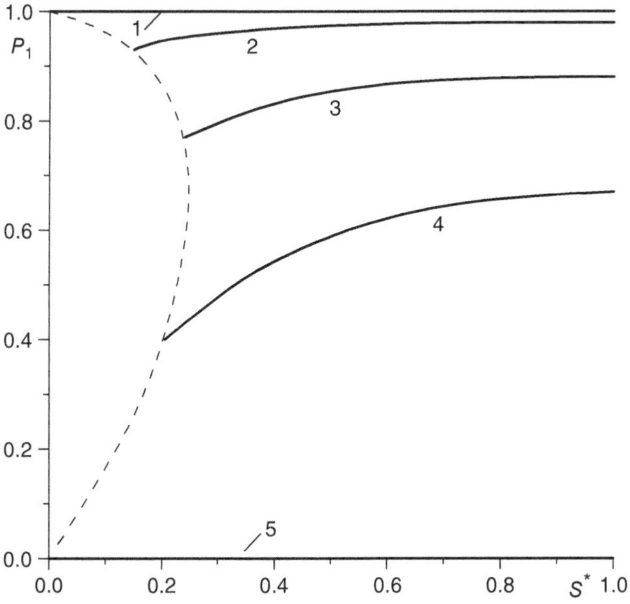

Figure 7.8 Extinction curves for the case of linear dependence of burning rate on initial temperature. 1, $\beta = 1.0$; 2, $\beta = 0.7$; 3, $\beta = 0.5$; 4, $\beta = 0.3$; 5, $\beta = 0.0$.

Asymptotically at $s^* \gg 1$

$$P_1 = \left\{ 1 - \frac{k\exp(1-k)}{s^*} \left[(1+k)\ln\left(\frac{1}{k}\right) - 2(1-k) \right] \right\} k\exp(1-k) \tag{7.109}$$

The point where the extinction curve ends is determined from the conditions

$$\Lambda = \Lambda^* = k - 1, \quad d\Lambda/d\tau = 0 \tag{7.110}$$

Eq. (7.43) gives $P_1^*(k-1) + s_{\min}^* = 0$, therefore the coordinates of the extinction curve termination points are

$$P_1^* = \exp\left[-\frac{(1-k)^2}{k} \right], \qquad s_{\min}^* = (1-k)\exp\left[-\frac{(1-k)^2}{k} \right] \tag{7.111}$$

Figure 7.9 illustrates dependencies provided by the relationship (7.108). The meaning of the dashed curve is the same as in Figure 7.8.

7.7 Propellant Combustion and Extinction Under Depressurization: Variable Surface Temperature

Qualitative considerations that explain propellant extinction on a rapid and deep pressure drop in the case of variable surface temperature are as follows. First of all, for the dependence $u(f)$ pertaining, for example, to ballistite N (Figure 3.19) the extinction mechanism described in Section 7.6 does not exist. Indeed, in this case temperature gradient, as a function of burning rate, does not have a maximum. The dependence $T_s(f)$ has a similar form.

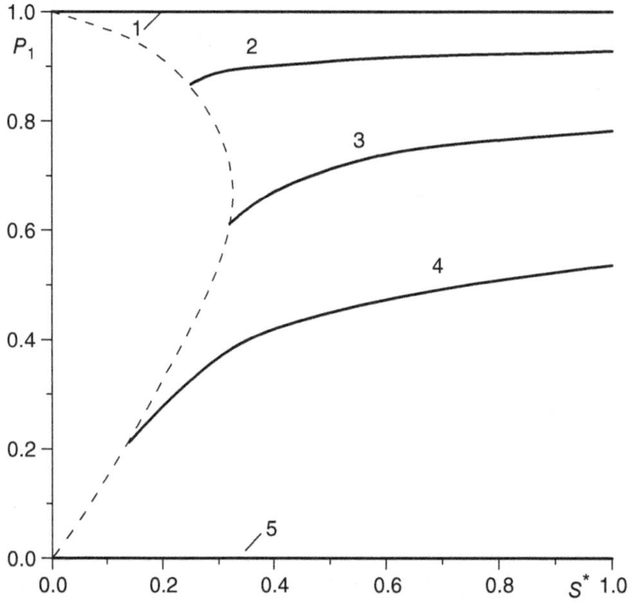

Figure 7.9 Extinction curves for the case of exponential dependence of burning rate on initial temperature. 1, $k = 1.0$; 2, $k = 0.7$; 3, $k = 0.5$; 4, $k = 0.3$; 5, $k = 0.0$.

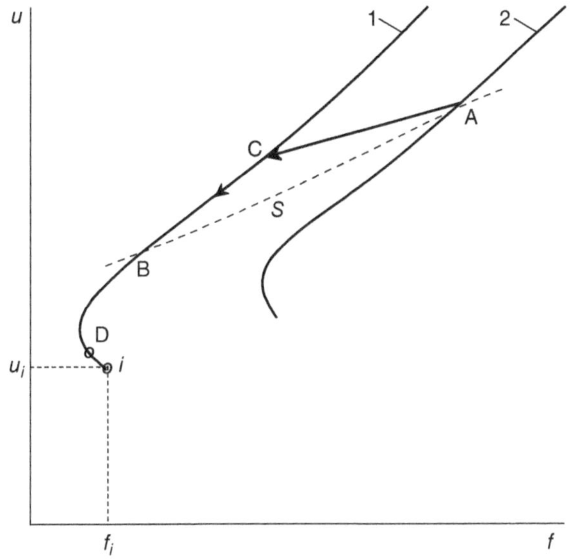

Figure 7.10 Extinction in the model with variable surface temperature. Dependence of the burning rate on temperature gradient for two different values of pressure.

Correct understanding of the extinction mechanism in the case of variable surface temperature must be based on information on the behaviour of the functions $u(f)$ and $T_s(f)$ for sufficiently small u and low T_s. Figure 3.19 shows data describing the steady-state combustion regime in the range of initial temperatures $-200°C < T_0 < 140°C$.

It may turn out that steady-state combustion cannot proceed under some sufficiently low initial temperature. In this case surface temperature would be too low to sustain continuous chemical reaction. The curves $u(f)$ and $T_s(f)$ would end at some points (f_i, u_i) and (f_i, T_{si}).

Another possibility is that steady-state combustion exists and may be observed down to the initial temperature of absolute zero. Then nonsteady burning laws $u(f, p)$ and $T_s(f, p)$ may be obtained from their steady-state counterparts $u^0(T_a, p)$ and $T_s^0(T_a, p)$ down to only some values of the burning rate and surface temperature which correspond to $T_0 = 0$ K. Nonsteady burning laws are still relevant below these values of the burning rate and surface temperature, but their quantification in this region would have to be based on experiments with nonsteady (e.g. ignition) combustion regimes. It is natural to expect that in this case combustion also may be observed only above a certain value of the surface temperature, that is the curve $T_s(f)$ ends at a certain point (f_i, T_{si}). This point has a corresponding point (f_i, u_i) where the burning rate curve ends.

Figures 7.10 and 7.11 show dependencies of burning rate and surface temperature on temperature gradient for two different values of pressure $p_1 < p_2$ (curves 1 and 2, respectively). These curves correspond to steady-state combustion regimes at a given propellant initial temperature and different values of pressure.

Consider, qualitatively, propellant behaviour if the pressure changes from p_2 to p_1. The initial state is denoted by A, and the final by B. The curve s corresponds to the transition process under slow pressure variation. In the case of rapid pressure change the surface temperature does not substantially change initially (and in the limit of sudden pressure drop does not change at all) while the temperature gradient at the surface and burning rate experience jumps. This is shown by the arrows AC. The temperature profile within

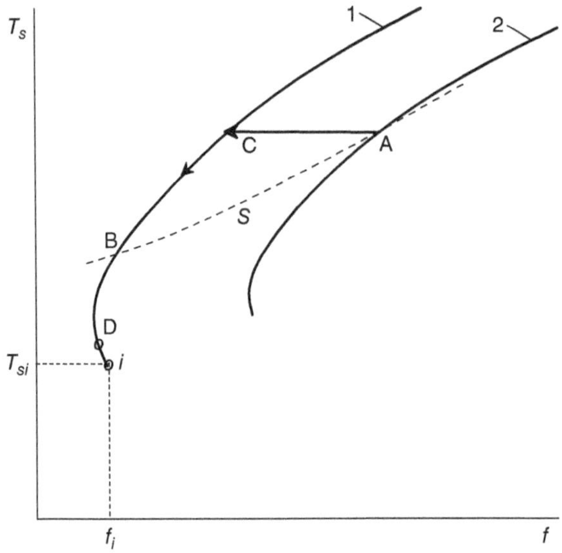

Figure 7.11 Extinction in the model with variable surface temperature. Dependence of the surface temperature on temperature gradient for two different values of pressure.

the propellant body changes only slightly during pressure variation. The solid curve A in Figure 7.12 corresponds to initial temperature distribution. The dashed curve C, distinguishable from A only near the propellant surface, shows temperature distribution just after pressure drop. Surface temperature changes very little, but the temperature gradient at the surface changes substantially. The steady-state profile C_1 with increased initial temperature corresponds to the new value of the gradient. If the pressure does not change any more and stays equal to p_1 then the state of the propellant would change along curve 1 (Figures 7.10 and 7.11) from C to B. Both surface temperature and burning rate decrease. Indeed, for the temperature distributions C and C_1 near the burning surface

$$\left(\frac{\partial T}{\partial t}\right)_{C_1}=0,\quad u\left(\frac{\partial T}{\partial x}\right)_C=u\left(\frac{\partial T}{\partial x}\right)_{C_1},\quad \kappa\left(\frac{\partial^2 T}{\partial x^2}\right)_C<\kappa\left(\frac{\partial^2 T}{\partial x^2}\right)_{C_1} \tag{7.112}$$

and according to the heat transfer equation $(\partial T/\partial t)_C < 0$.

In the case of rapid and deep pressure drop the temperature profile, as the system approaches point B, would be substantially different from the steady-state profile (curves B and B_1 in Figure 7.12). For the same reasons as outlined above, this would lead to a further decrease in the burning rate and surface temperature. The point denoting the state of the system would continue to move towards the end of nonsteady burning curves $u(f)$ and $T_s(f)$.

Furthermore, nonsteady process behaviour depends on the rate and magnitude of the pressure drop. For small ratios p_2/p_1, or for slow pressure variation, the temperature profile in section Bi of the curves (Figures 7.10 and 7.11) would be such that its second derivative near the surface coincides with the same derivative of the steady-state profile. This latter steady-state temperature profile corresponds to the existing instantaneous value of the gradient. Further surface temperature and burning rate decrease would stop at a certain point D, and the system would start to return to the steady-state regime B in an oscillatory manner. Larger values of the pressure drop rate and pressure drop magnitude may lead to the cessation of combustion at point *i*.

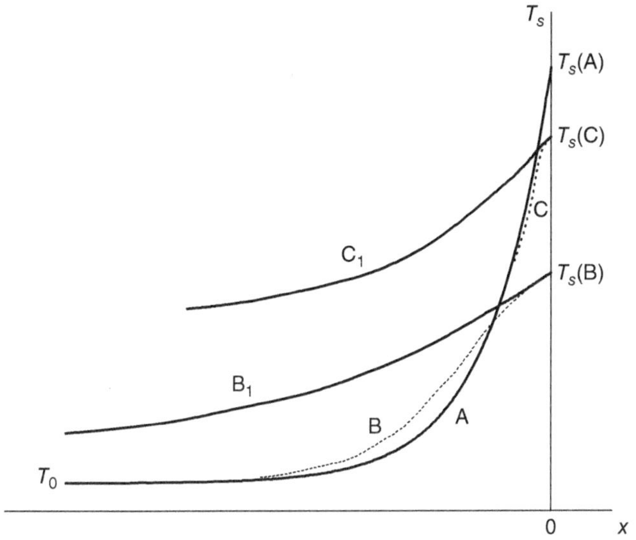

Figure 7.12 Temperature profile evolution under pressure drop.

This qualitative analysis shows that the principal reason for extinction in the case of variable surface temperature is the same as in the Zeldovich model (constant surface temperature), that is, the difference between temperature profiles in the initial and the final states. However, the details of the burning rate and temperature distribution behaviour in these two cases, as well as the extinction criterion, are very different. In the case of constant surface temperature extinction is determined by the point with infinite derivative on the curve $u(f)$. In the variable surface temperature case it is determined by the point i corresponding to the end of the nonsteady curves $u(f)$ and $T_s(f)$. In the absence of extinction relaxation of the burning rate and temperature distribution profiles to the steady-state regime occurs in a monotonic fashion in the case of constant surface temperature, and in an oscillatory manner if surface temperature is variable.

Both models give qualitatively identical relationships between the minimum magnitude of pressure drop and the pressure change rate required for extinction. The minimum pressure drop rate sufficient for extinction decreases in both models with increasing p_2/p_1 ratio. In the case of constant surface temperature the extinction curves, that is, the dependencies of the ratio p_2/p_1, sufficient for extinction, on the pressure decrease rate dp/dt may be obtained from the results in Section 7.6. In the case of variable surface temperature occurrence of extinction is determined by the distance between points C and B. The latter increases as pressure drop magnitude and pressure decrease rate become larger. Therefore the ratio p_2/p_1 sufficient for extinction would decrease with an increasing pressure drop rate dp/dt.

A proposed extinction scenario can be validated using detailed simulations within the formalism of the ZN theory and comparison with experimental data (Lidskii et al. 1983, 1985; Marshakov and Novozhilov 2011a).

Consider the transient combustion processes forced by the pressure drop law of the form

$$\eta(\tau) = \eta_k + (1 - \eta_k)\exp(-\tau/\tau_c), \quad \tau \geq 0 \tag{7.113}$$

where η_k is a specified final nondimensional pressure. The scaling of the variables is the same as in Section 2.3. In particular, $\tau_c = (u^0)^2 t_c/\kappa$ is nondimensional time corresponding to the time scale of pressure drop t_c. The initial temperature distribution is given by

$$\theta_i(\xi) = \exp(\xi), \quad \xi \leq 0, \tag{7.114}$$

The transient burning rate may be determined from the integral ZN theory formulations (2.49) and (2.50).

Lidskii et al. (1983) used the results of experiments by Zenin (1973), Pokhil et al. (1967), and Kowalskii et al. (1967). The dependencies of burning rate and surface temperature on pressure and initial temperature, provided by these studies, may be approximated by the following dependencies

$$u = \frac{Ap}{\left(1 - \frac{(1-\hat{\gamma})}{(1+p/\alpha)}\right)} \exp\left(\frac{CT_0^3}{1 - \frac{(1-\hat{\gamma})}{(1+p/\alpha)}}\right) \tag{7.115}$$

$$u = B\frac{T_s}{(T_s - T_a)^{1/2}} \exp\left[-\frac{E}{2RT_s}\right] \tag{7.116}$$

The specific parameters are $A = 6.08\times10^{-10}$ m^3/(N s), $C = 1.22\times10^{-8}$ K^{-3}, $\alpha = 3\times10^6$ N/m^2, $\hat{\gamma} = 0.33$, $E = 8.8\times10^4$ J/mol, $R = 8.31$ J/(mol $\cdot$ K), $B = 0.51$ m/s. (Burning rate and temperature units are m/s and K, respectively).

Initial temperature is excluded using the steady-state relationship $\kappa f = u(T_s - T_a)$. As a result, the following nondimensional implicit relationships, equivalent to (2.50), may be derived

$$\varphi = v\left\{\vartheta + \frac{1}{\hat{\mu}\vartheta_0} - \left(1 - \frac{(1-\hat{\gamma})}{(1+\eta_0\eta)}\right)^{1/3} \cdot \left[\frac{1}{\hat{\delta}}\ln\left(\frac{\left(1-\frac{(1-\hat{\gamma})}{(1+\eta_0\eta)}\right)}{\left(1-\frac{(1-\hat{\gamma})}{(1+\eta_0)}\right)}\frac{v}{\eta}\right) + \frac{\left(\frac{1}{\hat{\mu}\vartheta_0}\right)^3}{\left(1-\frac{(1-\hat{\gamma})}{(1+\eta_0)}\right)}\right]^{1/3}\right\} \tag{7.117}$$

$$\sqrt{v\varphi} = \frac{(1+\hat{\mu}\vartheta_0\vartheta)}{(1+\hat{\mu}\vartheta_0)} \exp\left(\frac{\vartheta_0}{(1+\hat{\mu}\vartheta_0)}\frac{(\vartheta-1)}{(1+\hat{\mu}\vartheta_0\vartheta)}\right) \tag{7.118}$$

Here the following nondimensional parameters are introduced

$$\hat{\mu} = \frac{2RT_0}{E}, \quad \vartheta_0 = \frac{E}{2RT_0^2}, \quad \hat{\delta} = C(T_s^0 - T_0)^3, \quad \eta_0 = \frac{p^0}{\alpha} \tag{7.119}$$

Initial pressure was set to $p^0 = 6\times10^6$ Pa and initial temperature to $T_a = 293$ K. These values imply $u^0 = 0.7\times10^{-2}$ m/s and $T_s^0 = 670$ K.

Eq. (2.49) was integrated numerically. Details of the numerical approach are provided by Lidskii et al. (1983).

Results are presented in Figures 7.13 and 7.14. Figure 7.13 shows that for pressure drop with the parameters $\eta_k = 0.4$ and $\tau_c = 1$ burning rate, after achieving a minimum, approaches asymptotically the value corresponding to the steady-state regime at final pressure (curve 1). Pressure drop with the parameters $\eta_k = 0.33$ and $\tau_c = 1$ causes extinction (curve 2).

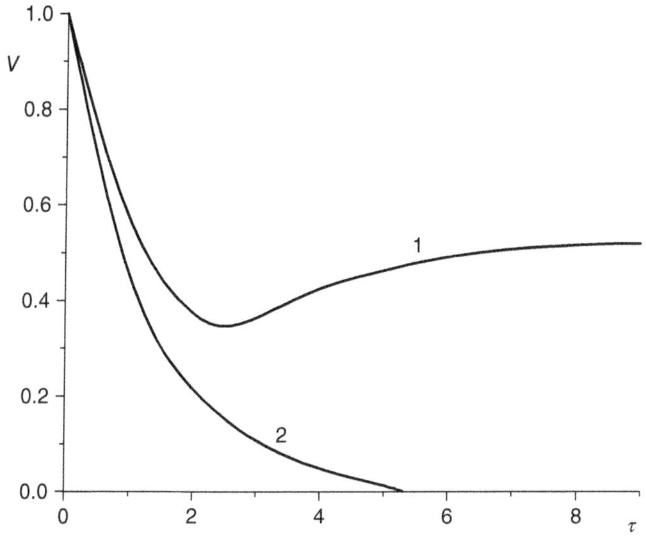

Figure 7.13 Burning rate evolution under pressure drop.

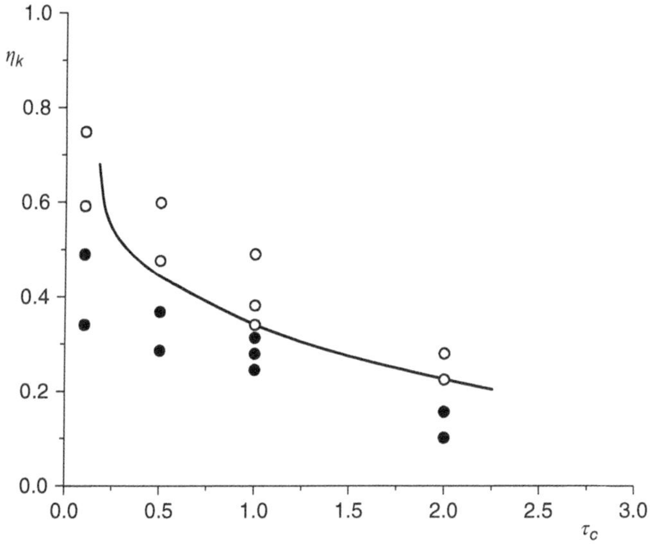

Figure 7.14 Extinction curve. ○, control parameters that do not cause extinction (simulations by Lidskii et al. 1983); •, control parameters forcing extinction (simulations by Lidskii et al. 1983). Solid curve, approximation of experimental data due to Marshakov and Leipunskii (1967).

Figure 7.14 shows the extinction curve that bounds, in the coordinates (τ_c, η_k), the regions of control parameters forcing extinction. Experimental data is due to Marshakov and Leipunskii (1967), who used the ballistite N with the addition of 2% MgO.

The results of the simulations are in a very good agreement with experimental data. It should be noted that the dependencies (7.115) and (7.116) were determined experimentally for limited ranges of pressure and temperature gradient variations. For the purpose of simulation they were extrapolated beyond the experimentally studied region. The results of the simulations demonstrate, however, that the choice of extrapolation has little effect on the transient burning rate behaviour and the position of the extinction curve. It should be noted that the dependencies (7.115) and (7.116) are derived for pure propellant N. Addition of 2% MgO may result in up to 10% burning rate variation compared to the pure propellant.

Marshakov and Novozhilov (2011a) conducted another set of experiments and simulations on the extinction of nitroglycerin-based propellant with the addition of 2% MgO. The following experimental dependencies for this particular propellant were obtained by Marshakov et al. (2010)

$$u^0(p^0) = 6.2 \times 10^{-2}(p^0)^n \exp[\beta(p^0)(T_a - T_k)]$$
$$u^0(T_s^0) = 1.12 \times 10^3 \exp\left[-\frac{\varepsilon_s}{T_s^0}\right]$$
$$\beta(p^0) = 10^{-3} \times [8.3 - 0.27(p^0)^n] \tag{7.120}$$

Here $n = 0.57$, $\varepsilon_s = 5 \times 10^3$ K, and $T_k = 273$ K.

On elimination of the initial temperature using the Michelson profile, these turn into the nonsteady burning laws

$$u = 6.2 \times 10^{-2} p^n \exp\left[\beta(p)\left(T_s - \frac{\kappa f}{u} - T_k\right)\right]$$
$$u = 1.12 \times 10^3 \exp\left[-\frac{\varepsilon_s}{T_s}\right]$$
$$\beta(p) = 10^{-3} \times [8.3 - 0.27p^n] \tag{7.121}$$

Pressure decays of the form

$$p = p_f + (p_i - p_f)\exp\left(-\frac{t}{t_p}\right) \tag{7.122}$$

are considered. Here, the timescale t_p is defined as the time required for pressure to drop to the level $p = p_f + (p_i - p_f)/e$.

The initial temperature distribution is in the form

$$T_i^0 = T_a + (T_{si}^0 - T_a)\exp\left(-\frac{u_i^0 x}{\kappa}\right) \tag{7.123}$$

The ZN theory equations are solved using a reduction of the infinite set of ODEs (Section 2.5).

A comparison with experimental data is presented in Figures 7.15–7.20. The following dimensionless time, relative velocity, and pressure drop are used in these figures

$$\tau = \frac{u_i^2}{\kappa}t, \quad U = \frac{u}{u_i}, \quad H = \frac{p_f}{p_i} \tag{7.124}$$

It should be noted that this sample is considered extinguished even in the cases where reignition followed occasionally after a period of time.

Different types of transition processes are illustrated in Figure 7.15. At a fixed rate of pressure, decay combustion regimes change dramatically under a small variation in the drop depth from transition to the new steady-state mode (curves 1 and 2) to complete extinction (curve 3).

A similar pattern is observed at a fixed pressure drop depth and variable drop rate (Figure 7.16).

To emphasize the nature of nonsteady combustion behaviour it is useful to compare the regimes in Figures 7.15 and 7.16 with the quasi-steady-state approximations obtained by substituting instantaneous pressures (7.122) into the burning laws (7.120). The latter approximations are illustrated by the curve 4 in Figures 7.15 and 7.16 and are completely different from genuinely nonsteady transient processes.

Figures 7.17–7.20 show extinction curves separating control parameters, causing a transition to new steady-state regimes from those forcing extinction. The domains of these two types of parameters are rather well defined. This is consistent with the observations in Figures 7.15 and 7.16 which show that the nature of transient processes is sensitive to the control parameters t_p and H. For the data presented in Figure 7.15, for example, at a fixed value of t_p the type of transition process changes within the range of values $H = 0.65 \pm 0.05$. In Figure 7.16, where H is fixed, the type of process changes at $t_p = 0.76 \pm 0.02$ ms. This circumstance allows theoretical extinction curves to be accurately defined.

Extinction curves separate fundamentally different transition processes. For the parameters above the curves the transient combustion process ends up approaching the steady-state regime at pressure p_f. For parameters below the curves the burning rate decreases to extremely low values interpreted as cessation of burning.

Calculations are performed using the conventional value of thermal diffusivity $\kappa = 10^{-3}\,\text{cm}^2/\text{s}$, although some data (Figure 7.17) are described better with the value $\kappa = 0.8 \times 10^{-3}\,\text{cm}^2/\text{s}$.

As can be seen from the studies by Lidskii et al. (1983, 1985) and Marshakov and Novozhilov (2011a), ZN theory successfully describes propellant extinction caused by

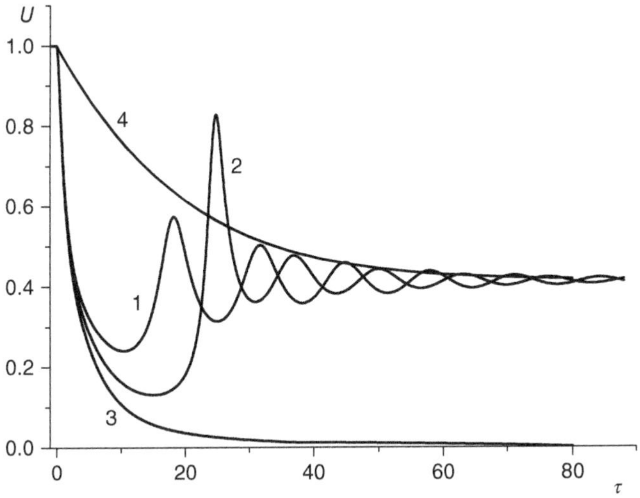

Figure 7.15 Burning rate as a function of time at $t_p = 0.5$ ms. 1, $H = 0.7$; 2, $H = 0.65$; 3, $H = 0.6$; 4, quasi-steady-state $H = 0.65$.

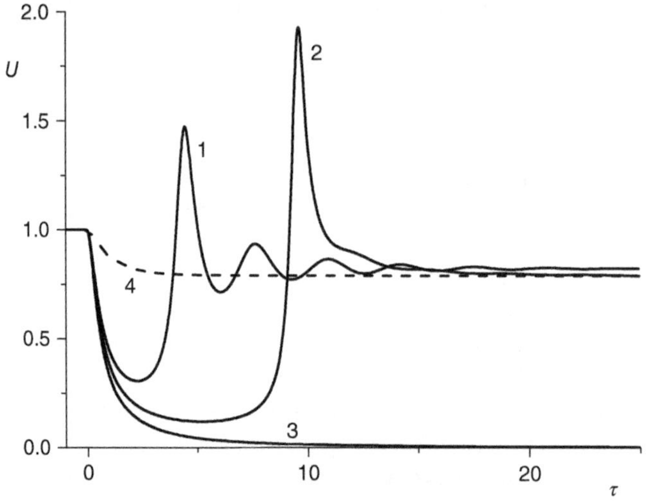

Figure 7.16 Burning rate as a function of time at $H = 0.2$. 1, $t_p = 0.78$ ms; 2, $t_p = 0.76$ ms; 3, $t_p = 0.74$ ms; 4, quasi-steady-state $t_p = 0.76$ ms.

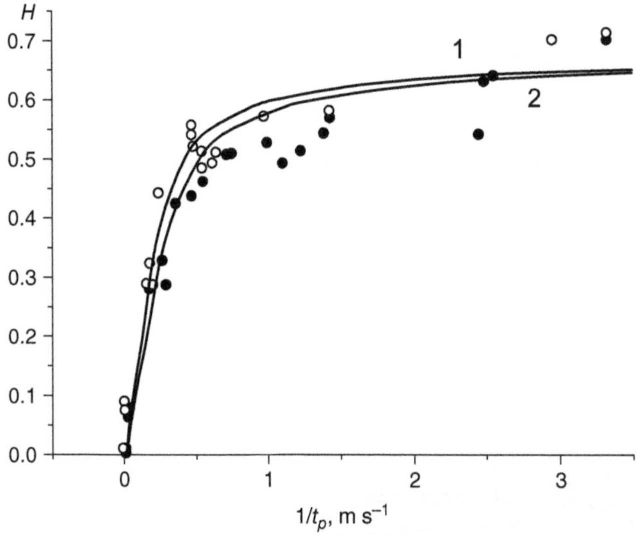

Figure 7.17 Experimental data and theoretical extinction curves at $p_i = 60$ atm. ○, burned out; •, extinguished. 1, $\kappa = 10^{-3}$ cm^2/s; 2, $\kappa = 0.8 \times 10^{-3}$ cm^2/s.

depressurization, and extinction curve predictions are in satisfactory agreement with experimental data. The largest deviations are observed at low initial pressures where significant scatter in experimental data occurs and reignition is rather frequent.

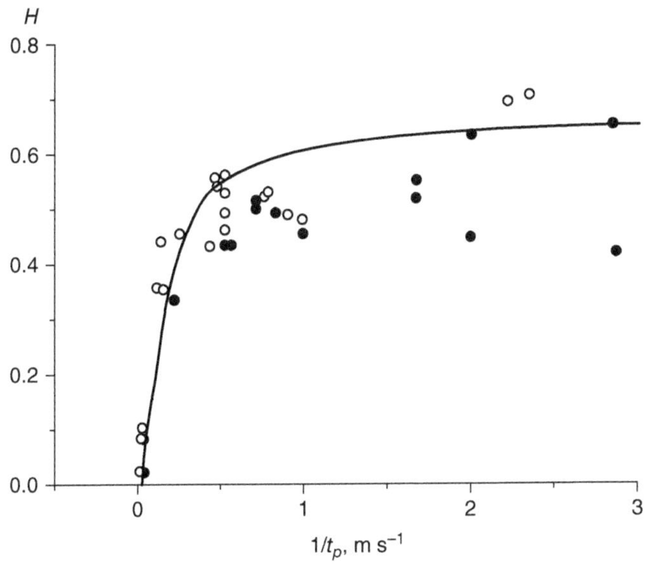

Figure 7.18 Experimental data and theoretical extinction curves for $\kappa = 10^{-3}$ cm^2/s at $p_i = 50$ atm. ○, burned out; •, extinguished.

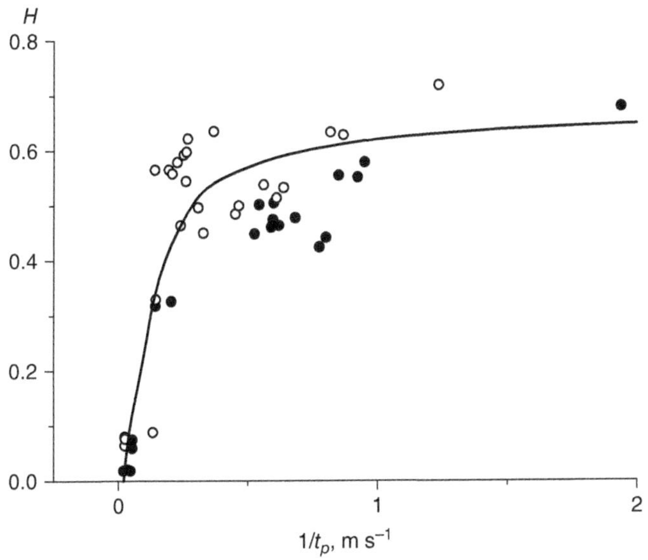

Figure 7.19 Experimental data and theoretical extinction curves for $\kappa = 10^{-3}$ cm^2/s at $p_i = 40$ atm. ○, burned out; •, extinguished.

The effects of various parameters on burning rate transitional behaviour and extinction may be illustrated further using a simple analytical form of the burning laws

$$v = \eta' \exp\left[k\left(\vartheta - \frac{\varphi}{v}\right)\right], \qquad v = \exp\left[\frac{k}{r}(\vartheta - 1)\right] \tag{7.125}$$

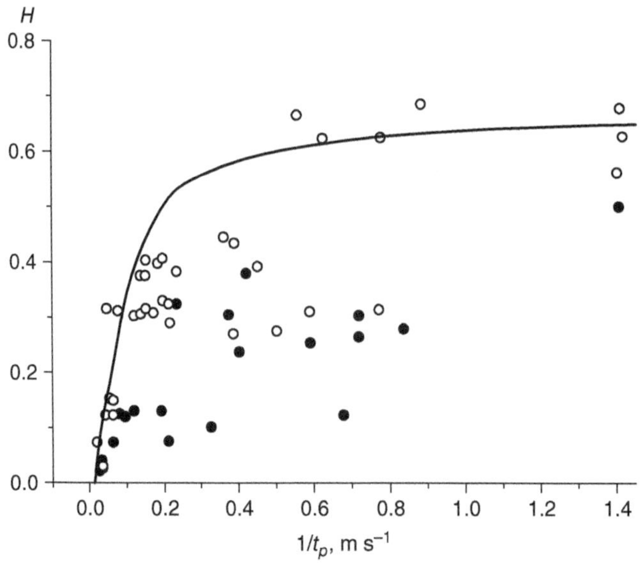

Figure 7.20 Experimental data and theoretical extinction curves for $\kappa = 10^{-3}$ cm²/s at $p_i = 30$ atm. ○, burned out; •, extinguished.

and exponential dependence of pressure on time

$$\eta = 1, \quad 0 \le \tau \le 1$$
$$\eta = \eta_f + (1 - \eta_f)\exp\left(^{-(\tau-1)}/_{\tau_p}\right), \quad \tau \ge 1 \tag{7.126}$$

Transitional process starts, therefore, at $\tau = 1$. At $\eta_f < 1$ (7.126) corresponds to an exponential decrease in nondimensional pressure from $\eta = 1$ to $\eta = \eta_f$. On the other hand, $\eta_f > 1$ implies an increase in pressure. Numerical analysis requires the simulation time interval $0 \le \tau \le \tau_{\max}$ to be fixed.

The following results may be obtained using different formulations of the ZN theory, namely PDE (Frost and Yumashev 1973), Novozhilov et al. (2010), integral (Lidskii et al. 1983, 1985), or the heat balance integral method (Marshakov and Novozhilov 2011a). Dashed curves in Figures 7.21–7.25 show pressure time history, while solid curves show burning rate history.

Asymptotically, as time goes to infinity, the pressure and burning rate approach the values η_f and $v_f = \eta_f^{\iota}$, respectively.

Some transitional regimes considered below exhibit burning rate oscillations. In the linear approximation, under constant pressure, nondimensional frequency is given by the expression (3.28)

$$\omega = \sqrt{\frac{k}{r^2} - \left(\frac{r(k+1) - (k-1)^2}{2r^2}\right)^2} \tag{7.127}$$

In transitional processes this expression applies only asymptotically when pressure can be considered as being essentially constant.

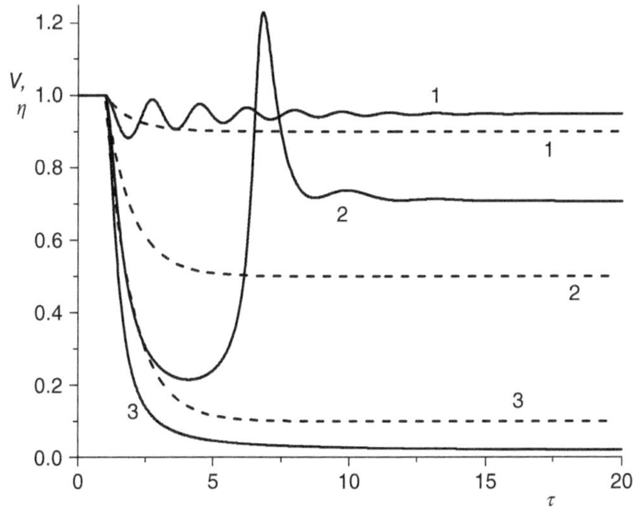

Figure 7.21 Effect of the magnitude of pressure drop on nonsteady burning rate. $\iota = 0.5$, $k = 2$, $r = 0.35$, $\tau_p = 1.0$. 1, $\eta_f = 0.9$; 2, $\eta_f = 0.5$; 3, $\eta_f = 0.1$.

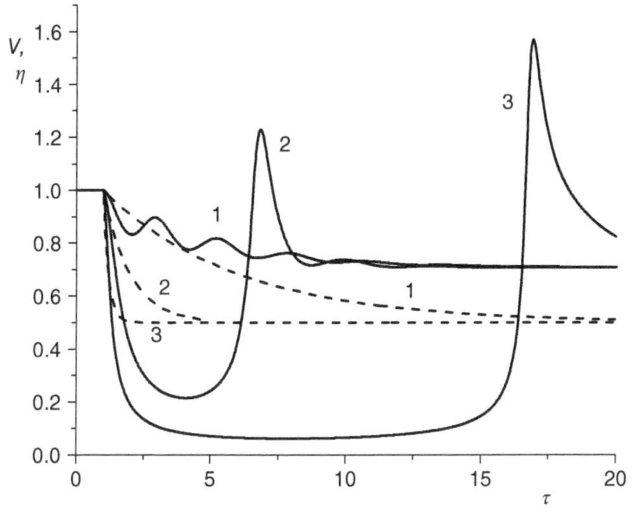

Figure 7.22 Effect of the pressure drop rate on nonsteady burning rate. $\iota = 0.5$, $k = 2$, $r = 0.35$, $\eta_f = 0.5$. 1, $\tau_p = 5$; 2, $\tau_p = 1.0$; 3, $\tau_p = 0.2$.

Figure 7.21 shows three transitional processes with the same pressure drop rate, but with different final values of pressure. Correspondingly, the final steady-state burning rate is different in all three cases. Chosen values of the parameters are in the vicinity of the stability boundary (at the boundary $r = 1/3$).

For small pressure drops (curve 1), the system behaviour is close to linear. The burning rate oscillates with the period $T = 1.73$ where $T = 2\pi/\omega$. This value can be found from the plot, as well as from the formula (7.127). Under deeper pressure drops, the burning rate

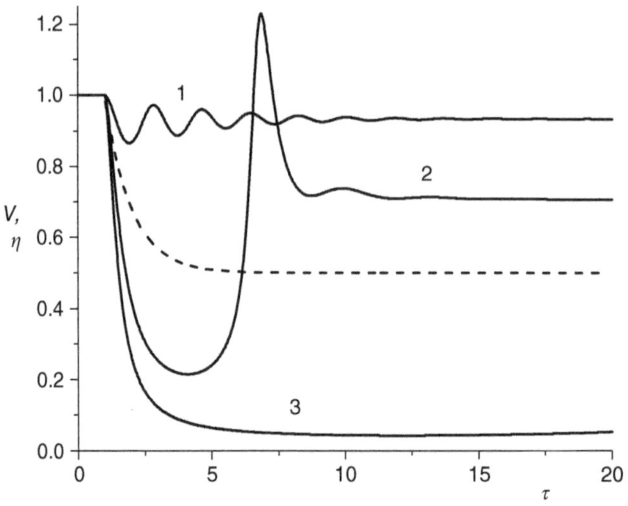

Figure 7.23 Effect of the parameter ι on nonsteady burning rate. $\eta_f = 0.5$, $k = 2$, $r = 0.35$, $\tau_p = 1.0$. 1, $\iota = 0.1$; 2, $\iota = 0.5$; 3, $\iota = 0.9$.

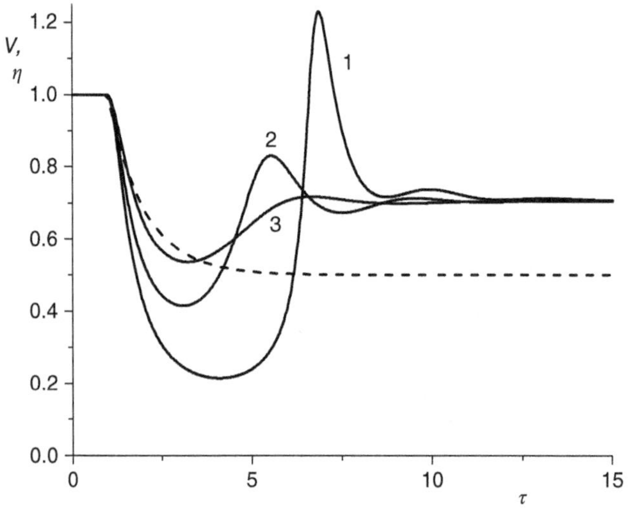

Figure 7.24 Effect of the parameter r on nonsteady burning rate. $\iota = 0.5$, $k = 2$, $\eta_f = 0.5$, $\tau_p = 1.0$ 1, $r = 0.35$; 2, $r = 0.4$; 3, $r = 0.5$.

decreases to values substantially lower than corresponding quasi-steady-state ones. Such burning rate depression lasts a certain period of time after which the burning rate, through oscillations, approaches a final steady-state value. For example, the pressure drops to half of its initial value (curve 2), resulting in a depression period of approximately four units of nondimensional time. When the final pressure is an order of magnitude lower than the initial pressure (curve 3) depression period is much longer. Depression may last for several hundred nondimensional time units. For this reason burning rate recovery and convergence

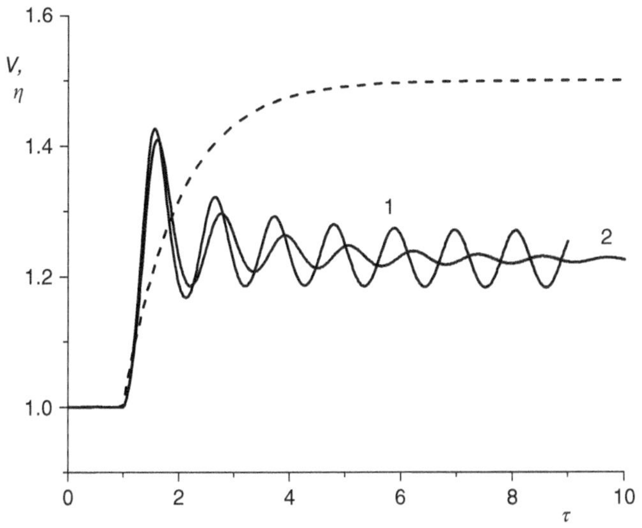

Figure 7.25 Nonsteady burning rate under pressure rise. $\iota = 0.5$, $k = 2$, $\eta_f = 1.5$, $\tau_p = 1.0$. 1, $r = 0.375$; 2, $r = 0.4$.

to the steady-state regime for this latter case are not shown. Similar features of transitional processes may also be seen in the other plots discussed below.

The case of different pressure drop rates with the same pressure drop magnitude is illustrated in Figure 7.22. An increase in the pressure drop rate (from case 1 to case 3) results in progressively more pronounced unsteadiness of combustion. It manifests through larger deviations of the burning rate from quasi-steady-state values, as well as increases in the periods of depression and the magnitudes of the later recovery.

Obviously, higher burning rate sensitivity to changes in pressure leads to more pronounced unsteadiness of the transition process. This is illustrated in Figure 7.23, where effect of the sensitivity parameter ι on burning rate behaviour in time is shown for fixed magnitude and rate of pressure drop.

For a fixed value of the parameter k, the system becomes more stable as surface temperature sensitivity to changes in initial temperature (i.e. the parameter r) increases. Figure 7.24 shows that nonsteady effects become less dramatic with increased system stability (the sequence of curves from 1 to 3).

Finally, Figure 7.25 illustrates the case of pressure rising with time. Curve 1 exhibits weakly diminishing oscillations. Their period may be estimated according to (7.127) to be $T = 1.12$. A similar value may be obtained from the plot itself.

8

Modelling Nonsteady Combustion in a Solid Rocket Motor

8.1 Introduction

This chapter considers the combustion of solid propellants in the formulation which couples together the Zeldovich–Novozhilov (ZN) theory with the description of processes occurring in a combustion chamber.

Explicit consideration of the chamber effects demonstrates a difference between acoustic and nonacoustic combustion regimes, discussed qualitatively and more precisely in Section 7.1.

There are three distinctive time scales associated with propellant combustion in a rocket engine chamber. These are the relaxation times of the preheat zone in the condensed phase t_c, the acoustic time t_a, and the time of combustion products efflux from the chamber t_{ch}.

In the acoustic combustion regimes, as investigated in Chapter 4, the acoustic time is of the same order of magnitude as the condensed phase relaxation time, that is $t_a \sim t_c$. This circumstance leads to the possibility of the development of acoustic (in general, nonlinear) oscillations within the engine. Such a development occurs in large-scale motors with high chamber pressure. Acoustic regimes are considered in Section 8.7.

A specific class of nonsteady nonacoustic phenomena is formed by problems where the condensed phase relaxation time and the chamber relaxation time are close, $t_c \sim t_{ch}$, and therefore problem time scales are much larger than the acoustic time. These conditions apply to smaller engines at low pressures. Following the pioneering work of Zeldovich (1942), this area of research may also be referred to as propellant combustion in a semi-enclosed volume. The number of theoretical studies devoted to such problems is much smaller than that carried out on acoustic investigations. In fact, only the problem of steady-state regime stability in the linear approximation has been considered in a fully consistent manner.

The acoustic time t_a may be defined as a period of gas acoustic oscillations in the combustion chamber. Taking L and R as the length and the radius of the cylindrical combustion chamber, respectively, and a as the speed of sound, the acoustic time is $t_a \sim L/a$ for the longitudinal and $t_a \sim R/a$ for the transversal combustion modes.

The condensed phase relaxation time is $t_c \sim \kappa/(u^0)^2$. Comparing t_a and t_c, the length of the combustion chamber L_a where longitudinal oscillations in combustion products may

Theory of Solid-Propellant Nonsteady Combustion, First Edition. Boris V. Novozhilov and Vasily B. Novozhilov.

Companion website: www.wiley.com/go/Novozhilov/solidpropellantnonsteadycombustion

occur is of the order of

$$L_a \sim \frac{a\kappa}{(u^0)^2} \tag{8.1}$$

For $\kappa \sim 10^{-3}$ cm^2/s, $a \sim 10^5$ cm/s, and $u^0 \sim 1$ cm/s this value is of the order of 1 m.

For nonacoustic combustion regimes the time t_c is comparable with the chamber relaxation time t_{ch}, which is estimated based on the time of gas efflux from the nozzle. In particular, mass conservation in the absence of combustion is

$$\frac{dM_g}{dt} = -Aps \tag{8.2}$$

Here M_g, p, s, and A are mass of the gas in the chamber, pressure, nozzle cross-sectional area, and nozzle discharge coefficient, respectively. The latter may be written following Landau and Lifshitz (1987) as

$$A = \frac{\Gamma(\gamma)}{a}, \Gamma(\gamma) = \sqrt{\gamma}\left(\frac{2}{\gamma+1}\right)^{\frac{\gamma+1}{2(\gamma-1)}} \tag{8.3}$$

For the ideal gas (8.2) takes the form

$$\frac{dp}{dt} = -\frac{p}{t_{ch}}, t_{ch} = \frac{V}{Asa^2} \tag{8.4}$$

For a cylindrical combustion chamber the volume is $V = SL$ where the chamber cross-sectional area S is much larger than the nozzle cross-sectional area $S \gg s$. Straightforward manipulations show that the length of the chamber resulting in $t_c = t_{ch}$ must be

$$L_{na} \sim \frac{s}{S}\frac{a\kappa}{(u^0)^2} \tag{8.5}$$

Comparison between (8.5) and (8.1) shows that $L_a \gg L_{na}$, that is, acoustic regimes develop in large-scale engines while nonacoustic regimes develop in small-scale engines.

8.2 Nonacoustic Regimes: Problem Formulation

The analysis of nonacoustic regimes in this chapter uses a number of approximations. Their introduction is justified in view of the large variety of existing propellant compositions and rocket engine designs. From a theoretical point of view this does not lead to a loss of generality as the described methodology remains valid, with relevant modifications, for more complicated combustion models and descriptions of the properties of combustion products.

1) The problem is considered as one-dimensional with L and S being the length and cross-sectional area of the cylindrical combustion chamber, respectively.
   The phase separation boundary is located at $x = 0$. The combustion chamber occupies the region $0 \le x \le L$ while the propellant fills the half-space $x \le 0$.
2) Sound waves are absent from nonacoustic regimes, therefore pressure is uniform over space and depends on time only.

3) Propellant thermal and physical properties such as density, specific heat, and thermal conductivity are considered to be constant.
4) The nonsteady combustion process is described by the ZN theory. Consequently, steady-state dependencies of the burning rate u^0, the surface temperature T_s^0, and the combustion temperature T_b^0 on the pressure p^0 and the initial temperature T_a must be known.
5) Combustion products obey the ideal gas law

$$\rho_g = \frac{\tilde{\mu} p}{RT_g} \tag{8.6}$$

where $\tilde{\mu}$ is molecular weight and R is the universal gas constant.
6) Variation of the combustion chamber volume $V = LS$ is negligible. Consideration of this effect leads to multiplication of the burning rate by the factor $1 - \rho_g/\rho$, which is practically equal to one. Here ρ_g and ρ are the densities of the gas and condensed phases, respectively.

Investigation of nonacoustic combustion regimes implies consideration of the two different nonsteady processes. The first is nonsteady combustion of the propellant ($x \leq 0$), considered in detail in preceding chapters. The second is nonsteady behaviour of the gas phase ($0 \leq x \leq L$).

It is clear that under a nonsteady combustion regime the combustion chamber admits gases with variable temperature. Therefore, the pressure inside the chamber depends on time whilst the chamber temperature varies with both chamber length and time. Certain simplifications are introduced to describe the behaviour of the gas inside the combustion chamber.

The simplest way to describe the behaviour of combustion products is the so-called isothermal approximation (the *I* approximation). This approximation was introduced in early studies on the stability of the steady-state combustion regime in a semi-enclosed volume by Zeldovich (1942) and Novozhilov (1967b). Zeldovich (1942) investigated an unrealistic model with constant surface temperature while Novozhilov (1967b) considered, within the framework of the ZN theory, a model with variable surface temperature.

The *I* approximation is based on the gas mass balance in the combustion chamber and on the assumption of uniformity of the gas temperature over the volume. The latter is assumed to be equal to the steady-state combustion temperature

$$\frac{d(\rho_g V)}{dt} = \rho u S - A(T_g) p s, \; T_g = T_b^0 \tag{8.7}$$

where V is the volume of the combustion chamber, u is the propellant linear burning rate, and A is the gas flow discharge coefficient through the nozzle with cross-sectional area s.

As a result of the ideal gas assumption

$$\frac{\tilde{\mu} V}{RT_b^0} \frac{dp}{dt} = \rho u S - A(T_b^0) p s \tag{8.8}$$

The *M* approximation corresponds to the instantaneous mixing of the new portion of gas entering the chamber with the whole mass of the gas present in the chamber already. In this case, the pressure and temperature of the gas depend on time, but are uniform over the space. The time histories of these quantities need to be found from the mass and energy conservation laws

$$\frac{d(\rho_g V)}{dt} = \rho uS - A(T_g)ps, \quad \frac{d(\rho_g VcT_g)}{dt} = c_p(\rho uT_bS - A(T_g)T_g ps) \tag{8.9}$$

For the ideal gas these take the form

$$\frac{\widetilde{\mu}V}{R}\frac{d}{dt}\left(\frac{p}{T_g}\right) = \rho uS - A(T_g)ps$$

$$\frac{\widetilde{\mu}V}{\gamma R}\frac{dp}{dt} = \rho uT_bS - A(T_g)T_g ps \tag{8.10}$$

where $\gamma = c_p/c_v$ is the specific heat ratio.

Finally, the W approximation describes the case when the new portion of gas does not either mix or exchange energy with the gas occupying the chamber. Pressure then depends on time only while temperature depends on the spatial coordinates as well. Conservation laws, assuming the gas behaves as ideal, take the form

$$\rho_g = \frac{\widetilde{\mu}p}{RT_g}, \quad \frac{\partial \rho_g}{\partial t} + \frac{\partial \rho_g u_g}{\partial x} = 0$$

$$\rho_g c_p\left(\frac{\partial T_g}{\partial t} + u_g\frac{\partial T_g}{\partial x}\right) = \frac{dp}{dt} \tag{8.11}$$

The following nondimensional variables will be used in the analysis in this chapter

$$\tau = \frac{(u^0)^2 t}{\kappa}, \ \xi = \frac{x}{L}, \nu = \frac{u}{u^0}, \nu_g = \frac{u_g}{u_g^0}$$

$$\eta = \frac{p}{p^0}, \theta_g = \frac{T_g}{T_b^0}, \rho = \frac{\rho_g}{\rho_g^0}, \chi = \frac{\widetilde{\mu}(u^0)^2 V}{A(T_b^0)RT_b^0 s\kappa} \tag{8.12}$$

The latter quantity is called an apparatus constant, and represents the ratio between the time of gas efflux from the chamber and the condensed phase relaxation time

$$\chi = \frac{t_{ch}}{t_c}, t_{ch} = \frac{\widetilde{\mu}V}{A(T_b^0)RT_b^0 s}, t_c = \frac{\kappa}{(u^0)^2} \tag{8.13}$$

A simpler expression for the chamber characteristic time $t_{ch} = L/u_g^0$ should be noted.

In the nondimensional variables the I approximation (8.8) assumes the form

$$\chi\frac{d\eta}{d\tau} = \nu - \eta \tag{8.14}$$

The M approximation (8.10) requires the two equations

$$\chi\frac{d}{d\tau}\left(\frac{\eta}{\theta_g}\right) = \nu - \frac{\eta}{\sqrt{\theta_g}}, \ \frac{\chi}{\gamma}\frac{d\eta}{d\tau} = \nu\theta_b - \eta\sqrt{\theta_g} \tag{8.15}$$

The W approximation (8.11) is governed by

$$\rho = \frac{\eta}{\theta_g}, \chi\frac{\partial \rho}{\partial \tau} + \frac{\partial \rho \nu_g}{\partial \xi} = 0, \chi\frac{\partial \theta_g}{\partial \tau} + \nu_g\frac{\partial \theta_g}{\partial \xi} = \frac{(\gamma - 1)\chi}{\gamma\rho}\frac{d\eta}{d\tau} \tag{8.16}$$

Eqs (8.14)–(8.16) must be solved in conjunction with the equations describing the nonsteady combustion of the condensed phase. It should be remembered that the spatial scales used to define the nondimensional coordinate ξ are different in the two phases. These are the Michelson thickness of the preheat zone for the condensed phase and the chamber length for the gas phase. The same remark is applicable to the temperature. The scaling procedure uses $T_s^0 - T_a$ for the condensed phase and T_b^0 for the gas phase. In addition, specific problems must of course embrace appropriate initial and boundary conditions.

8.3 Stability of the Steady-state Regime in a Semi-enclosed Volume

Burning rate under steady-state conditions in a semi-enclosed volume conforms with flux from the nozzle

$$\rho u^0 S = A p^0 s \tag{8.17}$$

It can be easily shown (Zeldovich 1942) that in a quasi-steady-state approximation, that is, within the assumption of a steady-state relation between burning rate and pressure $u^0 \sim (p^0)^\iota$, the combustion regime would only be stable at $\iota < 1$. Otherwise any pressure perturbation leads to violation of the condition (8.17) in such a way that the perturbation grows in time.

Such an approach to the stability investigation is inconsistent. The correct description of the stability conditions must take the nonsteady nature of the relevant processes as well as the inertia of both the fuel and the combustion chamber into account.

The ability to predict nonacoustic instability is of paramount importance in the design of solid fuel rocket motors operating at low pressure. It is essential to know the minimal value of pressure supporting the steady-state regime. Attempts to start the engine at operating conditions not compatible with stable steady-state burning leads to ignition delays or intermittent combustion. The steady-state combustion period precedes the onset of intermittency but turns into an oscillating regime with nonacoustic frequencies and growing amplitude. For the engine operation to be stabilized its length or operating pressure must be increased. This may be achieved by reducing the nozzle cross-sectional area. Nonacoustic instability may also be observed during combustion of charges with the burning area progressively reducing in time. In this case combustion starts in the steady-state mode but as pressure decreases conditions for instability development are achieved. The combustion regime becomes unstable and low-frequency oscillations develop.

Qualitatively, the development of nonacoustic instability may be described as follows. Pressure fluctuation in the combustion chamber leads to the change in the burning rate that is the amount of combustion products admitted to the chamber per unit time changes. In the same way change in the rate of admitted gases affects the chamber pressure. Therefore, the system consisting of the chamber and the fuel possesses a feedback loop. It should be noted that the interactions in both ways are not instantaneous. There are certain thermal inertia of both the solid fuel and the combustion chamber.

Consider first the stability of combustion in the I approximation. This problem was first formulated and investigated by Zeldovich (1942) using an assumption of constant propellant surface temperature. Consistent analysis admitting variable surface temperature within the framework of the ZN theory may be performed using the methodology developed by Novozhilov (1967b).

Small perturbations of the steady-state regime may be described in the linear approximation analytically.

For this purpose the method of nondimensional complex amplitudes is used. Any time-dependent function is written as a sum of its steady-state value and a small oscillating component

$$\phi = \phi^0 + [\phi_1 \exp(i\omega_c\tau) + c.c] \tag{8.18}$$

where ω_c is complex frequency, $\omega_c = \omega + i\lambda$.

The frequency-containing factor takes the form

$$\exp(i\omega_c\tau) = \exp(i\omega\tau)\exp(-\lambda\tau) \tag{8.19}$$

Therefore perturbations would either decay ($\lambda > 0$) or grow ($\lambda < 0$) in time oscillations with frequency ω. The value $\lambda = 0$ separates these two regimes and describes the stability boundary. Frequency at the boundary has purely real value.

Furthermore, pressure and burning rate are expressed in the form

$$\eta = 1 + [\eta_1 \exp(i\omega_c\tau) + c.c], v = 1 + [v_1 \exp(i\omega_c\tau) + c.c] \tag{8.20}$$

Linearization of (8.14) gives the relation between complex amplitudes of burning rare and pressure

$$(1 + i\chi\omega_c)\eta_1 - v_1 = 0 \tag{8.21}$$

On the other hand, the relationship (4.27) between these values was obtained by consideration of the burning rate response to oscillating pressure

$$v_1 = \frac{\iota + \delta(z-1)}{1 + (z-1)(r - k/z)}\eta_1, z = \frac{1}{2}(1 + \sqrt{1 + 4i\omega_c}), \delta = \iota r - \mu k \tag{8.22}$$

Eqs (8.21) and (8.22) give the characteristic equation for frequency

$$1 + i\chi\omega = U,\ U = \frac{\iota + \delta(z-1)}{1 + (z-1)(r - k/z)} \tag{8.23}$$

For any fixed propellant parameters (ι, μ, k, and r) its solution provides the values of the apparatus constant χ and the frequency ω at the stability boundary.

Figure 8.1 shows stability boundary on the (k, χ) plane for fixed values of the other problem parameters. The regions above the curves correspond to stable steady-state regimes. The region of stable combustion widens with an increase in the parameter r. For $\chi \to \infty$ (constant pressure) the stability boundary is found from the relationship (3.29).

Frequency dependence on parameter k at the stability boundary is shown in Figure 8.2. Corresponding values of the apparatus constant may be found from Figure 8.1.

Contraction of the stability region with an increase in the parameter ι (burning rate sensitivity to pressure changes) is illustrated in Figure 8.3.

To consider the M approximation where combustion products entering the chamber mix instantaneously with the chamber gas, the gas temperature θ_g and the combustion temperature are represented in a form similar to (8.20).

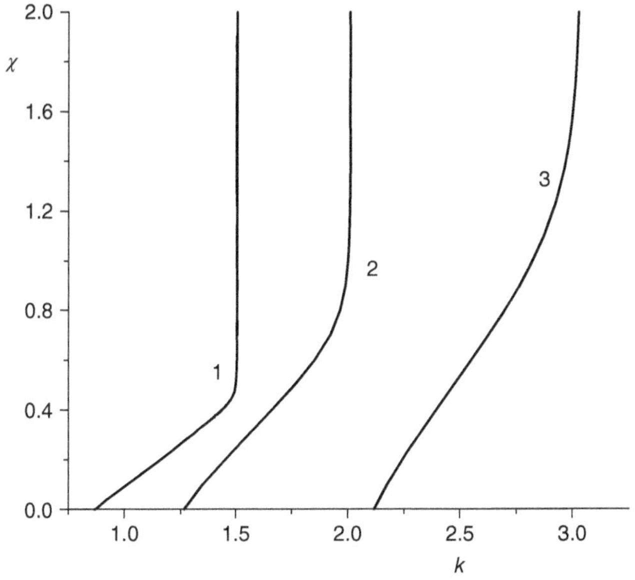

Figure 8.1 The stability boundary of the steady-state combustion regime in a semi-enclosed volume. $\iota = 0.5$, $\delta = 0.$ 1, $r = 0.1$; 2, $r = 1/3$; 3, $r = 1$.

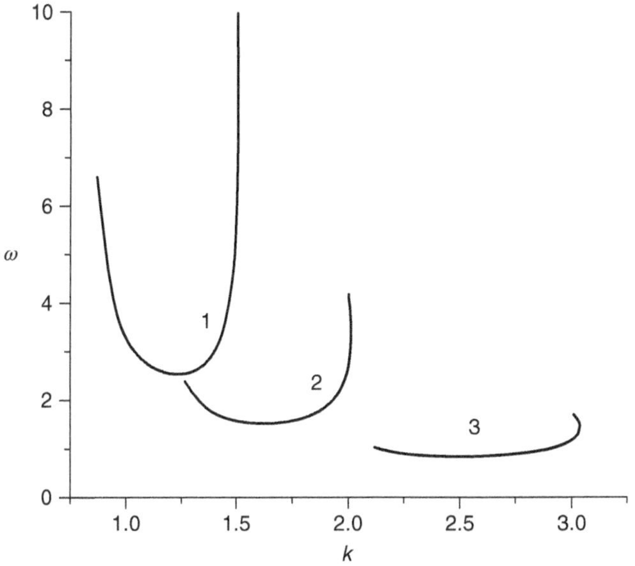

Figure 8.2 Frequency at the stability boundary of the steady-state combustion regime in a semi-enclosed volume. $\iota = 0.5$ and $\delta = 0.$ 1, $r = 0.1$; 2, $r = 1/3$; 3, $r = 1$.

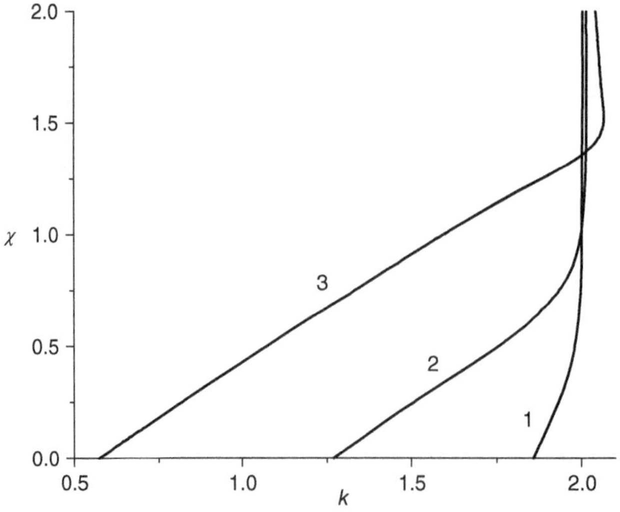

Figure 8.3 Effect of the parameter ι on the stability boundary of the steady-state combustion regime in a semi-enclosed volume. $r = 0.5$ and $\delta = 0$. 1, $\iota = 0.1$; 2, $\iota = 0.5$; 3, $\iota = 0.9$.

Linearization of Eq. (8.15) gives the two relationships

$$(1 + i\chi\omega) = v_1 + (1 + 2i\chi\omega)\frac{\theta_{g1}}{2}, \quad \frac{\theta_{g1}}{2} = v_1 + \theta_{b1} - \left(1 + \frac{i\chi\omega}{\gamma}\right)\eta_1 \tag{8.24}$$

Gas temperature perturbation may be eliminated from (8.24) to get the relation between complex amplitudes of burning rate and pressure

$$(1 + i\chi\omega)\eta_1 = v_1 + (1 + 2i\chi\omega)\left[v_1 + \theta_{b1} - \left(1 + \frac{i\chi\omega}{\gamma}\right)\eta_1\right] \tag{8.25}$$

This relation includes, however, an unknown amplitude of combustion temperature perturbation.

The latter may be found assuming the steady-state dependence of combustion temperature on initial temperature and pressure $T_b^0(T_a, p^0)$ is known. The nonsteady burning law $T_b\left(T_s - \frac{\kappa f}{u}, p\right)$ can then be derived following general formalism of the ZN theory.

Its linearization gives

$$\frac{T_{b1}}{T_s^0 - T_a} = r_b(\vartheta_1 - \varphi_1 + v_1) + \mu_b\eta_1 \tag{8.26}$$

where

$$r_b = \left(\frac{\partial T_b^0}{\partial T_a}\right)_{p^0}, \quad \mu_b = \frac{1}{T_s^0 - T_a}\left(\frac{\partial T_b^0}{\partial \ln p^0}\right)_{T_a} \tag{8.27}$$

are the linear coefficients of the combustion temperature sensitivity to variations of initial temperature and pressure. Using (4.26) the amplitude of combustion temperature perturbation may be expressed from (8.26) as a function of the perturbation amplitudes of burning rate and pressure

$$\theta_{b1} = \Delta\left(\frac{r_b}{k}\nu_1 + \left(\mu_b - \frac{\iota r_b}{k}\right)\eta_1\right), \quad \Delta = \frac{T_s^0 - T_a}{T_b^0} \tag{8.28}$$

Combining (8.25) and (8.28), the stability boundary in the M approximation may be described in the following form

$$1 + i\chi\omega + (1 + 2i\chi\omega)\left[1 + \frac{i\chi\omega}{\gamma} - \Delta\left(\mu_b - \frac{\iota r_b}{k}\right)\right] = U\left[1 + \left(1 + \frac{\Delta r_b}{k}\right)(1 + 2i\chi\omega)\right]$$

$$U = \frac{\iota + \delta(z-1)}{1 + \left(r - \frac{k}{z}\right)(z-1)} \tag{8.29}$$

The M approximation was considered, with different levels of detail, by Novozhilov (1967c) and Gostintsev and Sukhanov (1974). In particular, Novozhilov (1967c) demonstrated that accounting for conservation of energy (which is done within the M approximation) leads, compared to the I approximation, to corrections of the order Δ and $(\gamma - 1)/\gamma$. These corrections are small compared to unity. Indeed, setting $\Delta = 0$ and $\gamma = 1$ in (8.29) gives the characteristic Eq. (8.23), which is derived in the I approximation.

The W approximation assumes the adiabaticity of the gas dynamics, that is, the absence of heat exchange within the gas phase. This fact is reflected in Eqs (8.11). They must be supplemented by the boundary conditions

$$x = 0, \rho_g u_g = \rho u, T_g = T_b$$
$$x = L, S\rho_g u_g = A(T_g)ps \tag{8.30}$$

Linearization of Eqs (8.16) and the boundary conditions (8.30) leads to the following nondimensional boundary value problem (BVP)

$$\rho_1 = \eta_1 - \theta_1, i\chi\omega\rho_1 + \frac{\partial(\rho_1 + \nu_{g1})}{\partial\xi} = 0$$
$$i\chi\omega\theta_{g1} + \frac{\partial\theta_{g1}}{\partial\xi} = i\chi\omega\frac{\gamma - 1}{\gamma}\eta_1 \tag{8.31}$$

$$\xi = 0, \quad \rho_1 + \nu_{g1} = \nu_1, \quad \theta_{g1} = \theta_{b1}$$
$$\xi = 1, \quad \rho_1 + \nu_{g1} = \eta_1 - \frac{\theta_{g1}}{2}$$

Integration of the equation for the gas temperature complex amplitude θ_{g1}, taking into account the boundary condition at $\xi = 1$, gives

$$\theta_{g1}(\xi) = \left(\theta_{b1} - \frac{\gamma - 1}{\gamma}\eta_1\right)\exp(-i\chi\omega\xi) + \frac{\gamma - 1}{\gamma}\eta_1 \tag{8.32}$$

Integrating the continuity equation over the chamber length, the following equation is obtained

$$i\chi\omega\left(\eta_1 - \int_0^1 \theta_{g1}(\xi)d\xi\right) + \eta_1 - \frac{\theta_{g1}(1)}{2} - \nu_1 = 0 \tag{8.33}$$

Furthermore, on substitution of the expressions (8.32) and (8.28) for $\theta_{g1}(\xi)$ and θ_{b1}, respectively, the characteristic equation for frequency in the W approximation is

$$\frac{1+\gamma+2i\chi\omega}{2\gamma}+\left[\frac{\gamma-1}{\gamma}-\Delta\left(\mu_b-\frac{\iota r_b}{k}\right)\right]\left(1-\frac{1}{2}\exp(-i\chi\omega)\right)=$$

$$=U\left[1+\frac{\Delta r_b}{k}\left(1-\frac{1}{2}\exp(-i\chi\omega)\right)\right] \tag{8.34}$$

$$U=\frac{\iota+\delta(z-1)}{1+\left(r-\frac{k}{z}\right)(z-1)}$$

In the limit $\Delta\to 0$ and $\gamma\to 1$ this equation transforms into Eq. (8.23) for the isothermal approximation. The W approximation has also been studied by Assovskii and Rashkovsky (1998).

The simplest numerical procedure that can be used to find the stability boundary is as follows. The characteristic equation can be solved with respect to the parameter $k=K(\chi,\omega)$. By definition, this value is real and positive, that is, $\mathrm{Im}K(\chi,\omega)=0$. The latter restriction gives the relation between the frequency and the apparatus constant. Substituting such a relation into $k=\mathrm{Re}[K(\chi,\omega)]$, the three numbers ω, χ, and k at the stability boundary can be found, with all the other problem parameters fixed.

Figure 8.4 shows the stability boundaries for the three considered approximations. The three curves are close to each other, which is a consequence of the inequalities $\Delta\ll 1$ and $\gamma-1\ll\gamma$.

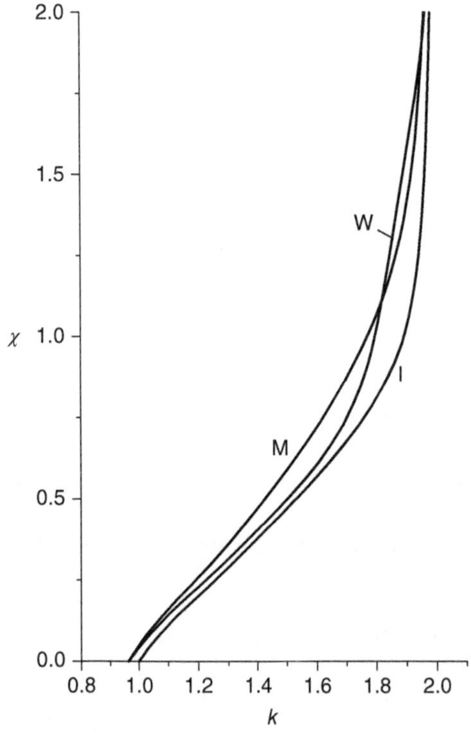

Figure 8.4 Stability boundaries for the steady-state combustion regime in a semi-enclosed volume for the I, M, and W approximations. $r=1/3$, $\iota=0.5$, $\delta=0$.

The obtained stability criterion (8.23) may be tested experimentally. In particular, Marshakov and Novozhilov (2011b) demonstrated an agreement of the criterion with their experimental data extracted from transient combustion tests of a double-base propellant.

8.4 Transient Regimes

Transient regimes in the rocket motor may be studied using a specific propellant combustion model. Consider, for example, the steady-state burning laws (7.120) (Marshakov et al. 2010). Using the standard formalism of the ZN theory these may be transformed into the following nondimensional nonsteady burning laws

$$v = \exp\left(\frac{\varepsilon_S}{T_s^i}\frac{(\vartheta - 1)}{(\vartheta + a)}\right), a = \frac{T_a}{(T_s^i - T_a)}$$
$$\frac{\varphi}{v} = \vartheta - \frac{1}{\beta(p^i\eta)(T_s^i - T_a)} \times \left\{\ln\frac{v}{\eta^n} - (T_a - T_K)[\beta(p^i\eta) - \beta(p^i)]\right\} \tag{8.35}$$

Here, the superscript is i instead of 0, and it is used to denote initial pressure at steady-state conditions in order to emphasize that (8.35) is applicable to the nonsteady combustion process. Similarly, the initial surface temperature is denoted T_s^i. The apparatus constant is defined using the initial (steady-state) pressure and the initial (steady-state) burning rate u^i.

Processes in the combustion chamber may be described by the I approximation. If the nozzle cross-sectional area varies during the combustion process then an additional factor $\sigma = s/s^i$, the ratio of the instantaneous cross-sectional area to its initial value, appears in the nondimensional mass balance (8.14)

$$\chi^i\frac{d\eta}{d\tau} = v - \sigma\eta \tag{8.36}$$

The conventional formulation (2.33)–(2.34) of the ZN theory, coupled with the nonsteady burning laws (8.35) and the chamber mass balance (8.36), allow transient regimes to be followed. The functional dependence $\sigma = \sigma(\tau)$ must also be provided. The initial conditions

$$\theta = \exp(\xi), \eta = 1, \vartheta = 1, \varphi = 1, v = 1 \tag{8.37}$$

correspond to the steady-state regime with the constant nozzle cross-sectional area $\sigma = 1$.

A solution may be obtained by either of the methods discussed in Chapter 2, for example by the integral balance method (Marshakov and Novozhilov 2011b).

Experimentally, transient regimes are obtained by abruptly opening the secondary nozzle in the combustion chamber shown in Figure 8.5. The secondary nozzle is opened by exploding a small auxiliary charge. A steady-state combustion regime is achieved just before the second nozzle opening. In such a case, the parameter σ is constant during the transient process. Full details of the apparatus and experiment have been reported by Marshakov and Novozhilov (2011b).

Following these experiments the function $\sigma = \sigma(\tau)$ is set as changing instantaneously from $\sigma = 1$ at $\tau = 0$ (the start of the transient process) to another constant value $\sigma = \text{constant} > 1$ maintained during instantaneous combustion.

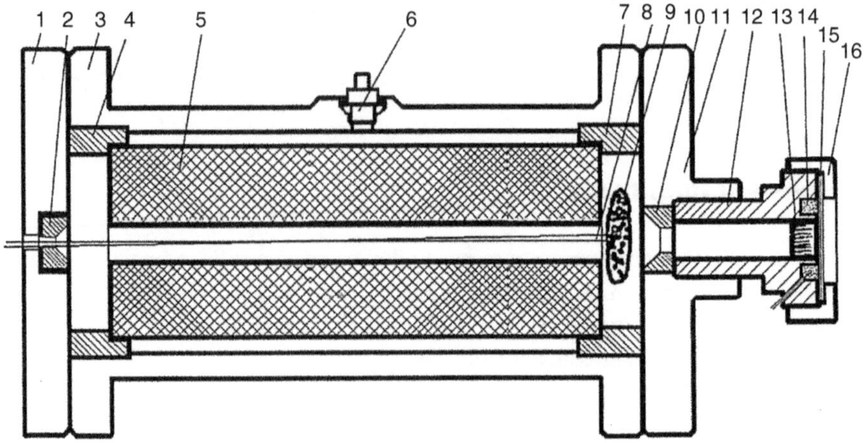

Figure 8.5 Sketch of the experimental rocket motor chamber. 1, front cap; 2, primary nozzle; 3, body of the chamber; 4, front fixing ring; 5, propellant charge; 6, pressure gauge; 7, back fixing ring; 8, electric ignitor; 9, black powder charge; 10, auxiliary nozzle; 11, back cap; 12, body of the ejection device; 13, cork; 14, ring channel; 15, membrane; 16, cap of the ejection device. Source: Reproduced with permission from Marshakov and Novozhilov (2011b).

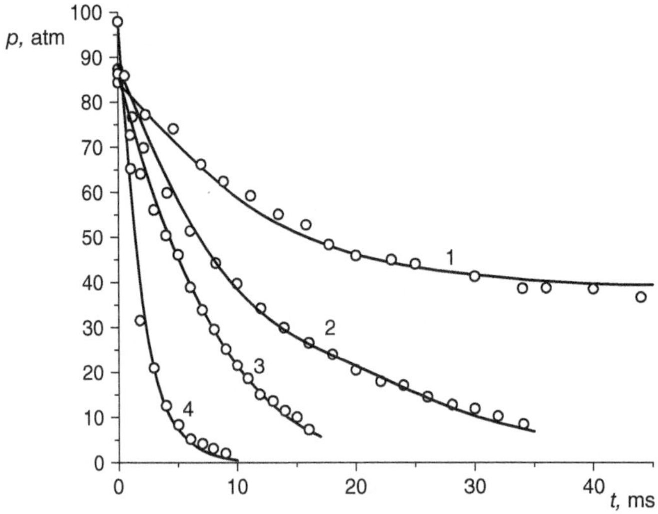

Figure 8.6 Pressure transient behaviour for different cases (Table 8.1). Lines, predictions by the ZN theory; circles, experimental data (Marshakov and Novozhilov 2011b).

The parameters σ and t_{ch} can only be determined in experiments to within a certain accuracy (Marshakov and Novozhilov 2011b). The relative error in measuring the experimental value σ_{exp} is about 10%. The chamber relaxation time (and consequently the apparatus constant) is measured with much worse accuracy. This parameter depends on the void volume of the chamber existing during the transition process. Since a certain burning time is required to achieve the steady-state regime, the latter may generally change from its initial value to full chamber volume. This leads to a large uncertainty in the determination of the relaxation time, namely $3 < t_{ch} < 10$ ms (Marshakov and Novozhilov 2011b).

In view of such experimental uncertainties the question which needs be addressed is whether the ZN theory may be reconciled with experimental data for the sets of parameters which fall within the uncertainty regions.

The results reported in Figure 8.6 for four different cases provide the affirmative answer. The simulation parameters for each of the cases are presented in Table 8.1.

Theoretical predictions for all four cases agree with the experiments for the sets of parameters, which are within the uncertainty ranges discussed above. For example, the assumed values of the parameter σ are different from the measured values σ_{exp} (also shown in Table 8.1) by not more than 10%. The adopted values of the chamber relaxation time are within the indicated uncertainty range between 3 and 10 ms, but higher than measured in each of the particular tests.

Evolution of the nondimensional pressure $\eta = p/p^i$ and the burning rate $v = u/u^i$ is shown in Figures 8.7 and 8.8. It should be noted first that qualitatively pressure and burning rate

Table 8.1 Parameters used in the transient combustion regime simulations and experiments.

Case no.	p^i (atm)	u^i (cm/s)	σ_{exp}	σ	$\frac{\sigma}{\sigma_{exp}}$	t_c (ms)	t_{ch} (ms)	χ^i
1	85	0.860	1.34	1.41	1.05	1.35	9.0	6.66
2	88	0.876	1.66	1.67	1.01	1.30	7.8	6.00
3	86	0.866	2.15	1.94	0.90	1.33	8.0	6.00
4	98	0.938	5.28	4.78	0.91	1.16	8.0	6.89

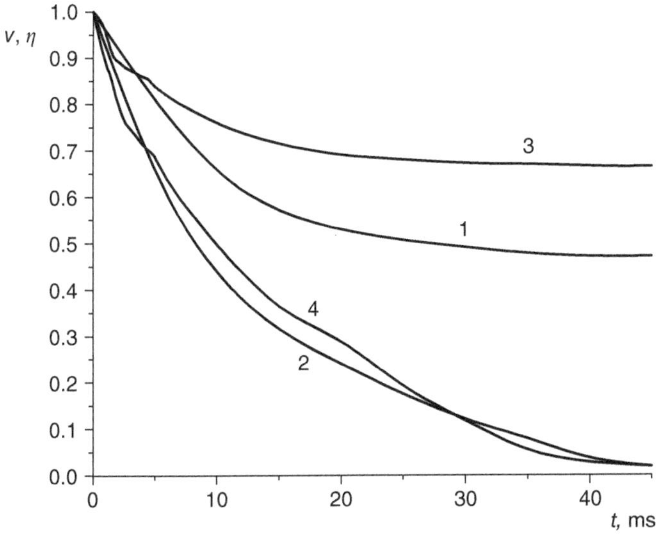

Figure 8.7 Transient behaviour of nondimensional pressure and burning rate. 1, η, case 1; 2, η, case 2, $\sigma = 1.67$; 3, v, case 1; 4, v, case 2, $\sigma = 1.67$.

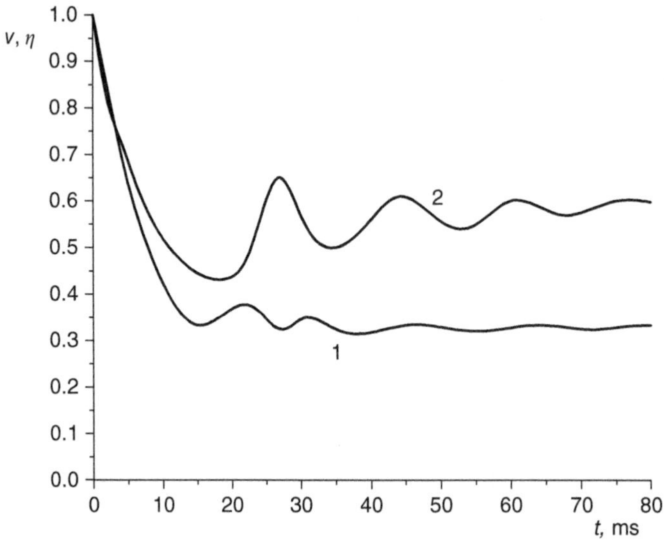

Figure 8.8 Transient behaviour of nondimensional pressure and burning rate. Case 2, $\sigma = 1.64$. 1, η; 2, ν.

Table 8.2 Values of pressure, burning rate, and apparatus constant at final steady-state regimes.

Case no.	p^f (atm)	u^f (cm/s)	χ^f
1	40.4	0.576	2.11
2	28.8	0.479	1.07
3	20.1	0.393	0.640
4	2.99	0.135	0.0307

behaviour are well correlated with each other. Transition to the new steady-state regime is evident for case 1 in Figure 8.7. This new regime occurs at final pressure p^f with burning velocity u^f, which may be found from the set of equations

$$u^f = u^0(p^f), \quad \frac{u^f}{u^i} = \sigma\frac{p^f}{p^i} \tag{8.38}$$

Results following from (8.38) for all the cases are listed in Table 8.2.

On the other hand, case 2 in Figure 8.7 illustrates extinction at the relative nozzle expansion $\sigma = 1.67$.

If the degree of the process unsteadiness is reduced slightly ($\sigma = 1.64$) then the behaviour of the pressure and burning rate changes dramatically. This is shown in Figure 8.8. The combustion process approaches the steady-state regime. The shape of the presented curves reflects the oscillating nature of the loss of stability on traversing the boundary separating the stable and unstable steady-state regimes.

It is instructive to compare these solutions to the problem with those obtained under quasi-steady-state approximation. The latter is obtained by assuming that the instantaneous burning rate is equal to its steady-state value at a given instantaneous pressure. In other words, the burning rate would follow the first of the dependencies (7.120) with the steady-state pressure p^0 being replaced by the instantaneous pressure p. With such a dependence $u^0(p) = u^0(\eta)$, and also using the steady-state mass balance identity

$$\rho u^0(p^i)S = Ap^i s^i \tag{8.39}$$

Eq. (8.36) is modified to

$$\chi^i \frac{d\eta}{d\tau} = \frac{u^0(\eta)}{u^0(1)} - \sigma\eta \tag{8.40}$$

Using pressure time history as an example, deviation of the solutions obtained under the quasi-steady-state approximation from the nonsteady ones is presented in Figure 8.9. This relative deviation is defined as $\Delta(t) = (p^{qs}(t) - p(t))/p(t)$, where $p^{qs}(t)$ is the quasi-steady-state solution, obtained from (8.40), and $p(t)$ is the unsteady solution obtained using the ZN theory equations and (8.36). It may be seen as a quantitative measure of the degree of unsteadiness of the process.

As seen in Figure 8.9, the relative deviation is positive for all cases, that is, the quasi-steady-state approximation consistently overpredicts the chamber pressure. Furthermore, it is in agreement with the monotonically decreasing nature of the pressure–time dependencies. Quantitatively, a small variation of the nozzle cross-sectional ratio σ (case 1) leads to a small difference between the unsteady and quasi-steady-state solutions. With increasing σ, the quasi-steady-state solution fails to agree quantitatively with the nonsteady one (and with the experimental data) due to the highly unsteady nature of the transition process.

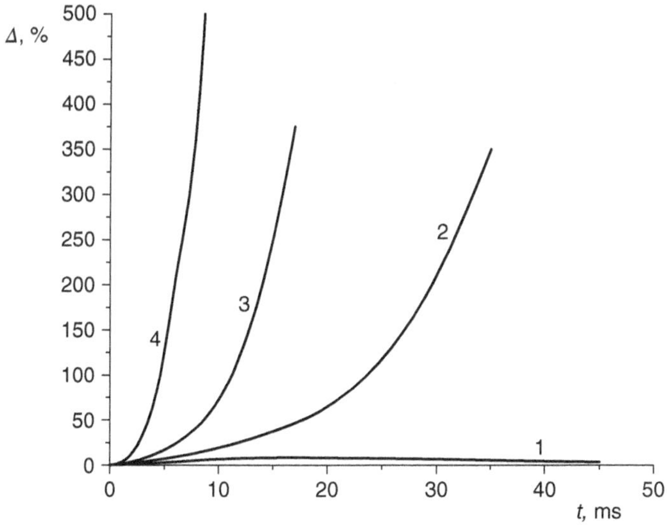

Figure 8.9 Relative deviation of the solutions in the quasi-steady-state approximation from unsteady solutions. Curve numbering corresponds to cases 1 to 4 (Table 8.1).

8.5 Unstable and Chaotic Regimes

There are unsteady regimes more complicated than those discussed in Section 8.4. These may be identified by systematically varying relevant problem parameters beyond the region of steady-state solution stability.

One possibility, in particular, is the chaotic behaviour of the unsteady burning rate. The existence of chaotic regimes at constant pressure has been demonstrated in Section 3.4. The same problem within the semi-enclosed volume is investigated in this section.

The nondimensional problem formulation is a natural modification of that used in Section 3.4. The problem is formulated as follows

$$\frac{\partial\theta}{\partial\tau} = \frac{\partial^2\theta}{\partial\xi^2} - v\frac{\partial\theta}{\partial\xi}, \quad -\infty < \xi \le 0$$

$$\xi = 0, \theta = \vartheta(\tau), \xi \to -\infty, \theta = 0$$

$$v = \eta^{\iota}\exp\left[k\left(\vartheta - \frac{\varphi}{v}\right)\right], v = \exp\left[\frac{k}{r}(\vartheta - 1)\right]$$

$$\chi\frac{\mathrm{d}\eta}{\mathrm{d}\tau} = v - \eta \tag{8.41}$$

The nonsteady burning laws used here are derivatives of the following steady-state dependencies

$$u^0 = A(p^0)^{\iota}\exp(\beta T_a), u^0 = B\exp(\beta_s T_s^0) \tag{8.42}$$

where A, B, β, β_s, and ι are constants.

The second of these is identical to (3.130), the first is modified to allow for variable pressure.

The last of the equations in (8.41) is a combustion chamber mass balance in the isothermal approximation (Eq. (8.14)).

The set of equations in (8.41) contains unknown functions $v(\tau)$, $\vartheta(\tau)$, $\varphi(\tau)$, $\eta(\tau)$, $\theta(\xi, \tau)$, and the parameters ι, k, r, χ.

The sensitivity coefficients with respect to changes in the initial temperature are the same as in Section 3.4

$$k = \beta(T_s^0 - T_a), r = \beta/\beta_s \tag{8.43}$$

At the steady-state regime

$$\eta^0 = 1, \theta^0 = e^{\xi}, \varphi^0 = 1, v^0 = 1, \vartheta^0 = 1 \tag{8.44}$$

The initial propellant temperature profile for the equations in (8.41) may be imposed as a weakly perturbed steady-state profile, that is in the form (2.54) where the change $\xi \to -\xi$ is being made for the solution on the left half-space and the moments of the distribution are

$$y_0(0) = 1 + \varepsilon, \varepsilon \ll 1$$

$$y_n(0) = 0, n = 1, 2, \ldots, N \tag{8.45}$$

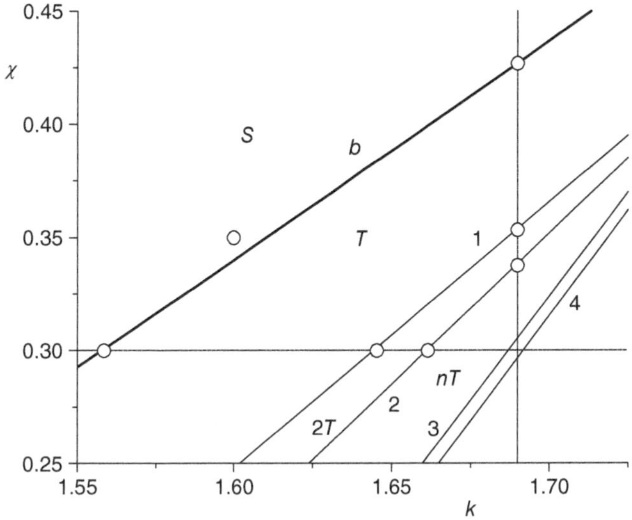

Figure 8.10 Regions of the existence of different nonsteady regimes on the plane (k, χ).

It should be noted that the majority of solutions considered in the present section are either periodic or chaotic. In these cases, regime characteristics do not depend on the specific form of initial conditions.

The solution of the set of equations in (8.41) may be obtained (Novozhilov et al. 2009a) by either one of the conventional finite volume/finite difference methods (Novozhilov and Posvyanskii 1991; Belyaev et al. 2004), or by the integral balance method (Section 2.5).

Two of the four problem parameters are fixed as $\iota = 0.5$ and $r = 1/3$. Qualitatively, various possible combustion regimes may be illustrated on the plane of the two remaining parameters (k, χ).

Curve b in Figure 8.10 is a stability boundary of the steady-state regimes, obtained in Section 8.3 in linear approximation. Region S above curve b is a set of parameters which produce stable steady-state regimes so that small burning rate perturbations decay with time. The point corresponding to one of the steady-state regimes is also shown in Figure 8.10 while the corresponding time history of the decaying burning rate is presented in Figure 8.11.

Below the boundary of stability of steady-state regimes, there are regions where periodic combustion regimes occur. They may be identified by moving, for example along the line $\chi = 0.3$ from the region of stability S into the unstable part of the diagram in Figure 8.10. The parameter k serves as a bifurcation parameter in this process. Analytical investigation in linear approximation gives at $\chi = 0.3$ the position of the stability boundary at $k = 1.5579$, with natural frequency $\omega = 1.5236$ and corresponding period of oscillations $T = 2\pi/\omega$ equal to 4.1239.

Numerical simulation shows that the first bifurcation occurs at $k_1 = 1.5586$. At this value the steady-state regime is replaced, by means of the Andronov–Hopf bifurcation, by the oscillating regime with constant amplitude. This regime, called the T regime, is

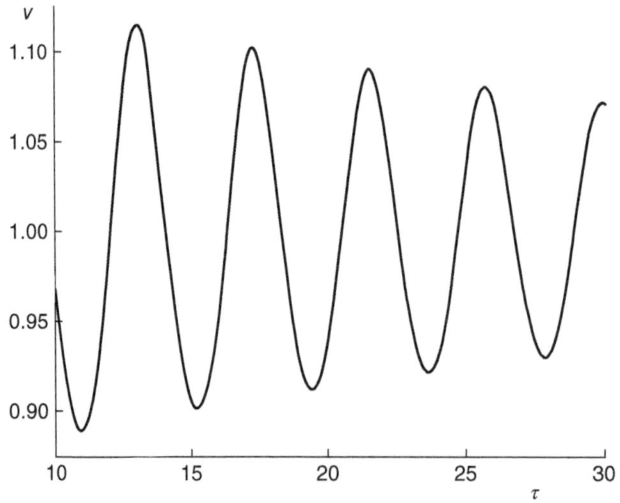

Figure 8.11 Stable steady-state combustion regime at $k = 1.6$, $\chi = 0.35$. Source: Reproduced with permission from Novozhilov et al. (2009a).

illustrated in Figure 8.12. Figure 8.12b shows a two-dimensional projection of the system phase trajectory on the plane (v, y_0) where y_0 is the zeroth moment of the temperature distribution (the heat content of the condensed phase)

$$y_0(\tau) = \int_{-\infty}^{0} \theta(\xi, \tau)\mathrm{d}\xi \tag{8.46}$$

Further increase of the parameter k leads to a cascade of bifurcations, with the period doubling at each of these. Figures 8.12–8.15 illustrate, using selected combustion regimes, the system evolution towards chaotic behaviour. The period of oscillations doubles after each bifurcation, and according to established terminology (Landau and Lifshitz 1987) the regimes are called the T regime, the $2T$ regime, the $4T$ regime, the $8T$ regime, etc. It should be noted that exact period doubling occurs at the bifurcation point. Variation of the bifurcation parameter between the two successive bifurcations causes a slight change in the period. Hence, the periods given in the captions of Figures 8.12–8.14 differ by a factor of 4 only approximately.

Numerical errors prevent the quantification of a full infinite series of bifurcation points. The bifurcation parameter values for the first five bifurcations are

$$k_1 = 1.5586, k_2 = 1.6454, k_3 = 1.6618, k_4 = 1.6656, k_5 = 1.6666 \tag{8.47}$$

The difference between the two successive values decreases quite rapidly as the bifurcation number grows, therefore only the points corresponding to the first three bifurcations are shown in Figure 8.10.

By analogy, simulations are also done to fix the parameter $k = 1.69$ and using the apparatus constant χ as a bifurcation parameter. As the latter decreases the system undergoes a series of successive period doubling bifurcations at the following values of the apparatus constant

$$\chi_1 = 0.4255, \chi_2 = 0.3532, \chi_3 = 0.3377, \chi_4 = 0.3337, \chi_5 = 0.3327 \tag{8.48}$$

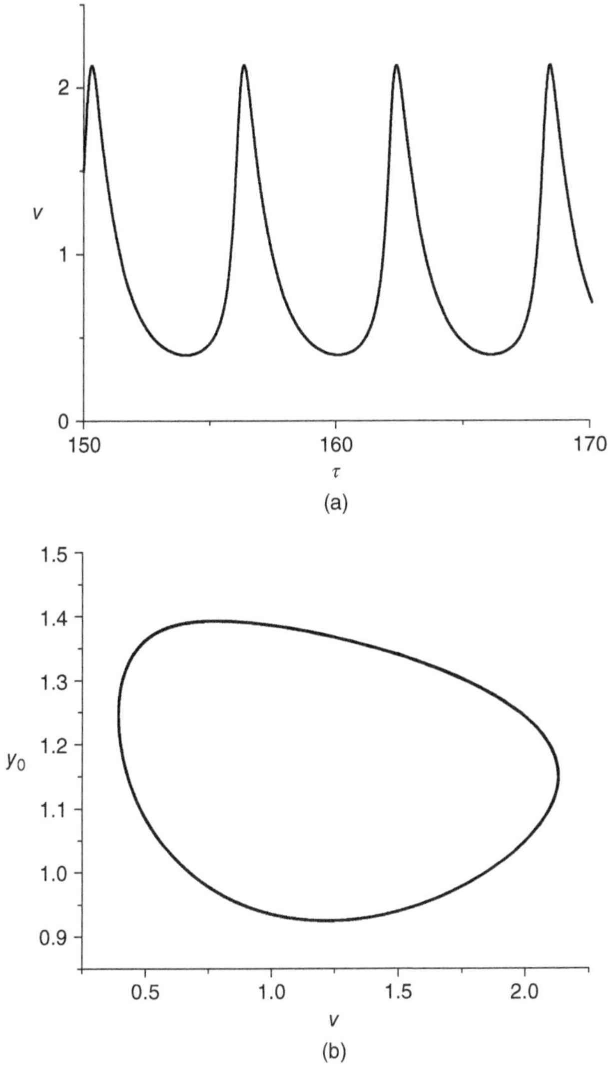

Figure 8.12 T regime. $k_1 < k < k_2$, $k = 1.64$, $T = 6.04$.

Figure 8.10 shows the points corresponding to the first three of these bifurcations.

The region between the stability boundary b and curve 1 in Figure 8.10 corresponds to T regimes. In the region between curves 1 and 2 the $2T$ regimes occur. Furthermore, there are regions where (between lines 2 and 3) the nT regimes occur successively.

A chaotic regime is observed after the fifth bifurcation on further small change of the bifurcation parameter (either k or χ). The corresponding phase plane is shown in Figure 8.15. Phase trajectories fill the bounded region in the phase plane almost uniformly.

The presented numerical analysis shows that transition from the stable steady-state regime to thechaotic regime follows the Feigenbaum cascading bifurcations scenario

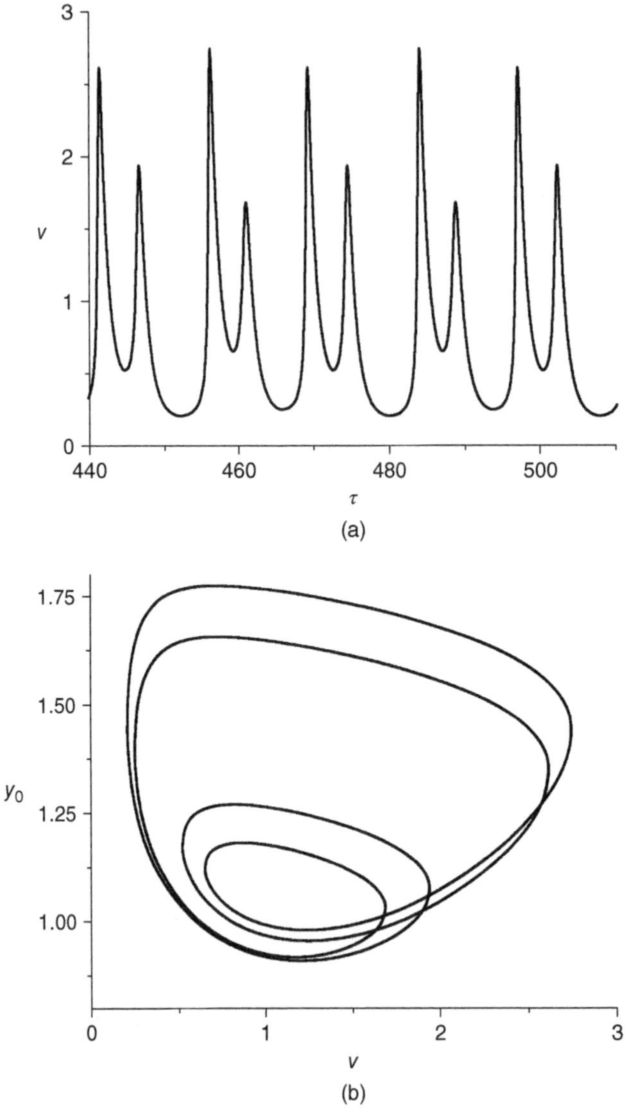

Figure 8.13 4T regime. $k_3 < k < k_4$, $k = 1.664$, $T = 27.8$.

(Landau and Lifshitz 1987; Arnold 1988). In this scenario, the successive bifurcation values k_m must satisfy the following law

$$\lim_{m\to\infty} \delta_m = \delta, \delta_m = \frac{k_m - k_{m-1}}{k_{m+1} - k_m} \tag{8.49}$$

where $\delta = 4.669\ldots$ is the universal Feigenbaum constant. The values (8.47) of the bifurcation parameter k provide the following estimations

$$\delta_2 = 5.3 \pm 0.1, \delta_3 = 4.3 \pm 0.3, \delta_4 = 4.5 \pm 1.2 \tag{8.50}$$

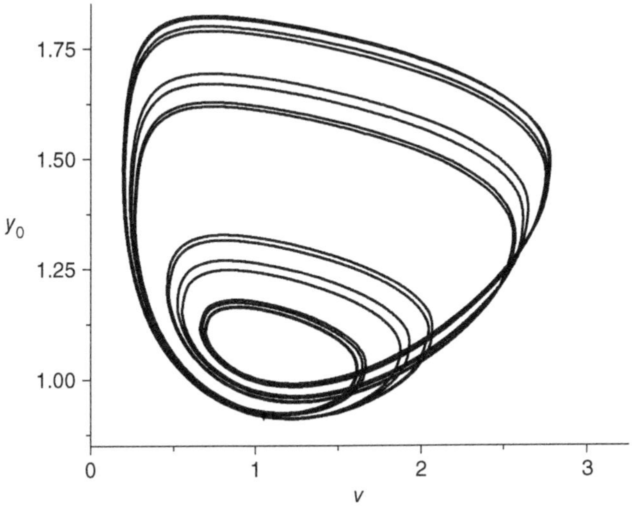

Figure 8.14 $16T$ regime. $k_5 < k$, $k = 1.66665$, $T = 113$.

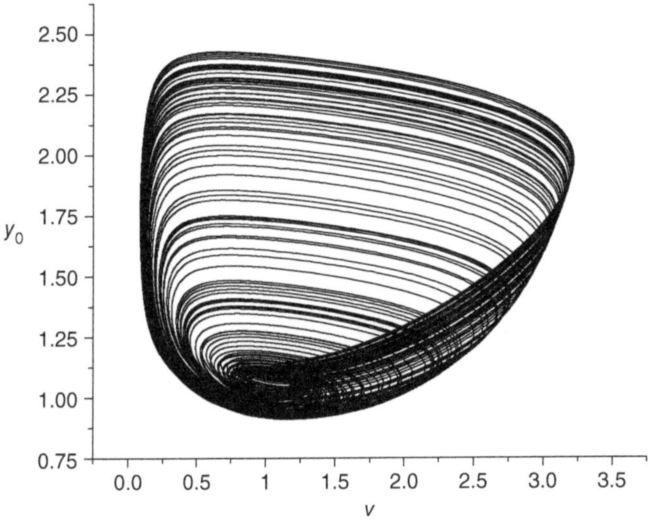

Figure 8.15 Chaotic combustion regime. $k = 1.6685$.

Similarly, defining Δ_m as

$$\Delta_m = \frac{\chi_m - \chi_{m-1}}{\chi_{m+1} - \chi_m} \tag{8.51}$$

and using (8.48)

$$\Delta_2 = 4.67 \pm 0.07,\ \Delta_3 = 3.9 \pm 0.3,\ \Delta_4 = 4.2 \pm 1.0 \tag{8.52}$$

Therefore, even the first elements of both sequences are close to the limiting value of δ, as is well known from other examples of the Feigenbaum scenario.

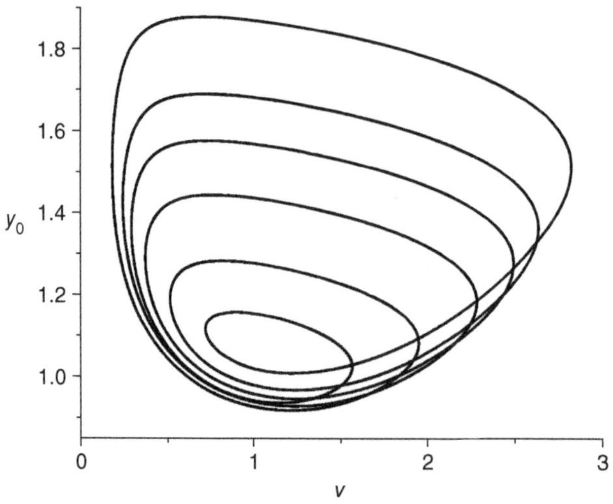

Figure 8.16 $6T$ regime. $k = 1.6695$, $T = 42.6$.

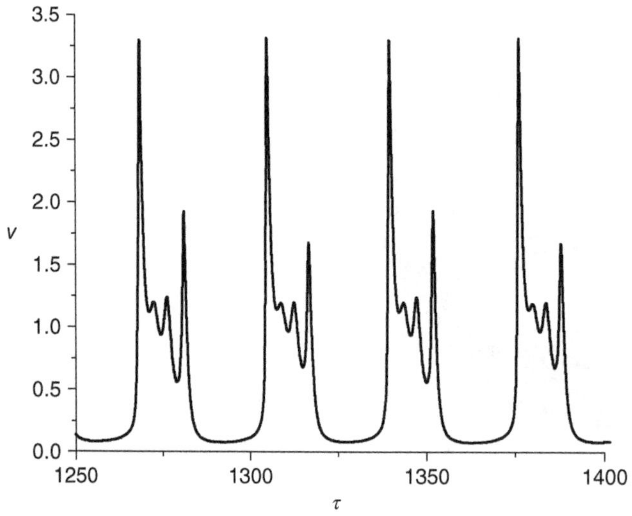

Figure 8.17 'Sneezing' regime. $k = 1.7$, $T = 711$.

The chaotic regime occupies the region between curves 3 and 4 in Figure 8.10. These boundaries are determined to within a certain accuracy due to numerical errors. For the value $\chi = 0.3$, chaos is observed for $1.6685 < k < 1.69$. It should be noted that sometimes simple periodic regimes emerge in the same region. As an example, Figure 8.16 shows the $6T$ regime observed for $k = 1.6695$.

The post-chaos region is located below curve 4. On moving away from this curve, the combustion regimes simplify. Figure 8.17 shows a periodic pulsating regime known as 'sneezing'. It consists of successive combustion bursts interchanging with periods of depression where the burning rate falls much below its steady-state value.

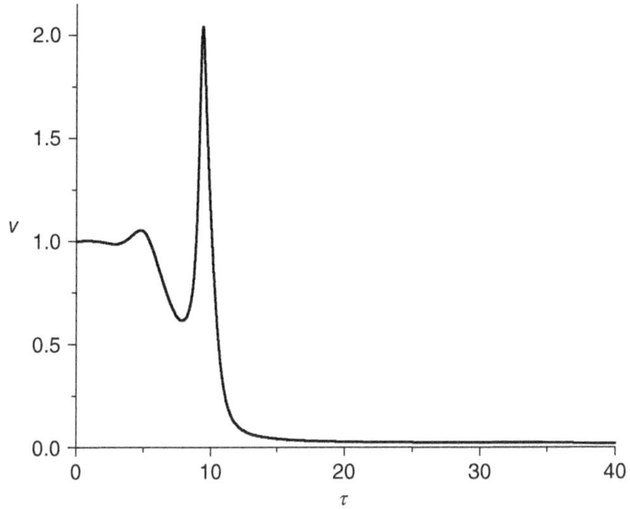

Figure 8.18 Extinction. $k = 1.8$.

On further movement into the post-chaos region duration of depression, periods grow which must eventually lead to the termination of combustion. Such an extinction regime is demonstrated in Figure 8.18.

It should be noted that strict extinction cannot be realized within the adopted propellant combustion model as the latter does not possess the steady-state regime corresponding to a zero burning rate. The claim of the existence of the extinction regime is based on the fact that the considered model may produce solutions having an arbitrary large depression period with an arbitrary small burning rate.

8.6 Experimental Data

This section discusses the available experimental data which, in principal, can be compared with predictions of nonacoustic regimes in rocket combustion chambers. The first experimental studies on nonacoustic instability of solid propellant combustion in a semi-enclosed volume were conducted in the second half of the twentieth century (Akiba and Tanno 1959; Eisel et al. 1964; Beckstead et al. 1966; Beckstead and Price 1967; Svetlichny et al. 1971; Simonenko et al. 1980). A comprehensive review of these results is provided by Price (1992).

Nonacoustic instability is often called L^* instability. This terminology originates from the property

$$L^* = \frac{V}{s} \tag{8.53}$$

having units of length (V is combustion chamber volume, s is the cross-sectional area of the nozzle) and representing a certain effective length of the chamber. Following from Section 8.2, it is proportional to the apparatus constant χ, and the steady-state combustion regime loses stability at a certain value of this parameter.

Combustion chambers with end burner grain geometry are most commonly used for studying nonacoustic instability. This arrangement eliminates the effect of erosion. Beckstead and Price (1967) provide a detailed description of the chamber design. Propellant samples were chosen as disks having diameter equal to the diameter of a cylindrical chamber. Burns were conducted in two arrangements, with either one or two disks. In the latter case a hole, concentric with the nozzle, was drilled in one of these disks. A burning surface area is constant in such a system, therefore combustion occurs with constant background pressure. The chamber side walls were made of pyrex, a material which provides better experiment repeatability compared to metallic tubes, allowing the radiation to be controlled.

This design allows the pressure to be varied from 3.5 to 14 atm, and the effective chamber length L^* to be varied from 50 to 250 cm. Beckstead and Price (1967) provide typical pressure oscillograms obtained under nonacoustically unstable combustion conditions. Amplifying pressure oscillations develop after propellant ignition, leading to periodic flame extinction. Oscillograms of these types allow values of oscillation frequency, oscillation exponential amplitude growth constant, and the effective length L^* to be extracted.

Most investigations of low-frequency oscillations are conducted using composite propellants. Comprehensive data on nonacoustic instability are provided by Beckstead et al. (1966). They used composite propellant U-TF containing 75% ammonium perchlorate (with average particle diameter 225 μm), 18% copolymer of polybutadiene and acrylic acid, and additions of aluminium and copper chromate. Steady-state burning rate dependence on pressure may be expressed as $u^0 = 0.32p^{0.46}$ where pressure units must be atm and burning rate units cm/s. The propellant thermal diffusivity is $\kappa = 1.96 \times 10^{-3}$ cm²/s.

Particle size in this particular case is bigger than the Michelson length (the latter is about 10^{-3} cm at a pressure of 10 atm). Although the ZN theory is strictly valid under the reverse relationship between these two characteristic lengths, an attempt nevertheless can be made to analyse the nonacoustic instability of this particular fuel. The parameters k and r of the composite propellant may be regarded in such an analysis as having certain effective values.

Beckstead et al. (1966) measured the frequency and damping decrement of oscillations as functions of the apparatus constant at several values of pressure.

At a certain value of the apparatus constant χ^* (damping decrement λ is zero; the critical regime) the steady-state combustion regime loses stability. Combustion is unstable for $\chi < \chi^*$ and the pressure amplitude in the chamber grows exponentially. Conversely, for $\chi > \chi^*$ the combustion regime is stable. Being disturbed (e.g. by injecting an additional amount of gas), the chamber pressure returns to its steady-state value in oscillatory manner ($\lambda > 0$).

Experimental data for the pressure $p = 8.4$ atm are plotted as crosses in Figures 8.19 and 8.20. The critical apparatus constant turns out to be $\chi^* = 0.12$ and the frequency at the stability boundary $\omega^* = 2.9$.

In the I approximation the relation between oscillation frequency and chamber and propellant parameters at the stability boundary is given by (8.22) and (8.23)

$$1 + i\chi^*\omega^* = \frac{\iota + \delta(z^* - 1)}{1 + (z^* - 1)(r - k/z^*)}, z^* = \frac{1}{2}(1 + \sqrt{1 + 4i\omega^*}) \tag{8.54}$$

The coefficient of linear sensitivity of burning rate with respect to pressure is known for the tested propellant to be $\iota = 0.5$. Sufficiently high accuracy of measurements is required to

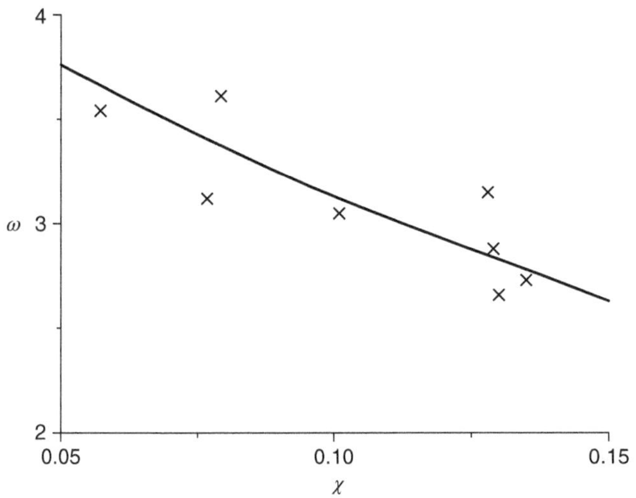

Figure 8.19 Pressure oscillation frequency as a function of apparatus constant.

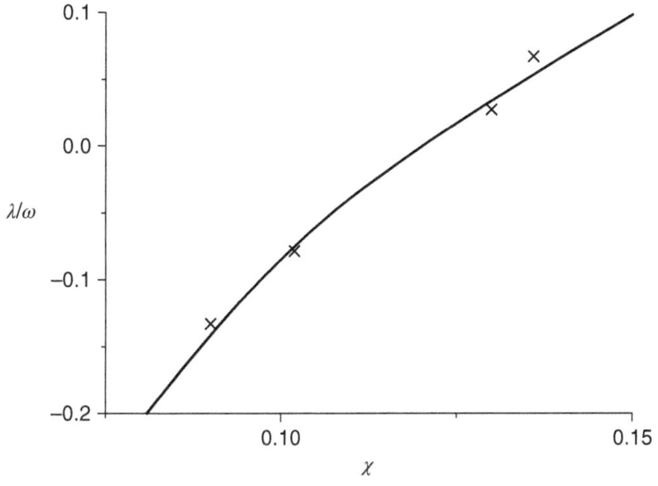

Figure 8.20 Damping decrement as a function of apparatus constant.

determine the parameter δ. The data of Beckstead et al. (1966) lacks the required accuracy, but it is known that normally $\delta \ll 1$. In the following calculations this value is set to zero. The first of the equations in (8.54) is complex, and is equivalent to two real equations. It allows the values of the two unknowns $k = 1.05$ and $r = 0.103$ to be found.

The relations (8.54) are valid at the stability boundary of the steady-state regime. Beyond this boundary the damping decrement λ is different from zero. Therefore, replacing χ^* by χ and ω^* by $\omega + i\lambda$ in (8.54), in the general case gives

$$1 + \chi(i\omega - \lambda) = \frac{\iota + \delta(z-1)}{1 + (z-1)(r - k/z)}, z = \frac{1}{2}(1 + \sqrt{1 + 4(i\omega - \lambda)}) \tag{8.55}$$

For the known propellant parameters ι, δ, k and r these relations allow the frequency and damping decrement dependencies $\omega(\chi)$ and $\lambda(\chi)$ to be obtained. These are the dependencies measured by Beckstead et al. (1966).

Comparison of calculations and experimental data, presented in Figures 8.19 and 8.20, shows their agreement. Using the values of the parameters k and r, obtained from critical conditions, the dependencies $\omega(\chi)$ and $\lambda(\chi)/\omega(\chi)$ may be reproduced in a whole range of the apparatus constant. The values k and r thus obtained from experimental data processing have quite reasonable values. For $k = 1.05$ and $T_s^0 - T_a \sim 400$ K it turns out that $\beta \sim 3 \times 10^{-3}$ K^{-1}. This value conforms by an order of magnitude with the values of temperature coefficient of the burning rate, common for composite systems of the considered type.

Despite such an agreement, the presented comparison should be considered with caution. One reason for this has already been indicated: this is a high nonhomogeneity state of the fuel. The other reason is a lack of information on energy losses in the combustion chamber. It is obvious that such losses contribute to a certain degree to damping the oscillations.

Examination of the body of research on nonacoustic instability suggests the absence of exhaustive experimental data on this phenomenon. In particular, information indicating unequivocal discrepancies between theory and experiment would be highly valuable. This would stimulate further development of the theory which is, by its very nature, approximate.

The results discussed in this chapter are evidence of the very complicated nature of the propellant combustion process in a semi-enclosed volume. Even in the simplest theoretical framework (the ZN theory) a large variety of possible regimes is predicted. Price (1992) provides classification of the observed combustion regimes. It is important to note that such observable regimes include the sneezing and extinction regimes predicted theoretically and demonstrated in Figures 8.17 and 8.18, respectively. A cascade of bifurcating nT regimes has not been detected so far. The reason for this is a quite narrow parameters region where such regimes occur, and very small changes in bifurcation parameter from one bifurcation to another.

8.7 Acoustic Regimes

As pointed out in Section 8.1, acoustic instabilities may develop if the acoustic time is of the same order of magnitude as the condensed phase relaxation time. The parameters leading to such a relation, that is, $t_a \sim t_c$, are generally those of large-scale rocket motors with high chamber pressure.

A long history of the design and operation of solid fuel rocket engines confirms this conclusion. It turns out that various unsteady phenomena, for example soft and hard excitation of burning rate and pressure oscillations, may occur instead of the designed steady-state regime. Several studies on these effects are collated, for example by De Luca et al. (1992). It has also been pointed out that achievements in understanding and theoretical explanation of various unsteady phenomena inherent in laboratory and practical operation of solid fuel engines are very modest (De Luca et al. 1992). This is despite a considerable body of studies concerned with general issues of interaction between the acoustic field and the solid propellant combustion process (Novozhilov 2005).

Attempts to develop a consistent theory of nonsteady processes in the combustion chamber of a solid propellant rocket engine have taken place for over 50 years. Most studies use an oversimplified approach to the interaction between acoustics and combustion. A review of these is presented by Culick and Yang (1992). All the attempts based on such an approach have been unsuccessful.

The major shortcomings of the discussed approach are as follows. First, only the linear burning rate response to varying pressure is considered. Such linear analysis may be used (in conjunction with a linear acoustics approximation) for establishing conditions of stability of the steady-state combustion regime. However, a consideration of unsteady processes with finite pressure variation amplitude require analysis beyond the restrictions of a linear approximation.

Second, in investigation of engines with circular bore grain geometry the erosion effect (i.e. the effect of tangential gas stream on burning rate) is ignored. This effect has been shown, however, to have paramount importance for unsteady combustion dynamics in such an engine configuration (Novozhilov 2007; Novozhilov et al. 2007).

These circumstances, combined with some additional assumptions that are hard to justify, have led to a conclusion (Culick 2000) that the results obtained over past several decades cannot be considered to be a satisfactory solution to the problem of predicting nonsteady combustion phenomena in engines with circular bore grain geometry. In fact, these studies have not even presented a consistent problem formulation.

The present section discusses the problem for engines with end burner charge configuration (Novozhilov et al. 2009b). Under such conditions the problem becomes one-dimensional, with no erosion effect present. A nonlinear burning rate response to varying pressure may be found within the framework of the ZN theory.

Mathematical problem formulation includes nonlinear equations of acoustics and nonsteady burning laws for a specific propellant combustion model with a minimum number of parameters. The problem has two vastly different time scales accounting for 'quick' acoustic and 'slow' oscillation amplitude variation times. These differ by about three orders of magnitude, setting a high benchmark for the numerical accuracy of the solution.

Nevertheless, a less demanding but still satisfactory solution may be proposed based on consideration of the quadratic approximation with respect to pressure oscillation amplitude. Using such an approach, the set of ordinary differential equations for the amplitudes of various oscillation modes can be derived. Effects related to 'quick' acoustic time are absent from the solutions of this set of equations. As a result, the numerical solution procedure is significantly simplified.

Consider solid fuel rocket engine with end burner grain geometry. The problem is one-dimensional: the coordinate $x = 0$ corresponds to the fuel surface and $x = L$ to position of the nozzle. It is assumed that the fuel surface includes a thin zone where both condensed phase and gas phase chemical reactions occur.

The set of equations describing unsteady acoustic phenomena in the gas phase ($0 \le x \le L$) includes the continuity equation, the Euler equation, and the condition of adiabaticity

$$\frac{\partial \rho_g}{\partial t} + \frac{\partial(\rho_g u_g)}{\partial x} = 0, \rho_g \left(\frac{\partial u_g}{\partial t} + u_g \frac{\partial u_g}{\partial x} \right) = -\frac{\partial p}{\partial x}$$

$$\rho_g = \frac{\gamma p^0}{a^2} \left(\frac{p}{p^0} \right)^{\frac{1}{\gamma}}, a^2 = \frac{\gamma p^0}{\rho_g^0} \tag{8.56}$$

Here ρ_g, p, and u_g are density, pressure, and gas velocity, respectively, γ is specific heat ratio and a is the speed of sound. It is assumed that the combustion product temperature T_b, just after the termination of gas phase reactions, obeys the adiabaticity condition

$$x = 0, \frac{T_b}{T_b^0} = \left(\frac{p}{p^0}\right)^{\frac{(\gamma-1)}{\gamma}} \tag{8.57}$$

This assumption filters out entropic waves and simplifies the problem significantly as the heat conduction equation in the gas phase does not need be considered. The error introduced by this assumption is of the same order of magnitude as the ratio of the maximum temperature difference across the condensed phase to the maximum temperature difference within the combustion wave (Novozhilov 1973a).

Combustion is treated within the framework of the ZN theory. The steady-state burning law dependencies are assumed to have the form (8.42), and the corresponding nonsteady burning laws are

$$u = Ap^{\iota} \exp\left[\beta\left(T_s - \frac{\kappa f}{u}\right)\right], u = B\exp(\beta_s T_s) \tag{8.58}$$

The boundary conditions essentially enforce continuity of temperature and mass flux

$$T = T_a, x \to -\infty$$
$$T = T_s(t), \rho u = \rho_g u_g, x = 0 \tag{8.59}$$

where ρ is density of the condensed phase.

The remaining required condition at the entrance to the nozzle takes the form (Landau and Lifshitz 1987)

$$g = \Gamma\sqrt{\gamma p \rho_g}, \Gamma = \left(\frac{2}{\gamma+1}\right)^{\frac{(\gamma+1)}{2(\gamma-1)}} \frac{S_{\min}}{S}; x = L \tag{8.60}$$

Here $S_{\min}$ and S are the cross-sectional areas of the nozzle and the combustion chamber, respectively.

The problem is reformulated in nondimensional form. It is convenient to use variables which represent deviations from the respective steady-state values.

Spatial coordinates are nondimensionalized using the κ/u^0 scale for the condensed phase and L for the gas phase. Time scale is taken as that of the gas phase L/a

$$\xi = \frac{u^0}{\kappa}x; -\infty < x \le 0, -\infty \le \xi \le 0$$
$$\xi = \frac{x}{L}; 0 < x \le L, 0 \le \xi \le 1$$
$$\tau = \frac{a}{L}t; t, \tau \ge 0 \tag{8.61}$$

The following nondimensional variables and parameters are used

$$\theta = \frac{T - T^0}{T^0 - T_a}, \vartheta = \frac{T_s - T_s^0}{T_s^0 - T_a}, \varphi = \frac{f}{f^0} - 1$$
$$v = \frac{u}{u^0} - 1; \zeta = \gamma\frac{(u_g - u_g^0)}{a}, \eta = \frac{p}{p^0} - 1$$
$$M = \frac{u_g^0}{a}; \chi = \frac{(u^0)^2 L}{\kappa a} \tag{8.62}$$

Here M is the Mach number for combustion products and χ is the apparatus constant, which is the ratio of acoustic time scale L/a to the condensed phase time scale $\kappa/(u^0)^2$.

Using these variables, Eqs (2.28) and (8.56) can be written in the form ($\tau \geq 0$)

$$\frac{1}{\chi}\frac{\partial\theta}{\partial\tau} = \frac{\partial^2\theta}{\partial\xi^2} - (1+v)\frac{\partial\theta}{\partial\xi} - v\exp(\xi), \quad -\infty < \xi \leq 0$$
$$\gamma\frac{\partial}{\partial\tau}(1+\eta)^{\frac{1}{\gamma}} + \frac{\partial}{\partial\xi}(1+\eta)^{\frac{1}{\gamma}}(\varsigma+\gamma M) = 0, \quad 0 \leq \xi \leq 1$$
$$(1+\eta)^{\frac{1}{\gamma}}\left[\frac{\partial\varsigma}{\partial\tau} + \left(\frac{\varsigma}{\gamma} + M\right)\frac{\partial\varsigma}{\partial\xi}\right] = -\frac{\partial\eta}{\partial\xi}, \quad 0 \leq \xi \leq 1 \tag{8.63}$$

The nonsteady burning laws (8.58) take the form

$$v = (1+p)^{\iota}\exp\left(k\frac{(\vartheta + v - \varphi + \vartheta v)}{(1+v)}\right) - 1$$
$$v = \exp\left(\frac{k}{r}\vartheta\right) - 1 \tag{8.64}$$

where $k = \beta(T_s^0 - T_a)$ and $r = \beta/\beta_s$.

Finally, the boundary conditions scaled from (8.59), (8.60) are

$$\theta = 0, \xi \to -\infty$$
$$\theta = \vartheta, (1+\eta)^{\frac{1}{\gamma}}(\varsigma + \gamma M) = \gamma M(1+v), \xi = 0$$
$$\varsigma = \gamma M\left[(1+\eta)^{\frac{(\gamma-1)}{2\gamma}} - 1\right], \xi = 1 \tag{8.65}$$

The set of Eqs (8.63)–(8.65) includes the unknown functions $v(\tau)$, $\vartheta(\tau)$, $\varphi(\tau)$, $\theta(\xi, \tau)$, $\eta(\xi, \tau)$, $\varsigma(\xi, \tau)$, and the parameters ι, k, r, γ, M, χ. The solutions of these equations are oscillations of pressure and other variables with the period of the order of acoustic time. On the other hand, the amplitude of these oscillations changes very slowly. The time scale of this change is of the order of acoustic time divided by Mach number (which is $\sim 10^{-3}$ for considered systems). The large discrepancy between the two time scales leads to difficulties in numerical investigation of Eqs (8.63)–(8.65).

This section discusses the approximate solution, capturing only slow amplitude variations.

In the acoustic approximation (i.e. in the absence of shock waves) the amplitudes of gas velocity and pressure oscillations are small. Therefore, the lowest nonlinear approximation is considered, namely the quadratic approximation with respect to the amplitude of the gas pressure oscillations.

Accounting for medium movement leads to corrections to the complex oscillation frequency of the order of the square of the Mach number. These corrections are neglected, and terms accounting for medium movement are dropped from the hydrodynamic equations.

First equations describing acoustic movement may be considered. In the linear approximation, the equations in (8.63) give the continuity equation and the Euler equation in the form

$$\frac{\partial\eta}{\partial\tau} + \frac{\partial\varsigma}{\partial\xi} = 0, \frac{\partial\varsigma}{\partial\tau} + \frac{\partial\eta}{\partial\xi} = 0 \tag{8.66}$$

On the other hand, in the quadratic approximation

$$\frac{\partial\eta}{\partial\tau}+\frac{\partial\varsigma}{\partial\xi}=\frac{(\gamma-1)}{2\gamma}\frac{\partial\eta^2}{\partial\tau}-\frac{1}{\gamma}\frac{\partial(\eta\varsigma)}{\partial\xi}$$
$$\frac{\partial\varsigma}{\partial\tau}+\frac{\partial\eta}{\partial\xi}=-\frac{1}{\gamma}\eta\frac{\partial\varsigma}{\partial\tau}-\frac{1}{2\gamma}\frac{\partial\varsigma^2}{\partial\xi} \tag{8.67}$$

Using the linear relationships (8.66), the equations in (8.67) may be transformed to the form

$$\frac{\partial\eta}{\partial\tau}+\frac{\partial\varsigma}{\partial\xi}=\frac{1}{2\gamma}\frac{\partial}{\partial\tau}(\gamma\eta^2+\varsigma^2)$$
$$\frac{\partial\varsigma}{\partial\tau}+\frac{\partial\eta}{\partial\xi}=\frac{1}{2\gamma}\frac{\partial}{\partial\xi}(\eta^2-\varsigma^2) \tag{8.68}$$

In the simplest case of classical acoustics where equations are linear and gas velocity perturbations are zero at the boundaries, the solution is a sum of the harmonics

$$\eta^{(cl)}=\sum_{n=1}^{\infty}(P_n\exp(i\omega_n\tau)+c.c.)\cos(k_n\xi)$$
$$\varsigma^{(cl)}=\sum_{n=1}^{\infty}(-iP_n\exp(i\omega_n\tau)+c.c.)\sin(k_n\xi) \tag{8.69}$$

Here *c.c.* stands for the complex conjugate and P_n are constant amplitudes, the values of which are undetermined in a linear approximation. Frequency and wave vectors for any of the harmonics are equal in the chosen nondimensional variables

$$\omega_n=n\pi,\ k_n=n\pi \tag{8.70}$$

The nonlinearity of the acoustics equations (8.68) results in the dependence of the pressure and gas velocity oscillation amplitudes on time. The solution is sought in the form

$$\eta=\sum_{n=1}^{\infty}(\eta_n(\tau)\exp(i\omega_n\tau)+c.c.)\cos(k_n\xi)+Mf_\eta(\tau)\sin(k_n\xi)$$
$$\varsigma=\sum_{n=1}^{\infty}(-i\eta_n(\tau)\exp(i\omega_n\tau)+c.c.)\sin(k_n\xi)+Mf_\varsigma(\tau)\cos(k_n\xi) \tag{8.71}$$

where $\eta_n(\tau)$ are amplitudes which change with the scale of slow time, and $f_{\eta,\varsigma}(\tau)$ are functions of the order of η_n containing both slow and fast times. The explicit form of the latter is not needed for further analysis.

An equation for the complex oscillation amplitudes $\eta_n(\tau)$ may be derived in the following way. The continuity equation (8.68) is multiplied by $\cos(k_n\xi)$ and integrated from zero to one using the expansions (8.71). The integral containing the derivative with respect to the coordinate is integrated by parts. By dividing the result by the common factor $\exp(i\omega_n\tau)$ the equations describing slow amplitude variations can be obtained. The result of these transformations is

$$\frac{1}{2}\frac{d\eta_n}{d\tau}=\left[-\varsigma\cos(k_n\xi)|_{\xi=1}+\varsigma|_{\xi=0}+\frac{i\omega_n}{2\gamma}\int_0^1(\gamma\eta^2+\varsigma^2)\cos(k_n\xi)d\xi\right]_{\omega_n} \tag{8.72}$$

where expression with subscript ω_n is a coefficient in front of the factor $\exp(i\omega_n\tau)$.

The application of the same procedure to the Euler equation (8.68) gives

$$\frac{1}{2}\frac{d\eta_n}{d\tau} = -\frac{ik_n}{2\gamma}\left[\int_0^1 (\eta^2 - \varsigma^2)\cos(k_n\xi)d\xi\right]_{\omega_n} \tag{8.73}$$

Addition of (8.72) and (8.73) provides the equation for the amplitude of oscillations

$$\frac{1}{2}\frac{d\eta_n}{d\tau} = \left\{-\varsigma\cos(k_n\xi)|_{\xi=1} + \varsigma|_{\xi=0} + \frac{i\omega_n}{2\gamma}\int_0^1 [(\gamma-1)\eta^2 + 2\varsigma^2]\cos(k_n\xi)d\xi\right\}_{\omega_n} \tag{8.74}$$

The three terms on the right-hand side correspond to contributions from the nozzle, the burning propellant surface, and the acoustic nonlinearity.

The problem contains two small parameters: Mach number and oscillation amplitude. In further analysis terms of the order of $M\eta_k$ and $\eta_k\eta_i$ are retained.

Using the boundary condition at the nozzle (8.65) the first term on the right-hand side of Eq. (8.74) can be found

$$\{-\varsigma\cos(k_n\xi)|_{\xi=1}\}_{\omega_n} = -M\frac{(\gamma-1)}{2}\eta_n \tag{8.75}$$

The third term can be calculated using the terms of the expansion (8.71) not containing the Mach number. Simple calculations provide the following result

$$\left\{\frac{i\omega_n}{2\gamma}\int_0^1 [(\gamma-1)\eta^2 + 2\varsigma^2]\cos(k_n\xi)d\xi\right\}_{\omega_n} = \frac{in\pi(\gamma+1)}{2\gamma}\left(\sum_{s=1}^{n-1}\eta_s\eta_{n-s} + 2\sum_{s=1}^{\infty}\overline{\eta}_s\eta_{n+s}\right) \tag{8.76}$$

where the over-bar on the right-hand side denotes the complex conjugate.

Terms of the order of $MU_{k,s}\eta_k\eta_s$ appear in the course of calculation of the second term on the right-hand side of Eq. (8.74). They contain the nonlinear response function of the burning rate to oscillating pressure $U_{k,s}$. The magnitude of this function may be sufficiently large (Novozhilov 2006) and therefore such the terms must also be retained in further calculations.

The following relation may be found (in the adopted approximation) from the conditions (8.65) between variations of the gas velocity at the fuel surface ς, the linear burning rate v, and the pressure η

$$\varsigma = \gamma M\left(v - \frac{\eta}{\gamma}\right) \tag{8.77}$$

Novozhilov (2006) demonstrated that

$$v_n = U_n\eta_n + \sum_{s=1}^{n-1} U_{s,n-s}\eta_s\eta_{n-s} + \sum_{s=1}^{\infty} U_{n+s,-s}\overline{\eta}_s\eta_{n+s} \tag{8.78}$$

where U_s and $U_{k,s}$ are the linear and quadratic response functions of burning rate to harmonically varying pressure (Sections 4.1 and 4.3).

Using expression (8.78)

$$\{\varsigma|_{\xi=0}\}_{\omega_n} = M\left[(\gamma U_n - 1)\eta_n + \gamma\left(\sum_{s=1}^{n-1} U_{s,n-s}\eta_s\eta_{n-s} + \sum_{s=1}^{\infty} U_{n+s,-s}\overline{\eta}_s\eta_{n+s}\right)\right] \tag{8.79}$$

The results (8.75), (8.76), and (8.79) give the final form of the equation for the complex amplitude of pressure oscillations (m is the number of the considered harmonic)

$$\frac{d\eta_n}{d\tau} = L_n\eta_n + \sum_{s=1}^{n-1} H_{n,s}\eta_s\eta_{n-s} + \sum_{s=1}^{m-n} G_{n,s}\overline{\eta}_s\eta_{n+s}$$

$$n = 1, 2, \ldots, m \qquad L_n = M\left(\gamma U_n - \frac{(\gamma+1)}{2}\right)$$

$$H_{n,s} = \frac{in\pi(\gamma+1)}{8\gamma} + M\gamma U_{n-s,s}$$

$$G_{n,s} = \frac{in\pi(\gamma+1)}{4\gamma} + M\gamma U_{n+s,-s} \tag{8.80}$$

Linear and quadratic response functions for the combustion model (8.42) may be expressed as (Novozhilov 2006)

$$U_{s,l} = \left(1 - \frac{\delta_{s,l}}{2}\right) U_{s+l} \times \left[\frac{U_s U_l}{\iota}(kK_{s,l} + rR_{s,l} - 1) + U_s + U_l - 1\right]$$

$$K_{s,l} = \frac{(s+l)}{z_{s+l}sl}\left(\frac{(s+l)}{z_{s+l}} - \frac{s}{z_s} - \frac{l}{z_l}\right)$$

$$R_{s,l} = \frac{(s+l)^2}{sl} - \frac{(s+l)}{z_{s+l}sl}(sz_s + lz_l) + z_{s+l} - 1 \tag{8.81}$$

where $\delta_{s,\,l}$ is Kronecker delta, and the characteristic roots, found by Novozhilov (1965a) are expressed as

$$U_n = \frac{\iota}{1 + (z_n - 1)(r - k/z_n)}$$

$$z_n = \frac{1}{2}\left(1 + \sqrt{1 + 4i\frac{n\pi}{\chi}}\right) \tag{8.82}$$

The real part of the coefficient L_n controls the oscillation dynamics in a linear approximation. If it is positive, then oscillations will amplify with time; in the opposite case oscillations are being damped.

Figure 8.21 shows the ratio of damping decrement to Mach number for the first four harmonics as a function of apparatus constant. For $\chi = 0.25$ (point A) all the harmonics are stable in the linear approximation. For $\chi = 2/3$ (point B) the first harmonic is unstable in the linear approximation.

In order to obtain the solution of Eq. (8.80) the number of harmonics m must be fixed.

The initial condition is set as a perturbation of the first harmonic. For a numerical solution of the full set of equations (8.63)–(8.65), this condition is written in the form

$$\varsigma(\xi, 0) = Z\sin(\pi\xi) \tag{8.83}$$

where Z is perturbation amplitude. Using the approximate method, the initial condition is written as

$$\eta_1(0) = \frac{i}{2}Z \tag{8.84}$$

The results for the two cases are discussed below. In both cases $\iota = 0.7$, $k = 1.8$, $r = 0.35$, and $\gamma = 1.25$. Furthermore, in case A $\chi = 0.25$, $M = 4\times10^{-3}$, and $Z = 4\times10^{-3}$, while in case B $\chi = 2/3$, $M = 2\times10^{-3}$, and $Z = 5\times10^{-3}$.

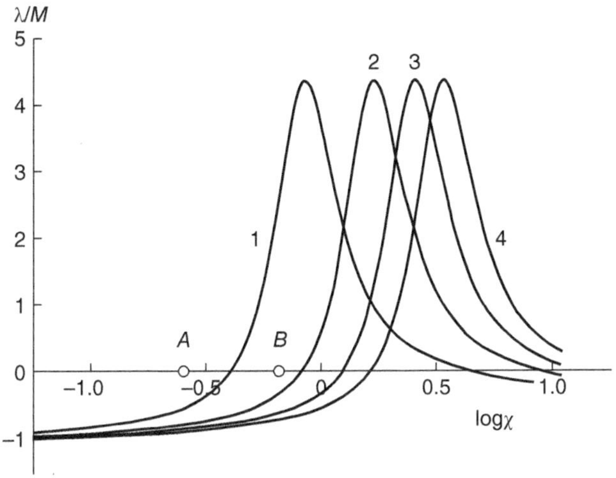

Figure 8.21 Dependence of damping decrement of oscillations on apparatus constant. $\iota = 0.7$, $k = 1.8$, $r = 0.35$, $\gamma = 1.25$. Source: Reproduced with permission from Novozhilov et al. (2009b).

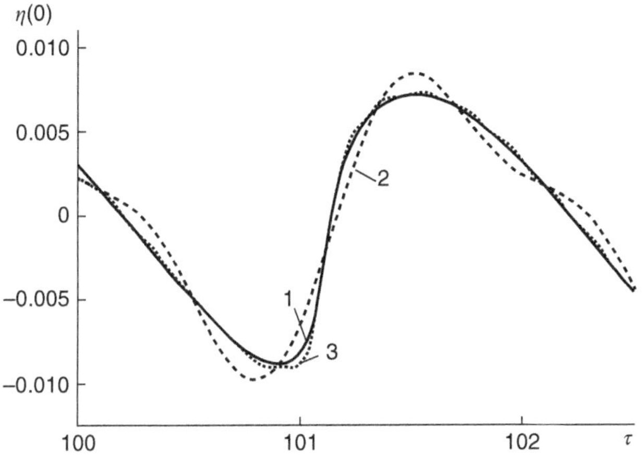

Figure 8.22 Comparison of approximate method with the numerical solution of the PDE set (8.63)–(8.65). Case *B*. 1, numerical PDE solution; 2, approximate method with two harmonics; 3, approximate method with 10 harmonics. Source: Reproduced with permission from Novozhilov et al. (2009b).

Figure 8.22 compares the results of numerical integration of the full set of equations (8.63)–(8.65) with results obtained by the approximate method. More details on the first (PDE integration) approach are given by Novozhilov et al. (2009b). It is apparent that the use of just two harmonics already gives an approximate solution with deviation from the full numerical solution within about 20%. Application of 10 harmonics produces an almost identical result.

The next two figures, 8.23 and 8.24, illustrate the time history of pressure perturbations for case A. They are plotted in different time scales. Figure 8.23 shows the full picture of individual oscillation periods which cannot be resolved. Figure 8.24 presents oscillation

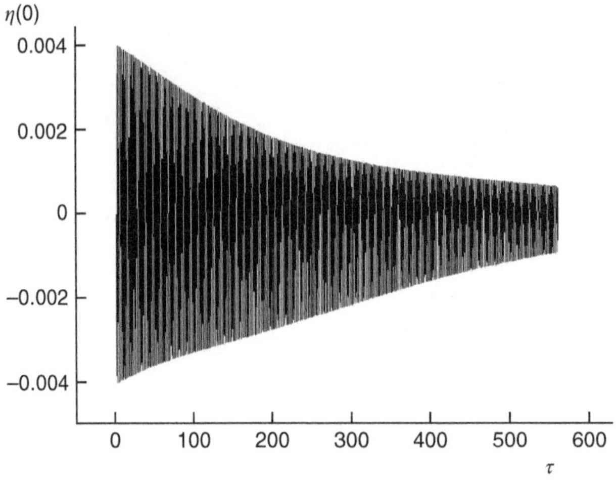

Figure 8.23 Variation of pressure oscillation amplitude with time. Case *A*. Source: Reproduced with permission from Novozhilov et al. (2009b).

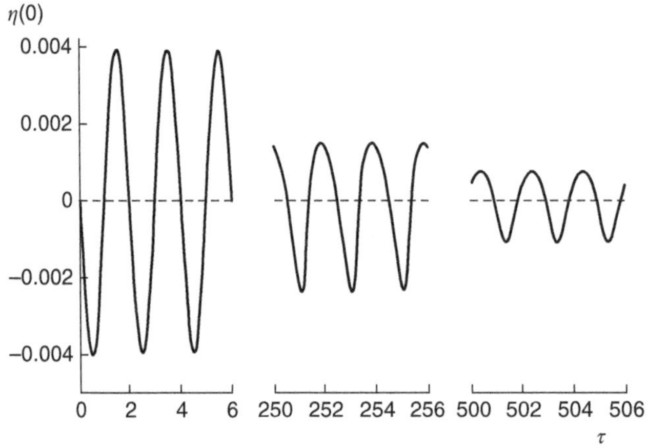

Figure 8.24 Evolution of oscillation profile with time. Case *A*. Source: Reproduced with permission from Novozhilov et al. (2009b).

evolution in different time intervals. The nonlinear nature of the oscillations can be clearly observed.

Similar plots for case B are presented in Figures 8.25 and 8.26. In this case oscillation amplitude grows with time, and sharp fronts of pressure perturbations appear. Shock waves may develop as a result of this process. The development of sharp fronts results in reduced accuracy of numerical PDE integration. In the approximate method this phenomenon is manifested by the rising prominence of higher harmonics.

The development of sharp pressure perturbation fronts is illustrated further in Figure 8.27. The upper part shows the reflection of the pressure wave from the nozzle; the bottom part shows the reflection from the burning surface of the propellant.

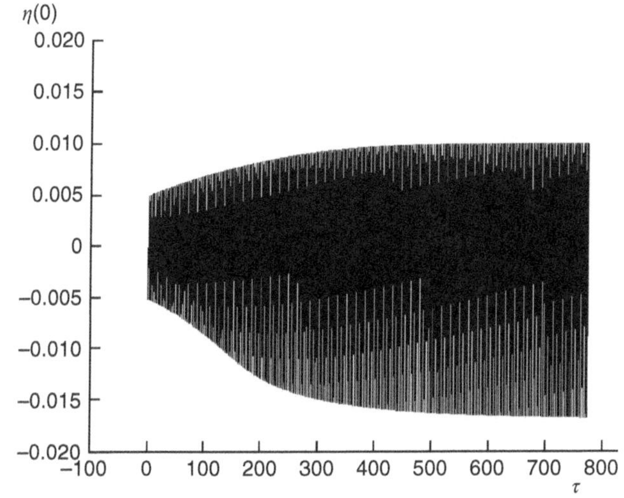

Figure 8.25 Variation of pressure oscillation amplitude with time. Case *B*. Source: Reproduced with permission from Novozhilov et al. (2009b).

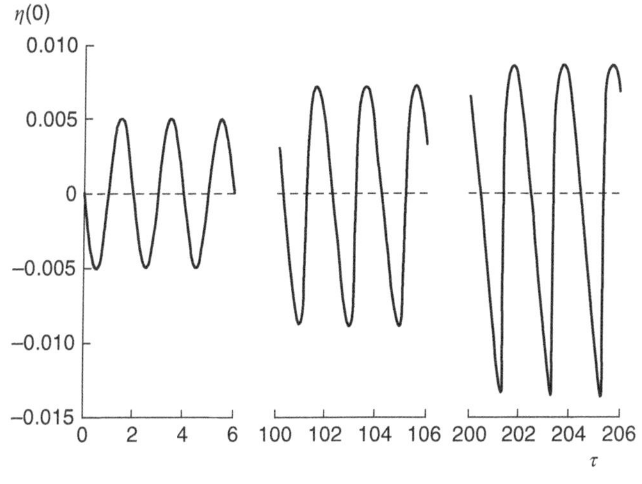

Figure 8.26 Evolution of oscillation profile with time. Case *B*. Source: Reproduced with permission from Novozhilov et al. (2009b).

8.8 Automatic Control of Propellant Combustion Stability in a Semi-enclosed Volume

This section discusses the possibility of the automatic control of nonsteady propellant combustion processes in a semi-enclosed volume. In particular, the position of the stability boundary of the steady-state regime may be changed in a controlled meaner. This boundary was determined in Section 8.3, and it was shown that steady-state regime becomes unstable for small values of the apparatus constant $\chi = t_{ch}/t_c$. As in Sections 8.2–8.5, this section

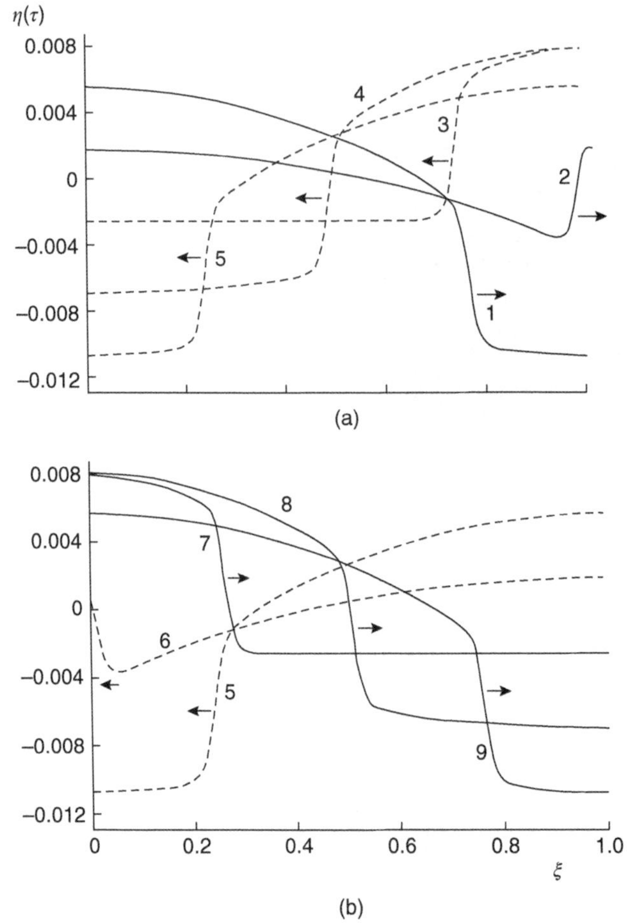

Figure 8.27 Dynamics of the pressure perturbation front. Case *B*. τ: 1, 170.00; 2, 170.25; 3, 170.50; 4, 170.75; 5, 171.00; 6, 171.25; 7, 171.50; 8, 171.75; 9, 172.00. Source: Reproduced with permission from Novozhilov et al. (2009b).

shows the condensed phase, and the chamber relaxation times are assumed to be comparable, i.e. nonacoustic regimes are being considered.

The importance of investigations of methods of automatic control of chamber pressure is underpinned by the need to ensure stable engine operations during quasi-steady-state and transient combustion regimes.

The most straightforward approach to altering the combustion regime in a semi-enclosed volume is to vary the cross-sectional area of the nozzle. In the process of automatic control dependence of the cross-sectional nozzle area on time must be determined by a control signal. The latter is determined by a control law. From a design point of view, the simplest control signal is derived from comparison of the steady-state value of pressure with a certain parameter of the process (at a current or earlier time instant) which is related to variable chamber pressure.

Existing studies on the automatic control of solid propellant engines are based on using the so-called transfer functions for a description of the combustion process, for example Prisnyakov (1984), Bobylev (1992), and Ivanov and Tsukanov (2000). Such an approach is not justified from a fundamental point of view and has not been particularly successful.

A consistent approach is based on the combination of methods of automatic control with consideration of nonsteady propellant combustion within the framework of the ZN theory.

The following analysis presents the theory of linear automatic control of the stability boundary of the steady-state propellant combustion regime in a semi-enclosed volume for different control laws.

Along with the ZN theory, the isothermal approximation (Section 8.2) is used to describe the behaviour of combustion products (Novozhilov 2010). In contrast to the numerous problems of classical control theory related to systems with a finite number of degrees of freedom, the present analysis considers the dynamical system with distributed parameters.

As mentioned earlier, the simplest method of automatic combustion regime control in a semi-enclosed volume is variation of the cross-sectional area of the nozzle.

It is assumed that apart from the constant component of the nozzle cross-sectional area s_0 there is also a time-dependent variable component $s_1(t)$. In such a case, the mass balance (8.7) transforms into

$$\frac{\mathrm{d}(\rho_g V)}{\mathrm{d}t} = \rho_c uS - A(T_b^0)p[s_0 + s_1(t)] \tag{8.85}$$

and its nondimensional form (8.14) transforms into

$$\chi\frac{\mathrm{d}\eta}{\mathrm{d}\tau} = \nu - \eta[1 + s(\tau)] \tag{8.86}$$

where the relative change in the nozzle cross-sectional area is given by

$$s(\tau) = \frac{s_1(\tau)}{s_0} \tag{8.87}$$

In the process of automatic control the function $s(\tau)$ must be determined by the control signal, which is determined in turn by the type of control law. The following analysis considers the three types of simple control laws, namely the proportional, the proportional-integral, and the proportional-differential (Popov 2014).

The results are compared with the case of constant nozzle cross-sectional area (Section 8.3). The characteristic equation for frequency in the latter case has the form (8.23).

Proportional control law assumes the comparison of the steady-state pressure value p^0 with the chamber pressure at some earlier time instant $p(t - t_d)$. The corresponding relative change in the nozzle cross-sectional area is written in the form

$$\frac{s_0 + s_1(t)}{s_0} = \hat{\zeta}\frac{p(t - t_d) - p^0}{p^0} \tag{8.88}$$

Here $\hat{\zeta}$ is a nondimensional proportionality coefficient. This property may be called a degree of the nozzle response to varying pressure or the regulator (i.e. nozzle) gain coefficient.

By writing (8.88) in nondimensional form and substituting into (8.86), the following equation describing the proportional control law is obtained

$$\chi\frac{\mathrm{d}\eta}{\mathrm{d}\tau} = \nu - \eta\{1 + \hat{\zeta}[\eta(\tau - \tau_d) - 1]\} \tag{8.89}$$

In a linear approximation when all the time-dependent variables are of the form

$$\phi = 1 + [\phi_1 \exp(i\omega\tau) + c.c.] \tag{8.90}$$

the latter equation gives

$$(1 + i\chi\omega)\eta_1 - \nu_1 = -\hat{\zeta} \exp(-i\omega\tau_d)\eta_1 \tag{8.91}$$

Under nonsteady combustion conditions, the burning rate response to varying pressure $U = \nu_1/\eta_1$ is given by

$$U = \frac{\iota + \delta(z-1)}{1 + (z-1)(r - k/z)}, z = \frac{1}{2}(1 + \sqrt{1 + 4i\omega}) \tag{8.92}$$

and therefore the characteristic equation for frequency is

$$1 + i\chi\omega + \hat{\zeta} \exp(-i\omega\tau_d) = \frac{\iota + \delta(z-1)}{1 + (z-1)(r - k/z)} \tag{8.93}$$

This equation is a generalization of Eq. (8.23) for the case of proportional control law.

In a particular case the time delay of the control signal, t_d, may be equal to zero. Then Eq. (8.93) simplifies to

$$1 + i\chi\omega + \hat{\zeta} = \frac{\iota + \delta(z-1)}{1 + (z-1)(r - k/z)} \tag{8.94}$$

Figure 8.28 shows stability boundaries for different values of the nozzle gain coefficient $\hat{\zeta}$. It is evident that sign of the parameter $\hat{\zeta}$ does not determine unambiguously whether the regime becomes more stable or less stable. For example, for $\hat{\zeta} > 0$ (curves 2 and 3) and for small values of the parameter k, automatic control widens the region of stability as these curves are below curve 1, which corresponds to the absence of any control. For $k \sim 1.8$ the situation is the opposite. For $k > 1.8$, the region of stability narrows as curves 2 and 3 appear in this case above curve 1.

This behaviour is related to the fact that chamber pressure changes for the two reasons: first, due to changes in the cross-sectional area of the nozzle, and, second, due to the response of the burning rate to varying pressure. In the considered example, $\hat{\zeta} > 0$, pressure growth above the steady-state value forces the nozzle cross-sectional area to increase. If the burning rate does not change under increased pressure, the chamber pressure drops. In reality, the burning rate may either increase or decrease. In the latter case (increased nozzle cross-sectional area and reduced burning rate), the chamber pressure would drop. However, the effect of increased burning rate may become stronger than the effect of increased nozzle area. In this case, automatic control fails to stabilize the process and the region of stability narrows ($k > 1.8$). It is natural to expect that the dynamics of the considered system with distributed parameters under control conditions is more complicated compared to systems with finite numbers of degrees of freedom (considered normally in classical control theory).

It was noted in Section 8.3 that the area of stability of the steady-state regime in a semi-enclosed volume reduces with growth of the exponent ι describing the power dependence of the burning rate on pressure. For $\iota \geq 1$ the combustion regime is only stable at $\chi \to \infty$ (i.e. at constant pressure).

Using the proportional control law stable steady-state regimes may also be achieved at $\iota \geq 1$ for finite values of the apparatus constant χ. Examples of stability boundaries for $\iota \geq 1$

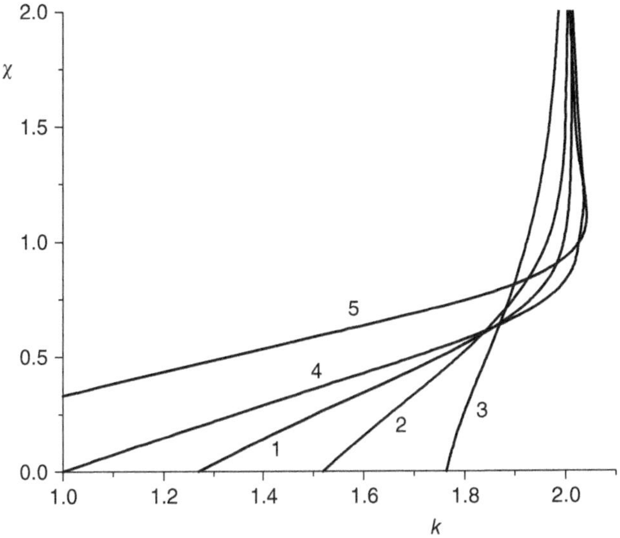

Figure 8.28 Stability boundaries of the steady-state combustion regime in a semi-enclosed volume under the proportional control law. $\tau_d = 0$, $r = 1/3$, $\iota = 0.5$, $\delta = 0.\,1$, $\zeta = 0$; 2, $\zeta = 0.5$; 3, $\zeta = 2.0$; 4, $\zeta = -0.25$; 5, $\zeta = -0.499$.

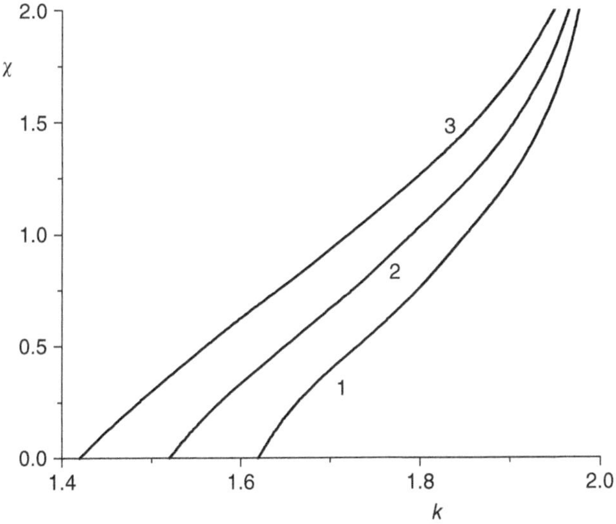

Figure 8.29 Stability boundaries for large values of the exponent ι. $\tau_d = 0$, $r = 1/3$, $\delta = 0.\,1$, $\iota = 0.8$; 2, $\iota = 1.0$; 3, $\iota = 1.2$.

are given in Figure 8.29. For comparison, the stability boundary for the case when $\iota < 1$ is also shown.

Under the proportional-integral control law the steady-state pressure value p^0 is compared with pressure averaged over a certain time interval t_i.

Consequently, the relative change in the nozzle cross-sectional area is

$$\frac{s_1(t)}{s_0} = \frac{\hat{\zeta}}{t_i}\int_{t-t_d-t_i}^{t-t_d} \frac{p(t') - p^0}{p^0} dt' \tag{8.95}$$

and, in nondimensional form,

$$s(\tau) = \hat{\zeta}\int_{\tau-\tau_d-\tau_i}^{\tau-\tau_d} [\eta(\tau') - 1]d\tau', \tau_i = \frac{t_i(u^0)^2}{\kappa} \tag{8.96}$$

The simplest consideration corresponds to the case of a no signal delay, i.e. $\tau_d = 0$. Applying (8.96), Eq. (8.86) is written in the form

$$\chi\frac{d\eta}{d\tau} = \nu - \eta\left\{1 + \hat{\zeta}\int_{\tau-\tau_i}^{\tau} [\eta(\tau') - 1]d\tau'\right\} \tag{8.97}$$

In the linear approximation (8.90) the latter equation gives

$$(1 + i\chi\omega)\eta_1 - \nu_1 = -\hat{\zeta}\frac{(1 - \exp(-i\omega\tau_i))}{i\omega\tau_i}\exp(-i\omega\tau_d)\eta_1 \tag{8.98}$$

This result is combined with the burning rate response to varying pressure (8.92) to provide the following characteristic equation for complex frequency in the case of the proportional-integral control law

$$1 + i\chi\omega + \hat{\zeta}\frac{(1 - \exp(-i\omega\tau_i))}{i\omega\tau_i}\exp(-i\omega\tau_d) = \frac{\iota + \delta(z - 1)}{1 + (z - 1)(r - k/z)} \tag{8.99}$$

If the averaging time interval is zero, $\tau_i = 0$, then this equation transforms into Eq. (8.94) corresponding to the proportional control law.

Figure 8.30 shows the stability boundaries for different values of the time averaging interval. For comparison, the stability boundary corresponding to $\tau_i = 0$ is also presented. It is completely understandable that the proportional-integral control law gives worse results compared to the proportional law.

The proportional-differential control law assumes the relation of cross-sectional nozzle area variation with pressure time derivate taken at some earlier time instant $t - t_d$

$$\frac{s_1(t)}{s_0} = \frac{\tilde{\zeta}}{p^0}\left.\frac{dp(t')}{dt'}\right|_{t'=t-t_d} \tag{8.100}$$

where $\tilde{\zeta}$ is a constant having units of time. The nondimensional form of (8.100) is

$$s(\tau) = \hat{\zeta}_1\left.\frac{d\eta(\tau')}{d\tau'}\right|_{\tau'=\tau-\tau_d}, \hat{\zeta}_1 = \tilde{\zeta}\frac{(u^0)^2}{\kappa} \tag{8.101}$$

In the linear approximation, applying exactly the same technique as in the two previous cases, the following equations are successively derived

$$[1 + i\omega(\chi + \hat{\zeta}_1\exp(-i\omega\tau_d))]\eta_1 - \nu_1 = 0 \tag{8.102}$$

$$1 + i\omega(\chi + \hat{\zeta}_1\exp(-i\omega\tau_d)) = \frac{\iota + \delta(z - 1)}{1 + (z - 1)(r - k/z)} \tag{8.103}$$

The characteristic equation (8.103) means that in the absence of signal delay, the apparatus constant is changed by an additive value that is equal to the nozzle gain coefficient. Figure 8.31 demonstrates the stability boundaries for different values of this coefficient.

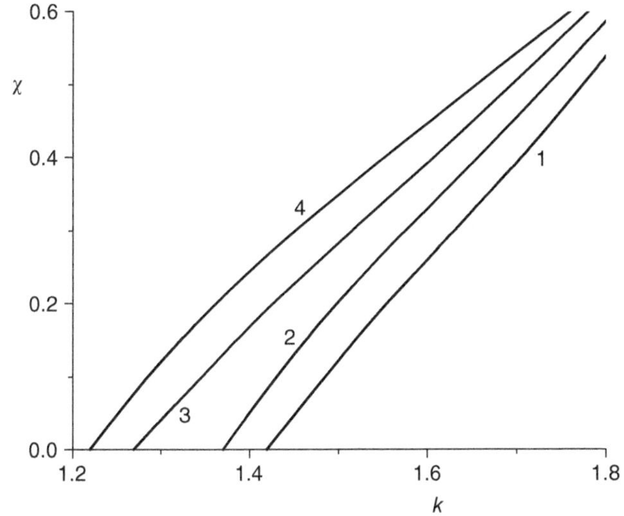

Figure 8.30 Stability boundaries for the proportional-integral control law. $\tau_d = 0$, $r = 1/3$, $\iota = 0.5$, $\hat{\zeta} = 0.25$. 1, $\tau_i = 0.0$; 2, $\tau_i = 0.5$; 3, $\tau_i = 1.0$; 4, $\tau_i = 2.0$.

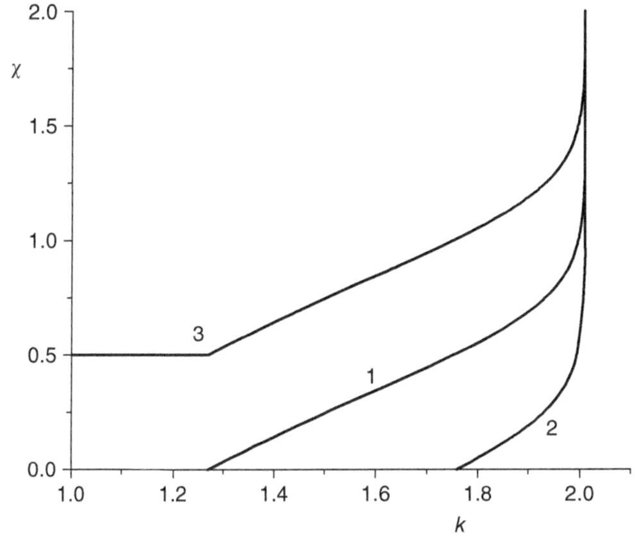

Figure 8.31 Stability boundaries for the proportional-differential control law. $\tau_d = 0$, $r = 1/3$, $\iota = 0.5$, $\delta = 0$. 1, $\hat{\zeta}_1 = 0.0$; 2, $\hat{\zeta}_1 = 0.5$; 3, $\hat{\zeta}_1 = -0.5$.

So far, stability boundaries have been plotted only for the simplest case of trivial time delay, $\tau_d = 0$. To find a stability boundary for $\tau_d \neq 0$ is more difficult, as the rapidly oscillating term $\exp(-i\omega\tau_d)$ appears in a characteristic equation for all three types of considered control laws.

For $\tau_d = 0$, the imaginary part of the expression for the apparatus constant has just two roots, and $\mathrm{Re}\chi > 0$ corresponds to one of those giving the position of the stability boundary.

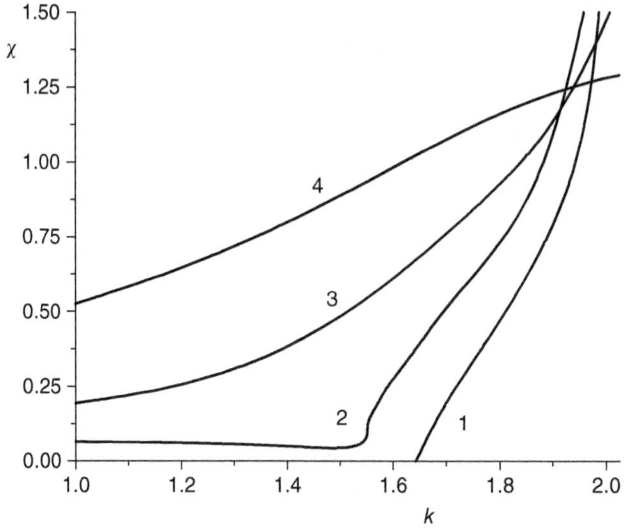

Figure 8.32 Influence of the parameter τ_d on the stability boundary under the proportional control law. $r = 1/3$, $\imath = 0.5$, $\delta = 0$, $\hat{\zeta} = 1.0$. 1, $\tau_d = 0$; 2, $\tau_d = 0.25$; 3, $\tau_d = 0.5$; 4, $\tau_d = 1.0$.

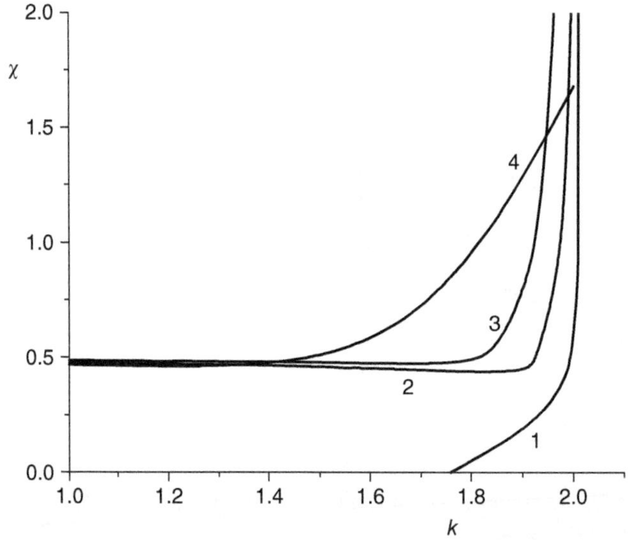

Figure 8.33 Influence of the parameter τ_d on the stability boundary under the proportional-differential control law. $r = 1/3$, $\imath = 0.5$, $\delta = 0$, $\hat{\zeta}_1 = 0.5$. 1, $\tau_d = 0$; 2, $\tau_d = 0.25$; 3, $\tau_d = 0.5$; 4, $\tau_d = 1.0$.

For $\tau_d \neq 0$, the imaginary part of the apparatus constant has many roots, and some may have Re $\chi > 0$. It is obvious that the largest value of the real part of the apparatus constant corresponds to the stability boundary. Therefore, different values of the real part of the apparatus constant must be compared. In the cases considered in the present study only two or three roots need to be compared as the real parts of the apparatus constant decrease quickly with an increased root number.

Examples accounting for the signal time delay in the proportional and proportional-differential laws are considered. For the proportional law, from (8.93)

$$\chi = \frac{1}{i\omega}\left(\frac{\imath + \delta(z-1)}{1 + (z-1)(r - k/z)} - 1 - \hat{\zeta}\exp(-i\omega\tau_d)\right) \tag{8.104}$$

Stability boundaries for different time delays are plotted in Figure 8.32.

For the proportional-differential law, from the characteristic equation (8.103)

$$\chi = \frac{1}{i\omega}\left(\frac{\imath + \delta(z-1)}{1 + (z-1)(r - k/z)} - 1 - i\omega\hat{\zeta}_1\exp(-i\omega\tau_d)\right) \tag{8.105}$$

The stability boundaries for this law for different time delays are shown in Figure 8.33.

Comparing Figures 8.32 and 8.33 with their counterparts Figures 8.28 and 8.31, it should be noted that accounting for time delay $\tau_d \neq 0$ leads to not only quantitative but also qualitative changes. For example, instead of smooth curves in Figure 8.28, curves with a sudden change of the first derivative appear (curve 2, Figure 8.32). In the case of the proportional-differential law, Figure 8.33, essentially constant values of the apparatus constant (within a certain range of variation of the parameter k) are followed by its rapid increase.

Such peculiarities are related to the fact that variation of the parameter k changes the roots of the function Im $\chi(\omega)$, and these roots actually determine the stability boundary.

9

Influence of Gas-phase Inertia on Nonsteady Combustion

9.1 Introduction

In this chapter, we present an extension of the theory allowing for the consideration of inertia of the gas phase. First, in this section, motivations for such an extension and the cases for which it may be necessary are discussed.

The development and applications of the Zeldovich–Novozhilov (ZN) theory within the t_c approximation, discussed in previous chapters, relied on the validity of the following inequalities (Section 2.1):

$$t_c \gg t_{cr}, t_c \gg t_g; t_p \gg t_{cr}, t_p \gg t_g \tag{9.1}$$

It is necessary to realize, however, that even the fulfilment of all of the conditions (9.1) does not completely rule out the possibility of obtaining contradictory results. This may be demonstrated using the example of a steady-state burning regime at constant pressure.

As discussed in Section 3.1 (formula (3.32)), in the vicinity of the stability boundary small perturbations of the burning rate behave as

$$v(\tau) = 1 + v_1 \exp(-\lambda\tau)\cos(\omega\tau + \psi) \tag{9.2}$$

that is, they oscillate with either increasing ($\lambda < 0$) or decreasing ($\lambda > 0$) amplitude, depending on whether the parameters of the system belong to the unstable or stable region.

Here (Section 3.1, formula (3.28))

$$\lambda = \frac{r(k+1) - (k-1)^2}{2r^2}, \qquad \omega = \sqrt{\frac{k}{r^2} - \lambda^2} \tag{9.3}$$

These results are obtained under the assumption that $t_{cr} = t_g = 0$. However, they cannot be considered correct at high decrements of damping or at high frequencies.

In fact, for $\lambda \gg 1$ or $\omega \gg 1$ it follows from (9.2) and (9.3) (recalling the time scaling (2.31)) that the intrinsic time scales of approaching the steady-state regime (system relaxation) or moving away from that regime, t_c/λ or t_c/ω, may become comparable to the relaxation times of the propellant reaction zone or the gas phase. The t_c approximation can also introduce errors for long flames, that is, for combustion regimes where chemical reactions in the gas phase terminate at a significant distance from the propellant surface.

Theory of Solid-Propellant Nonsteady Combustion, First Edition. Boris V. Novozhilov and Vasily B. Novozhilov.

Companion website: www.wiley.com/go/Novozhilov/solidpropellantnonsteadycombustion

In such situations, neglecting the thermal inertia of the latter zones is not justified. The inequalities (9.1) should be replace by more rigid ones

$$t_c/\lambda \gg t_{cr}, \quad t_c/\omega \gg t_{cr}; \quad t_c/\lambda \gg t_g, \quad t_c/\omega \gg t_g$$
$$t_p/\lambda \gg t_{cr}, \quad t_p/\omega \gg t_{cr}; \quad t_p/\lambda \gg t_g, \quad t_p/\omega \gg t_g \tag{9.4}$$

It is possible that the results of the stability analysis preformed in Section 3.1 would change significantly for large damping decrements and frequencies. Therefore, one has to realize the inherent inconsistency of the t_c approximation at $\lambda \gg 1$ and $\omega \gg 1$.

As follows from (9.2) and (9.3), the most problematic regions are those where at least one of the following inequalities holds

$$(k-1) \ll 1; r \ll 1 \tag{9.5}$$

The latter inequality is quite often fulfilled for real systems, therefore there is a need to extend the theory in such a way that relaxation times of small-inertia zones can be accounted for.

The number of studies considering the relaxation times of small-inertia zones has been very limited. Vilyunov and Rudnev (1973) and Romanov (1975) attempted to take into account the inertia of the propellant reaction layer. The influence of this effect on combustion in a semi-enclosed volume was studied by Vilyunov and Rudnev (1973), but no consistent results on burning stability under constant pressure were presented. The study by Romanov (1975) investigated the effect within the framework of numerous assumptions and numerical modelling of multiparametric problems. It is very difficult to derive any general relationships from this paper. Volkov and Medvedev (1969) took this one step further to include the thermal inertia of both the reaction layer in the condensed phase and the gas phase by introducing a phenomenological parameter, the delay time.

Very few studies have addressed the issue of gas-phase inertia by means of numerical analysis of relevant sets of time-dependent differential equations. Most (Hart and McClure 1959; T'ien 1972; Allison and Faeth 1975) focus on nonsteady propellant combustion driven by small-amplitude harmonically oscillating pressure. In particular, Hart and McClure (1959) investigated the problem in relation to acoustic admittance at high frequencies where, as just discussed, the effect of small-inertia zones has to be taken into account. T'ien (1972) and Allison and Faeth (1975) integrated numerically linearized equations governing the combustion process. In the paper by T'ien (1972), as well as in the later analytical study by Clavin and Lazimi (1992), a two-stage model of the combustion process was adopted. One stage described the chemical reaction in the condensed phase and the other described reaction in the gaseous phase. The relaxation time of the chemical reaction zone in the condensed phase was set to zero in order to simplify the problem. Such an approach is inconsistent as the effects produced by finite relaxation of both zones are comparable.

It is highly desirable to study multiscale aspects of combustion of systems with chemical reaction occurring in the gas phase. Such an attempt was made by Margolis and Williams (1988) using rather crude assumptions (constant gas density and one-dimensional geometry of flow).

Conditions for the stability of the steady-state burning regime at constant pressure, with an allowance for gas phase inertia, were first derived by Novozhilov (1988a). Essentially, the results of the analysis show (at small r) an expansion of the region of stable burning and lead to a qualitatively new expression for the natural frequency of the combustion system.

In the following sections, the effect of gas-phase inertia on combustion stability, the burning rate response to oscillating pressure, and the acoustic admittance of the propellant's surface are systematically investigated within the framework of the Belyaev model.

9.2 Steady-state Combustion Regime Stability

The first realistic model of solid propellants combustion is due to Belyaev (1938, 1940). This model was discussed in Section 1.8 in its steady-state form. The model takes into account processes occurring in the gas phase and therefore it is used in the present section (in non-steady form) to derive conditions of steady-state combustion stability with an allowance for the gas phase relaxation time.

The Belyaev model assumes that chemical transformation occurs in the vapours of liquid evaporated from the propellant surface. Evaporation is driven by heat feedback from the combustion zone; L is the latent heat of evaporation.

Assume the simplest reaction scheme $A \to P + Q_g$ with A and P being the initial propellant and reaction products, respectively, and Q_g the heat of reaction. We shall also assume that the molecular weights of the initial substance and products are close to each other so that the molecular weight $\tilde{\mu}$ of the mixture can be considered as constant.

Under the approximation of high activation energy, the thickness of the chemical reaction zone may be considered negligible in comparison with the gas preheat zone. This allows the chemical source terms in the governing equations for the gas phase to be dropped, and accounts for chemical transformation processes using appropriate boundary conditions at the flame surface. This approach is similar to the analysis of the thermo-diffusion stability of laminar gaseous flames (Barenblatt et al. 1962). Obviously, the inertia of the chemical transformation zone is neglected under this approach.

Let $x_s(t)$ and $x_f(t)$ be the positions of the liquid–gas interface and the flame front, respectively. The set of problems governing these equations includes the heat transfer equation of the condensed phase

$$\rho c \frac{\partial T}{\partial t} = \frac{\partial}{\partial x}\left(\lambda \frac{\partial T}{\partial x}\right), \quad -\infty < x < x_s(t) \tag{9.6}$$

as well as the continuity, the initial substance balance, and the heat transfer equations in the gas preheat zone.

$$\begin{aligned}
&\frac{\partial \rho_g}{\partial t} + \frac{\partial(\rho_g u_g)}{\partial x} = 0 \\
&\rho_g\left(\frac{\partial Y}{\partial t} + u_g \frac{\partial Y}{\partial x}\right) = \frac{\partial}{\partial x}\left(D\rho_g \frac{\partial Y}{\partial x}\right), \quad x_s(t) < x < x_f(t) \\
&\rho_g c_p\left(\frac{\partial T_g}{\partial t} + u_g \frac{\partial T_g}{\partial x}\right) = \frac{\partial}{\partial x}\left(\lambda_g \frac{\partial T_g}{\partial x}\right)
\end{aligned} \tag{9.7}$$

The half-space $x_f(t) < x < \infty$ is filled with combustion products, the initial substance concentration Y is equal to zero in this region, and the corresponding heat transfer equation is identical to (9.7) (with the obvious replacement of T_g and u_g by the temperature and the velocity of products T_p and u_p).

The notation in the governing equations is conventional for combustion theory. T stands for temperature, Y for mass fraction of reactant, ρ_g and u_g for density and velocity of gas, respectively, and c for specific heat. D and λ_g denote the gas diffusion coefficient and thermal conductivity, respectively.

Let us discuss the boundary conditions. The initial state of propellant is described by the initial temperature

$$x \to -\infty, \quad T = T_a \tag{9.8}$$

The following conditions must be imposed at the phase interface $x = x_s(t)$: overall mass balance and mass balance of the reactant, continuity of the temperature field, energy balance, and the condition of equilibrium evaporation, which relates temperature and concentration. These conditions have the following form

$$\begin{aligned} &-\rho\frac{\mathrm{d}x_s}{\mathrm{d}t} = -\rho_g\frac{\mathrm{d}x_s}{\mathrm{d}t} + \rho_g u_g, \quad -\rho\frac{\mathrm{d}x_s}{\mathrm{d}t} = -\rho_g Y\frac{\mathrm{d}x_s}{\mathrm{d}t} + \rho_g u_g Y - D\rho_g\frac{\partial Y}{\partial x} \\ &T_g = T, \quad \lambda\frac{\partial T}{\partial x} = \lambda_g\frac{\partial T_g}{\partial x} + \rho L\frac{\mathrm{d}x_s}{\mathrm{d}t} \\ &Y = \exp\left[-\frac{L\tilde{\mu}}{R}\left(\frac{1}{T} - \frac{1}{T_{bl}}\right)\right] \end{aligned} \tag{9.9}$$

Here T_{bl} is the boiling temperature at a given pressure and R is the universal gas constant.

At the flame front $x = x_f(t)$ the reactant concentration becomes zero, and velocity and temperature are continuous. In addition, there are conditions of temperature flux jump discontinuity and complete consumption of the original substance dictated by an infinitely fast chemical reaction

$$\begin{aligned} &Y = 0, \quad u_g = u_p, \quad T_g = T_p \\ &\lambda_g\frac{\partial T_g}{\partial x} = \lambda_g\frac{\partial T_p}{\partial x} - D\rho_g Q_g\frac{\partial Y}{\partial x} \\ &-D\rho_g\frac{\partial Y}{\partial x} = m \end{aligned} \tag{9.10}$$

where m is the mass burning rate, considered to be dependent on the flame temperature only. The gas is considered to be ideal, therefore $\rho_g T_g = \rho_{bl} T_{bl}$ where ρ_{bl} is the density at boiling temperature.

Two circumstances lead to the possibility of analytical investigation of the stability problem with an allowance for inertia of both the condensed and the gas phases (inertia of the chemical reaction zone, as already mentioned, is neglected). First, the transformation to Lagrangian variables decouples the hydrodynamic part of the problem from thermodiffusion. The advantages of the Lagrangian coordinates were demonstrated by Shkadinskii (1971) while solving numerically the problem of ignition of volatile substances and subsequent transition to a steady-state combustion regime.

Second, physically reasonable assumptions of the parameters of the condensed phase (i.e. ρ, c, λ and consequently thermal diffusivity κ) being constant, as well as independence of the product $D\rho_g^2$ of gas temperature ensure the linearity of the relevant equations in the considered approximation (infinitely thin chemical reaction zone). On the other hand, unperturbed steady-state solutions are of such a form that solutions of nonhomogeneous linear equations for perturbations may be found analytically.

To reduce the number of parameters, in the following analysis $c \equiv c_p$ and $Le = 1$ ($D\rho_g^2 = \lambda_g \rho_g / c_p$). These assumptions do not limit the generality of the analysis as the problem can be equally solved without these restrictions.

Along with transformation to Lagrangian coordinates, the following nondimensional variables and parameters are introduced

$$\theta = \frac{T}{T_{bl}}, \quad \theta_g = \frac{T_g}{T_{bl}}, \quad \theta_p = \frac{T_p}{T_{bl}}, \quad \theta_0 = \frac{T_a}{T_{bl}}$$
$$v = -\frac{\rho}{m^0}\frac{\mathrm{d}x_s}{\mathrm{d}t}, \quad v_g = \frac{\rho_b u_g}{m^0}, \quad v_p = \frac{\rho_b u_p}{m^0}$$
$$\sigma = \frac{D\rho_g^2}{\kappa \rho^2}, \quad q = \frac{Q_g}{cT_{bl}}, \quad l = \frac{L}{cT_{bl}}, \quad \Gamma = \frac{\tilde{\mu} c}{R} \tag{9.11}$$

Here v is the nondimensional burning rate, and the parameter $\sigma = t_g/t_c$ is the ratio of relaxation times of the gas and condensed phases ($t_g = D\rho_g^2/(m^0)^2$, $m^0 = \rho u^0$).

Nondimensional time and the new (Lagrangian) variables are introduced as follows:

$$\tau = \frac{(u^0)^2 t}{\kappa}, \quad -\infty < x < x_s(t), \quad \xi = \left(\frac{u^0}{\kappa}\right)(x - x_s(t)), \quad \xi < 0$$
$$x_s(t) < x < \infty, \quad \xi = \frac{1}{\rho\sigma}\frac{u^0}{\kappa}\int_{x_s(t)}^{x} \rho_g(y,t)\mathrm{d}y, \quad \xi > 0 \tag{9.12}$$

It is convenient to also introduce the operator

$$\hat{H} = \frac{\partial}{\partial \tau} + v(\tau)\frac{\partial}{\partial \xi} - \frac{\partial^2}{\partial \xi^2} \tag{9.13}$$

as well as the operator $\hat{H}_\sigma$, which differs from $\hat{H}$ by a factor σ in front of the time derivative.

In the new coordinates the problem is reduced to the following set of equations

$$-\infty < \xi < 0, \quad \hat{H}(\theta) = 0$$
$$0 < \xi < \xi_f(\tau), \quad \hat{H}_\sigma(\theta_g) = 0, \quad \hat{H}_\sigma(Y) = 0$$
$$\xi_f(\tau) < \xi < \infty, \quad \hat{H}_\sigma(\theta_p) = 0, \quad Y = 0 \tag{9.14}$$

with boundary conditions

$$\theta = \theta_0, \quad \xi \to -\infty$$
$$\theta = \theta_g, \quad \frac{\partial \theta}{\partial \xi} = \frac{\partial \theta_g}{\partial \xi} - lv, \quad \xi = 0$$
$$v(1 - Y) + \frac{\partial Y}{\partial \xi} = 0, \quad Y = \exp[l\Gamma(1 - \theta^{-1})]$$
$$\xi = \xi_f(\tau), \quad Y = 0, \quad \theta_g = \theta_p, -\frac{\partial Y}{\partial \xi} = \frac{m}{m^0}$$
$$\frac{\partial \theta}{\partial \xi} = \frac{\partial \theta_p}{\partial \xi} - q\frac{\partial Y}{\partial \xi}; \quad \theta_p < \infty, \quad \xi \to \infty \tag{9.15}$$

The set (9.14) contains four second-order equations. Apart from the functions $\theta(\xi, \tau)$, $\theta_g(\xi, \tau)$, $Y(\xi, \tau)$, and $\theta_p(\xi, \tau)$, the burning rate $v(\tau)$ and the position of the flame front $\xi_f(\tau)$ also have to be found. Therefore, (9.15) contains 10 boundary conditions.

Once the solution of (9.14) and (9.15) is known, gas velocity, if needed, may be obtained from the boundary value problem (BVP)

$$\sigma\frac{\partial \theta_g}{\partial \tau}+v\frac{\partial \theta_g}{\partial \xi}-\frac{\partial v_g}{\partial \xi}=0; \quad \sigma\frac{\partial \theta_p}{\partial \tau}+v\frac{\partial \theta_p}{\partial \xi}-\frac{\partial v_p}{\partial \xi}=0$$

$$\xi=0,\quad v_g=v\left(\theta-\frac{\rho_s}{\rho}\right),\quad \xi=\xi_f(\tau),\quad v_g=v_p \tag{9.16}$$

The steady-state solution of the set (9.14) and (9.15) has the form

$$-\infty<\xi<0,\quad \theta^0=\theta_0+(\theta_s^0-\theta_0)\exp(\xi)$$
$$0<\xi<\xi_f^0,\quad \theta_g^0=(\theta_0-l)+(\theta_s^0-\theta_0+l)\exp(\xi)\quad Y^0=1-(1-Y_s^0)\exp(\xi)$$
$$\xi_f^0<\xi<\infty,\quad \theta_p^0=\theta_0+q-l\quad Y^0=0$$
$$v^0=1,\quad \xi_f^0=-\ln(1-Y_s^0) \tag{9.17}$$

Nondimensional surface temperature and reactant mass fraction at the surface in the steady-state regime can be found from the equations

$$\theta_s^0=\theta_0+q(1-Y_s^0)-l,\quad Y_s^0=\exp\left[l\Gamma\left(1-\frac{1}{\theta_s^0}\right)\right] \tag{9.18}$$

Derivatives of the burning rate and surface temperature with respect to initial temperature, in the steady-state regime, have the following form in nondimensional variables

$$k=(\theta_s^0-\theta_0)\frac{\partial \ln m^0}{\partial \theta_0},\quad r=\frac{1}{\left[1+\frac{ql\Gamma Y_s^0}{(\theta_s^0)^2}\right]} \tag{9.19}$$

An investigation of stability of the steady-state solution (9.17) in the linear approximation may be carried out by the standard procedure, that is, by imposing small perturbations varying in time as $\exp(\Omega\tau)$ and obtaining the stability boundary as a condition, $\mathrm{Re}\Omega=0$. It should be noted that such a method gives a result identical to the one obtained by investigating stability as an unsteady problem with initial conditions (Novikov and Ryazantsev 1966).

In greater detail, the major steps of the derivation of the characteristic equation are as follows.

All the variables are represented in the form $f=f^0+f_1\exp(\Omega\tau)$ where f^0 is the steady-state solution and f_1 is the small perturbation. Corrections to θ, θ_g, θ_p, Y, v, and ξ_f are denoted as ϑ, ϑ_g, ϑ_p, y, b, and s, respectively.

The linearized set of equations for the perturbations is

$$\hat{h}(\vartheta)=b\Delta\exp(\xi),\quad \hat{h}_\sigma(y)=-ba\exp(\xi)$$
$$\hat{h}_\sigma(\vartheta_g)=b(\Delta+l)\exp(\xi),\quad \hat{h}_\sigma(\vartheta_p)=0 \tag{9.20}$$

where

$$\hat{h}=\frac{d^2}{d\xi^2}-\frac{d}{d\xi}-\Omega,\quad \Delta=\theta_s^0-\theta_0 \tag{9.21}$$

and $\hat{h}_\sigma$ differs from $\hat{h}$ by the multiplier σ in front of Ω.

The solutions of these equations are

$$\begin{aligned}
\vartheta &= Aq\exp(z\xi) - b\frac{\Delta}{\Omega}\exp(\xi)\\
y &= C\exp(z_1\xi) + D\exp(z_2\xi) + \frac{ba}{\sigma\Omega}\exp(\xi)\\
\vartheta_g &= Fq\exp(z_1\xi) + Gq\exp(z_2\xi) - \frac{b(\Delta + l)}{\sigma\Omega}\exp(\xi)\\
\vartheta_p &= Hq\exp(z_2\xi)
\end{aligned} \tag{9.22}$$

(here the boundary conditions at $\xi \to \pm\infty$ have been taken into account).

The remaining eight boundary conditions are linearized, and solutions for perturbations as well as expressions for the burning rate and the position of flame front are substituted into the resulting equations.

As a result, the following set of linear homogeneous algebraic equations for the unknowns A, C, D, F, G, H, b, and s is obtained

$$A - \frac{ab\delta}{\Omega} = F + G - \frac{ab}{\sigma\Omega} \tag{9.23}$$

$$Az - \frac{ab\delta}{\Omega} = Fz_1 + Gz_2 - \frac{ab}{\sigma\Omega} - ab(1-\delta) \tag{9.24}$$

$$ab = Cz_2 + Dz_1 \tag{9.25}$$

$$C + D + \frac{ab}{\sigma\Omega} = \left(\frac{1}{r} - 1\right)\left(A - \frac{ab\delta}{\Omega}\right) \tag{9.26}$$

$$F\exp(z_1\xi_f^0) + G\exp(z_2\xi_f^0) - \frac{b}{\sigma\Omega} + s = H\exp(z_2\xi_f^0) \tag{9.27}$$

$$C\exp(z_1\xi_f^0) + D\exp(z_2\xi_f^0) + \frac{b}{\sigma\Omega} - s = 0 \tag{9.28}$$

$$Fz_1\exp(z_1\xi_f^0) + Gz_2\exp(z_2\xi_f^0) - \frac{b}{\sigma\Omega} + s = \left(z_2 + \frac{k}{a\delta}\right)H\exp(z_2\xi_f^0) \tag{9.29}$$

$$s - \frac{b}{\sigma\Omega} - Cz_1\exp(z_1\xi_f^0) - Dz_2\exp(z_2\xi_f^0) = \frac{kH\exp(z_2\xi_f^0)}{a\delta} \tag{9.30}$$

By substituting A from (9.23), ab from (9.25), $s - \frac{b}{\sigma\Omega}$ from (9.27), and $H\exp(z_2\xi_f^0)$ from (9.30) into (9.24), (9.26), (9.28), and (9.29), the variables A, b, s, and H are eliminated.

The remaining four equations are

$$\begin{aligned}
&(Cz_2 + Dz_1)\left[\frac{(z-1)}{\sigma\Omega}(a\delta - 1) + 1 - \delta\right] + F(z - z_1) + G(z - z_2) = 0\\
&r(C + D + F + G) = F + G - \frac{(Cz_2 + Dz_1)}{\sigma\Omega}\\
&(F + C)\exp(z_1\xi_f^0) + (G + D)\exp(z_2\xi_f^0) =\\
&\frac{\delta a}{k}(Cz_2\exp(z_1\xi_f^0) + Dz_1\exp(z_2\xi_f^0))\\
&(F + C)z_2\exp(z_1\xi_f^0) + (G + D)z_1\exp(z_2\xi_f^0) =\\
&\frac{\delta a}{k}z_1(Cz_2\exp(z_1\xi_f^0) + Dz_1\exp(z_2\xi_f^0))
\end{aligned} \tag{9.31}$$

The last two equations imply $F = -C$, therefore just three equations for the unknowns C, D, and G remain.

$$C\left(z_2\left[\frac{(z-1)}{\sigma\Omega}(a\delta-1)+1-\delta\right]-z+z_1\right)$$
$$+Dz_1\left[\frac{(z-1)}{\sigma\Omega}(a\delta-1)+1-\delta\right]+G(z-z_2)=0$$
$$C\left(1+\frac{z_2}{\sigma\Omega}\right)+D\left(r+\frac{z_1}{\sigma\Omega}\right)+G(r-1)=0$$
$$Cz_2\exp[(z_1-z_2)\xi_f^0]+D\left(z_1-\frac{k}{\delta a}\right)-G\frac{k}{\delta a}=0 \tag{9.32}$$

The set of equations (9.32) has a nontrivial solution if the determinant of its matrix is equal to zero. This implies

$$r\Phi(\Omega)-\frac{k\varsigma}{a}=\Psi(\Omega)$$
$$\Phi(\Omega)=z[z_1a^{-\varsigma}-z_2]-\delta z_1z_2[a^{-\varsigma}-1]$$
$$\Psi(\Omega)=-\frac{z^2}{\Omega}z_1[z_1a^{-\varsigma}-z_2]-\delta z_1z_2[a^{-\varsigma}-1] \tag{9.33}$$

where

$$2z=1+(1+4\Omega)^{1/2},\quad 2z_{1,2}=1\pm(1+4\sigma\Omega)^{1/2}$$
$$\varsigma=z_1-z_2,\qquad a=\exp(-\xi_f^0),\qquad \delta=\frac{(\theta_s^0-\theta_0)}{(\theta_s^0-\theta_0+l)} \tag{9.34}$$

The parameter $a=1-Y_s^0$ is directly related to the thickness of the heating zone in gas phase ξ_f^0, while $\delta<1$ is the ratio of the amount of heat required for the substance to heat up to the surface temperature to the total amount of heat required for heating up and evaporation.

The characteristic equation (9.33) shows that the possibility of the existence of the steady-state regime is determined by the values of the following five parameters: k, r, σ, a, and δ. The parameter k is related to burning rate sensitivity and changes in flame temperature. The value of σ only varies if pressure is variable. The other three parameters, r, a, and δ, can also be considered as independent as they include many other problem parameters (θ_0, q, L, Γ).

For real systems, the surface temperature only depends – weakly – on initial temperature, that is, $r \ll 1$. This region, as already pointed out, is of most interest when considering the influence of the inertia of the gas phase. Formally, in the framework of the obtained characteristic equation, the case $r \sim 1$ may also be considered. This case may be used, for example, to show that for $\omega \sim 1$ neglecting the inertia of the gas phase is fully justified, and to find corrections associated with this effect. The same remark applies to the case $r=0$. Although probably never realized in real systems, this equality represents a sufficiently interesting limiting case.

An investigation of stability may be conducted on the conventional (k, r) plane, providing the parameters σ, a, and δ with specific constant values.

Since at the stability boundary the frequency Ω is purely imaginary ($\Omega = i\omega$), the characteristic equation (9.33) is equivalent to two real equations. Multiplying (9.33) once by $\overline{\varsigma}$ and once by $\overline{\Phi}$ (the overbar indicates the complex conjugate), and taking imaginary parts of the products provides

$$r = \frac{\mathrm{Im}(\Psi\overline{\varsigma})}{\mathrm{Im}(\Phi\overline{\varsigma})}, \quad k = a\frac{\mathrm{Im}(\Psi\overline{\Phi})}{\mathrm{Im}(\Phi\overline{\varsigma})} \tag{9.35}$$

that is, the parametric representation $(r(\omega), k(\omega))$ of the stability boundary $r(k)$.

Figure 9.1 shows the results of calculations of the critical boundary and the frequency at this boundary according to (9.33–9.35).

Figure 9.1 shows that the stability region widens on an increase in the parameter σ. It turns out that at $r = 0$ the value of the parameter k is not equal to unity, in contrast to the original ZN theory where the inertia of the gas phase is neglected. The discrepancy from unity is rather large, for example at $\sigma = 10^{-2}$ it is about 30%.

The dashed curves in Figure 9.1 show the frequency at the stability boundary as a function of the parameter r.

It is evident that the frequency at the stability boundary is always finite (including the case $r = 0$), and for the realistic values $r \sim 0.1 - 0.2$ may differ several times from the one predicted by the original ZN theory (inertia-free gas phase).

In the considered range of values of the parameter σ, the inequality $\sigma\omega \ll 1$ always holds. This allows the characteristic equation (9.33) and the functions (9.34) to be simplified. Expansions of the functions $z_1(\sigma\omega)$ and $z_2(\sigma\omega)$ with respect to the small parameter $\sigma\omega$ lead to the following expressions

$$\begin{aligned} r &= \frac{2}{p_\omega(p_\omega - 1)} \frac{(2 - \sigma p_\omega[h(p_\omega^2 - 1) - gp_\omega(p_\omega - 1)])}{(2 + \sigma p_\omega[(h-1)(p_\omega + 1) + 2g])} \\ k &= \frac{(p_\omega + 1)}{(p_\omega - 1)} \frac{(2 - \sigma(p_\omega - 1)[p_\omega - g(p_\omega + 2)])}{(2 + \sigma p_\omega[(h-1)(p_\omega + 1) + 2g])} \end{aligned} \tag{9.36}$$

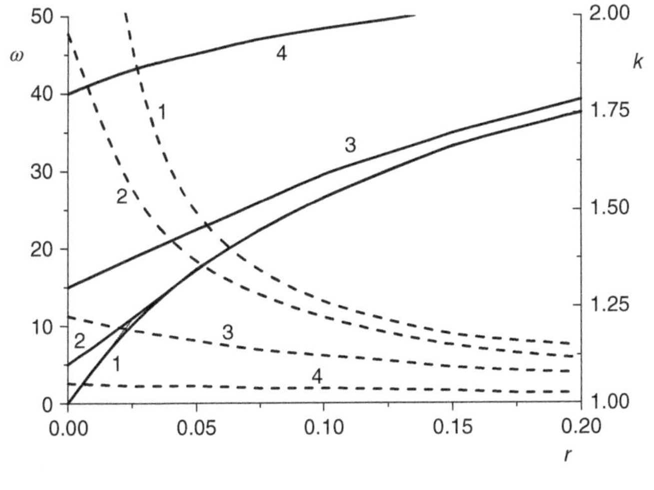

Figure 9.1 Stability boundary and frequency at the stability boundary. $a = 0.4$, $\delta = 0.15$. Solid lines, stability boundaries $k(r)$; dashed lines, frequencies at the stability boundaries as functions of the parameter r. 1, $\sigma = 0$; 2, $\sigma = 10^{-3}$; 3, $\sigma = 10^{-2}$; 4, $\sigma = 10^{-1}$.

Here, for convenience, the following new notation is introduced: $h = a - 2\ln a$, $g = \delta(1 - a)$.

Also, the function of the frequency $p_\omega = [1/2((1 + 16\omega^2)^{1/2} + 1)]^{1/2}$ is used instead of the frequency itself, so that $z = 1/2[p_\omega + 1 + i(p_\omega^2 - 1)^{1/2}]$.

The expression (9.36) is effectively parametric (where p_ω serves as a parameter) relation between the values of k, r, and ω at the stability boundary, at fixed values of the other variables h, g, and σ.

The relation (9.36) approximates well such dependencies for $\sigma \leq 10^{-1}$ (the error at $\sigma = 10^{-2}$ turns out to be a fraction of a percentage, and for $\sigma = 10^{-1}$ is about 10%). The formulae (9.36) show that the major contribution to the gas phase effects is provided (at fixed σ) by the parameter h, which is related to the thickness of the gas preheat zone. This is a result of the value g being small compared to h (their ratio has a maximum of 0.27 at $l = 0$, $a = 0.42$).

It is important to realize that the second term in the numerator in (9.36) for r may not always be considered as a correction to the first one. Expansion is made with respect to the small parameter $\sigma\omega \sim \sigma p_\omega^2$, while the second term is proportional to σp_ω^3 and therefore becomes an order of unity for large values of p_ω. The appearance of such a term in the numerator of r results from the fact the multiplier z^2/Ω in the function $\Psi(\Omega)$ is of the form $1 - i/p_\omega$ at $p_\omega \gg 1$. On its multiplication by $(1 + \alpha i\sigma p_\omega^2)$ (in the course of expansion with respect to the small parameter σp_ω^2) and further extraction of an imaginary part, the combination of the two terms of the same order $\alpha\sigma p_\omega^2 - 1/p_\omega$ appears.

These considerations result in the fact that r may become zero at some finite value of frequency, in contrast to the theory where the inertia of the gas phase is neglected. Such a strong effect of a small parameter is related to the fact that in the latter theory the time derivative in the equations describing gas phase is dropped. This omission is based purely on the fact that the equations contain the product of this derivative and the small parameter σ. This incorrect mathematical operation (i.e. dropping the time derivative) leads to $\omega \to \infty$ at $r \to 0$. The latter implies inconsistency of the theory for small values of r.

It is easy to demonstrate that the results of the stability analysis of the original ZN theory (3.29, 3.30, and 3.32) may be obtained from (9.36) by assigning $\sigma = 0$. For $\omega \sim 1$, that is for $p_\omega \sim 1$, the expressions (9.36) contain only small corrections, in the order of t_g/t_c, to the latter results.

Of most interest is, obviously, the behaviour of the functions $r(p_\omega)$, $k(p_\omega)$, and $\omega(p_\omega)$ at the large frequency values $\omega \gg 1$, that is, for $p_\omega \gg 1$. In this limit

$$r = \left(\frac{2}{p_\omega^2}\right) - \sigma(h - g)p_\omega, k - 1 = \left(\frac{2}{p_\omega}\right) - \frac{\sigma(h - g)p_\omega^2}{2}, \quad \omega = \frac{p_\omega^2}{2} \tag{9.37}$$

Maximum values of the parameter p_ω and the frequency are attained at $r = 0$

$$p_\omega = \left(\frac{2}{\sigma(h - g)}\right)^{1/3}, \quad \omega = (2^{1/2}\sigma(h - g))^{-2/3} \tag{9.38}$$

At this maximum, the parameter k differs from unity by the value $k - 1 = (\sigma(h - g)/2)^{1/3}$. The relations (9.37) and (9.38) hold true for sufficiently small σ. This follows from the condition $p_\omega \gg 1$, which means $(\sigma(h - g)/2)^{1/3} \ll 1$. Since frequency is being scaled by t_c (time constant of the condensed phase) then the dimensional frequency at this limit $\widetilde{\omega} = \omega/t_c$

is $\widetilde{\omega} = [2^{1/2}(h-g)t_g t_c^{1/2}]^{-2/3}$. In other words, under the condition $t_g \ll t_c$ the natural time scale of the system relaxation depends significantly on the inertia of the gas phase and is proportional to the product $t_g^{2/3} t_c^{1/3}$.

The results in this section for the combustion stability of volatile systems under constant pressure show that for small values of $r = dT_s^0/dT_a$, accounting for the inertia of the gas phase results in widening of the region of stable combustion as well as in a qualitatively new expression for the natural frequency of the system. This holds true even if the inertia of the gas phase is small compared to the inertia of the condensed phase. It is expected that similar conclusions regarding the influence and necessity of consideration of small inertia zones will be true for a wide spectrum of nonsteady combustion phenomena and for other combustion models.

9.3 Burning Rate Response to Harmonically Oscillating Pressure

Similar to the case of the steady-state burning regime under constant pressure, considered in Section 9.2, an analysis of propellant combustion under oscillating pressure shows that under certain conditions the influence of the inertia of reaction zones, as well as of preheat and combustion product zones in the gas phase, needs be taken into account (Novozhilov 1989).

Indeed, the problem of combustion under harmonically oscillating pressure in the t_c approximation was considered in Section 4.1. The principal result is Eq. (4.27). This expression allows the relative variation of the burning rate under harmonically varying pressure, as well as a phase shift between the burning rate and pressure oscillations, to be found.

This result is valid under sufficiently small frequencies of pressure oscillations $\widetilde{\gamma}$ and not too high pressures. Indeed, if the time t_r of relaxation of those zones whose inertia is neglected is such that $\widetilde{\gamma} t_r \gtrsim 1$, then the latter inertia should be taken into account. On the other hand, an increase in pressure leads to a corresponding increase in the time of relaxation of the gas phase t_g, which may be comparable with the relaxation time of the preheat zone in the condensed phase t_c.

Only a few publications (Hart and McClure 1959; T'ien 1972; Allison and Faeth 1975) have attempted to account for the inertia of the gas phase while considering the response of the burning rate to harmonically oscillating pressure. All three approaches are based on numerical analysis.

This section presents an analytical solution of the problem in the framework of the specific propellant combustion model, which is the Belyaev model discussed in Section 9.2.

It is easy to show that for reasonable pressure frequencies, the acoustic wavelength is much larger than the thickness of the gas preheat zone. Therefore, pressure can be considered uniform over space. Its variation in time is supposed to be harmonic

$$p = p^0(1 + \varepsilon \cos(\widetilde{\gamma} t)) \tag{9.39}$$

The discussion in Section 9.2 concerning model assumptions and the possibility of an analytical solution still holds.

Some modifications in governing equations and boundary conditions, compared to Section 9.2, are required due to gas compression effects. Taking these modifications into account, the nondimensional formulation of the problem is as follows.

Nondimensional variables and parameters are introduced as

$$\theta = \frac{T}{T_{bl}^0}, \quad \theta_g = \frac{T_g}{T_{bl}^0}, \quad \theta_p = \frac{T_p}{T_{bl}^0}, \quad \theta_0 = \frac{T_a}{T_{bl}^0}, \quad \theta_s = \frac{T_s}{T_{bl}^0}$$
$$q = \frac{Q_g}{cT_{bl}^0}, \quad l = \frac{L}{cT_{bl}^0}, \quad \Gamma = \frac{\widetilde{\mu}c}{R}, \quad \eta = \frac{p}{p^0}$$
$$v = -\frac{\rho}{m^0}\frac{dx_s}{dt}, \quad \sigma = \frac{(D\rho_g^2)^0}{\kappa\rho^2} \quad \gamma = \widetilde{\gamma}\frac{\kappa}{(u^0)^2} \tag{9.40}$$

Here the superscript 0 explicitly denotes boiling temperature in the steady-state regime (i.e. at pressure p^0).

The assumptions that allow an analytical solution to be obtained are the same as in Section 9.2. Note that the product $D\rho_g^2$ for the gas phase is proportional to pressure. Again, it is assumed that $c \equiv c_p$ and $Le = 1$.

Nondimensional time and the Lagrangian coordinates are given by (9.12).

The operator $\widehat{H}$ is defined in the same way as in Section 9.2 and the operator $\widehat{H}_\sigma$ is modified as

$$\widehat{H}_\sigma = \sigma\frac{\partial}{\partial\tau} + v(\tau)\frac{\partial}{\partial\xi} - \eta\frac{\partial^2}{\partial\xi^2} \tag{9.41}$$

In the Lagrangian variables (9.12) the considered thermodiffusion part of the problem is reduced to the following set of equations

$$-\infty < \xi < 0, \quad \widehat{H}(\theta) = 0 \quad 0 < \xi < \xi_f(\tau), \quad \widehat{H}_\sigma(\theta_g) = \frac{\sigma\theta}{\Gamma\eta}\frac{d\eta}{d\tau}, \quad \widehat{H}_\sigma(Y) = 0$$
$$\xi_f(\tau) < \xi < \infty, \quad \widehat{H}_\sigma(\theta_p) = \frac{\sigma\theta_p}{\Gamma\eta}\frac{d\eta}{d\tau}, \quad Y = 0 \tag{9.42}$$

with boundary conditions

$$\theta = \theta_0, \quad \xi \to -\infty$$
$$\theta = \theta_g, \quad \frac{\partial\theta}{\partial\xi} = \eta\frac{\partial\theta_g}{\partial\xi} - lv, \quad \xi = 0$$
$$v(1-Y) + \eta\frac{\partial Y}{\partial\xi} = 0, \quad Y = \frac{1}{\eta}\exp[l\Gamma(1-\theta^{-1})]$$
$$\xi = \xi_f(\tau) \; Y = 0, \quad \theta_g = \theta_p, \quad -\eta\frac{\partial Y}{\partial\xi} = \frac{m}{m^0}$$
$$\frac{\partial\theta_g}{\partial\xi} = \frac{\partial\theta_p}{\partial\xi} - q\frac{\partial Y}{\partial\xi}, \quad \theta_p < \infty, \quad \xi \to \infty \tag{9.43}$$

and given explicit pressure dependence on time

$$\eta = 1 + \varepsilon\cos(\gamma\tau) \tag{9.44}$$

The set of equations (9.42) contains four differential equations of second order. Apart from the functions $\theta(\xi,\tau)$, $\theta_g(\xi,\tau)$, $Y(\xi,\tau)$, and $\theta_p(\xi,\tau)$, the burning rate $v(\tau)$ and the flame front location $\xi_f(\tau)$ must be found. Correspondingly, (9.43) contains 10 boundary conditions.

The steady-state solution ($\eta = 1$) of the problem (9.42) and (9.43) is

$$\begin{aligned}
&v^0 = 1,\ \xi_f^0 = -\ln(1 - Y_s^0)\\
&-\infty < \xi < 0,\ \theta^0 = \theta_0 + (\theta_s^0 - \theta_0)\exp(\xi)\\
&0 < \xi < \xi_f^0,\ \theta_g^0 = (\theta_0 - l) + (\theta_s^0 - \theta_0 + l)\exp(\xi),\quad Y^0 = 1 - (1 - Y_s^0)\exp(\xi)\\
&\xi_f^0 < \xi < \infty,\ \theta_p^0 = \theta_0 + q - l\ \ Y^0 = 0
\end{aligned} \tag{9.45}$$

Nondimensional temperature and the mass fraction of reactant at the surface must be found by solving simultaneously

$$\theta_s^0 = \theta_0 + q(1 - Y_s^0) - l,\ \ Y_s^0 = \exp\left[l\Gamma\left(1 - \frac{1}{\theta_s^0}\right)\right] \tag{9.46}$$

The parameters characterizing the sensitivities of the burning rate and the surface temperature may be found from the relationships

$$\begin{aligned}
&k = (\theta_s^0 - \theta_0)\left(\frac{\partial \ln m^0}{\partial \theta_0}\right)_\eta, \qquad r = \frac{1}{\left[1 + \frac{ql\Gamma Y_s^0}{(\theta_s^0)^2}\right]}\\
&\iota = \left(\frac{\partial \ln m^0}{\partial \ln \eta}\right)_{\theta_0}, \qquad \mu = \frac{qY_s^0}{(\theta_s^0 - \theta_0)\left[1 + \frac{ql\Gamma Y_s^0}{(\theta_s^0)^2}\right]}
\end{aligned} \tag{9.47}$$

To solve the problem in the linear approximation, the method of complex amplitudes can be used again. All the variables are written in the form $f = f^0 + f_1 \exp(i\gamma\tau)$, where f^0 is the steady-state value and f_1 is a small correction (complex amplitude).

Specifically, the complex amplitudes of the functions θ, θ_g, Y, θ_p, v, and ξ_f are denoted as $\aleph$, $\aleph_g$, y, $\aleph_p$, b, and s, respectively.

Linearization of the set of equations (9.42) gives

$$\begin{aligned}
&\hat{h}(\aleph) = b\Delta\exp(\xi),\ \ \hat{h}_\sigma(\aleph_g) = (\Delta + l)\left[b - \varepsilon\left(1 + \frac{i\sigma\gamma}{\Gamma}\right)\right]\exp(\xi)\\
&\quad - \frac{i\sigma\gamma}{\Gamma}(\theta_0 - l)\varepsilon\\
&\hat{h}_\sigma(y) = a(\varepsilon - b)\exp(\xi),\ \ \hat{h}_\sigma(\aleph_p) = -\frac{i\sigma\gamma}{\Gamma}\theta_p^0\varepsilon
\end{aligned} \tag{9.48}$$

where

$$\hat{h} = \frac{d^2}{d\xi^2} - \frac{d}{d\xi} - i\gamma \tag{9.49}$$

and the operator $\hat{h}_\sigma$ differs from $\hat{h}$ by the multiplier σ in front of $i\gamma$.

The solutions of these equations, taking into account boundary conditions at $\xi = \pm\infty$, take the form

$$\begin{aligned}
&\aleph = Aq\exp(z\xi) + \frac{ib\Delta}{\gamma}\exp(\xi)\\
&\aleph_g = Fq\exp(z_1\xi) + Gq\exp(z_2\xi) + \frac{\varepsilon}{\Gamma}(\theta_p^0 - q)\\
&\quad + \frac{(\Delta + l)}{i\sigma\gamma}\left[b - \varepsilon\left(1 + \frac{i\sigma\gamma}{\Gamma}\right)\right]\exp(\xi)\\
&y = C\exp(z_1\xi) + D\exp(z_2\xi) - \frac{ia}{\sigma\gamma}(b - \varepsilon)\exp(\xi)\\
&\aleph_p = Hq\exp(z_2\xi) + \frac{\varepsilon\theta_p^0}{\Gamma}
\end{aligned} \tag{9.50}$$

where A, C, D, F, and G are constants.

The boundary conditions (9.43), on their linearization, give ($R = 1/r - 1$)

$$\xi = 0, \quad \aleph = \aleph_g, \quad \aleph' = \aleph_g' + (\Delta + l)\varepsilon - lb$$
$$ab - y + y' - a\varepsilon = 0, \quad \varepsilon(1 - a) + y = \frac{R\aleph}{q}$$
$$\xi = \xi_f, \quad qs + \aleph_g = \aleph_p, \quad \aleph_g' = \aleph_p' - qy'$$
$$\varepsilon + y - y' = \frac{k}{\Delta}\aleph_p, \quad y = s \tag{9.51}$$

The last of these relations allows the variable s to be eliminated. The remaining relations, on substitution of the solutions (9.50) of the linearized equations, provide the set of seven linear algebraic equations

$$\sum_{m=1}^{7} U_{mn} x_m = W_n \varepsilon \qquad m, n = 1, 2, \ldots, 7 \tag{9.52}$$

Here unknown x_m stands for b, A, C, D, F, G, and He_2, respectively (the notations $e_1 = \exp(z_1 \xi_f^0)$ and $e_2 = \exp(z_2 \xi_f^0)$) are adopted. The matrices U_{mn} and W_n are as follows

$$U_{mn} = \begin{vmatrix} \frac{i}{\gamma}\left(\frac{\Delta}{q} - \frac{a}{\sigma}\right) & 1 & 0 & 0 & -1 & -1 & 0 \\ \frac{i}{\gamma}\left(\frac{\Delta}{q} - \frac{a}{\sigma}\right) + \frac{l}{q} & z & 0 & 0 & -z_1 & -z_2 & 0 \\ a & 0 & -z_2 & -z_1 & 0 & 0 & 0 \\ \frac{i}{\gamma}\left(\frac{a}{\sigma} + R\frac{\Delta}{q}\right) & R & -1 & -1 & 0 & 0 & 0 \\ 0 & 0 & e_1 & e_2 & e_1 & e_2 & -1 \\ 0 & 0 & e_1 z_1 & e_2 z_2 & e_1 z_1 & e_2 z_2 & -z_2 \\ 0 & 0 & e_1 z_2 & e_2 z_1 & 0 & 0 & -\widetilde{\beta} \end{vmatrix}$$

$$W_n = \begin{vmatrix} \frac{\theta_s^0}{\Gamma q} - \frac{ia}{\sigma\gamma} \\ a\left(1 + \frac{1}{\Gamma} - \frac{i}{\sigma\gamma}\right) \\ a \\ 1 - a\left(1 - \frac{i}{\sigma\gamma}\right) \\ 0 \\ -\frac{1}{\Gamma} \\ \frac{\widetilde{\beta}\theta_p^0}{\Gamma q} - 1 + \imath \end{vmatrix} \tag{9.53}$$

The amplitude of the burning rate is expressed as

$$b = \left(\sum_{n=1}^{7} W_n D_{1n} \right) \left(\sum_{n=1}^{7} U_{1n} D_{1n} \right)^{-1} \varepsilon \tag{9.54}$$

where the D_{1n} values are minors corresponding to the elements of the first column of U_{mn}. The latter are calculated relatively easy thanks to the large number of zero elements:

$$\begin{aligned}
D_{1n} &= \frac{\varsigma d_n}{ar} \\
d_1 &= z_1 z_2^2 (1 - r)(a^{-\varsigma} - 1) - \tilde{\beta}\varsigma[1 + r(z - 1)] \\
d_2 &= \tilde{\beta}\varsigma - z_1 z_2 (1 - r)(a^{-\varsigma} - 1) \\
d_3 &= \varsigma[(1 - r)z_1 - \tilde{\beta}] - r(z - z_2)(z_2 a^{-\varsigma} - z_1) \\
d_4 &= r[z_1 z_2 (z - z_2)(a^{-\varsigma} - 1) - \tilde{\beta}\varsigma(z - 1)] \\
d_6 e_1 &= \tilde{\beta}(z_1 a^{-\varsigma} - z_2) + [\tilde{\beta} r(z - 1) - z_1 z_2 (1 - r)](a^{-\varsigma} - 1) \\
d_7 e_2 &= \varsigma[z_1 + r(z - 1)]
\end{aligned} \tag{9.55}$$

The value of D_{15} is not needed as $U_{15} = W_5 = 0$.

Further elementary calculations lead to the following result for the complex amplitude $U = b/\varepsilon$

$$\begin{aligned}
U &= [M_1 + M_2 + (N_1 + \tilde{\beta} N_2)/\Gamma](P_1 + P_2)^{-1} \\
M_1 &= \varsigma[z_1 \imath a^{z_2} + (\imath r a^{z_2} - \mu k)(z - 1)] \\
M_2 &= [z_1 + r(z - 1)][a(a^{-\varsigma} z_1 - z_2) - \varsigma a^{z_2}] \\
&\quad + i\sigma\gamma(1 - a)(z_2 - z)(a^{-\varsigma} - 1) \\
N_1 &= i\sigma\gamma(1 - r)(a^{-\varsigma} - 1)[a z_1 - a^{z_1} + z_2(1 - \theta_p^0/q)] \\
N_2 &= \varsigma\{ \theta_p^0(z_1 a^{z_2} - 1)/q + 1 + r(z - 1)[\theta_p^0(a^{z_2} - 1)/q + 1 - a]\} \\
&\quad + z_2 a^{z_1} - z_1 a^{z_2} + r(z - 1)(a^{z_1} - a^{z_2}) \\
P_1 &= a[z_1 + r(z - 1)](a^{-\varsigma} z_1 - z_2) - \tilde{\beta}\Delta\varsigma(z - 1)/qz \\
P_2 &= \Delta\sigma\gamma^2(a^{-\varsigma} - 1)(1 - r)/qz^2
\end{aligned} \tag{9.56}$$

where

$$\begin{aligned}
&2z = 1 + \sqrt{(1 + 4i\gamma)}, \quad 2z_{1,2} = 1 \pm \sqrt{(1 + 4i\sigma\gamma)}, \quad \varsigma = z_1 - z_2 \\
&a = 1 - Y_s^0, \quad \Delta = \theta_s^0 - \theta_0, \quad \tilde{\beta} = kq/\Delta
\end{aligned} \tag{9.57}$$

Note that $\tilde{\beta} = \beta Q_g/c$ where $\beta = (\partial \ln u^0/\partial T_0)_{p^0}$ is the temperature coefficient of the burning rate.

If $\sigma = 0$, the only terms that are different from zero in (9.56) are M_1 and P_1. These correspond to the numerator and denominator, respectively, of the formula (4.27) applicable to the inertia-free gas phase.

The combination $(N_1 + \tilde{\beta} N_2)/\Gamma$ describes gas temperature variations due to pressure oscillations, with the second term (containing the temperature coefficient of the burning rate) accounting for the fact that temperature is not constant at the flame front.

As an example of the calculation of the burning rate under harmonically varying pressure according to (9.56), consider a system with the parameters provided by Allison and Faeth (1975). These are $Q_g = 4860$ kJ/kg, $L = 1690$ kJ/kg, $\Gamma = 8.94$, $c = 3.05$ kJ/(kg × K), $\tilde{\mu} = 0.024$ kg/mol, $\lambda = 0.383$ W/(m × K), and $\rho = 10^3$ kg/m^3. For these values of the parameters and for pressure $p^0 = 10^5$ Pa, $T^0_{bl} = 384\,K$ and $D\rho_g^2 = 1.26 \times 10^{-5}$ kg^2/(m^4 × s). Calculations are performed for the value of the initial temperature $T_a = 298$ K which corresponds to the combustion products temperature $T_p^0 = 1340$ K. Allison and Faeth (1975) assume also that the chemical reaction is second order with respect to the concentration of the initial substance, which implies $\iota = 1$.

Figures 9.2 and 9.3 demonstrate the calculated results for the pressure values of $p^0 = 10^6$ Pa and $p^0 = 4 \times 10^6$ Pa, respectively. The corresponding values of the parameters involved in (9.56) are given in Table 9.1.

The value of the temperature coefficient of the burning rate β is chosen according to the following procedure. The plots of frequency dependencies of the modulus U and phase shift φ of burning rate oscillations relative to pressure oscillations are provided by Allison and Faeth (1975). These plots correspond to the absence of the gas-phase inertia ($\sigma = 0$) and are reproduced in Figure 9.2 (curves 1′ and 2′ respectively). These dependencies are determined at $\sigma = 0$ by the four sensitivity parameters (4.21), of which only three (r, μ, and ι) are known. Therefore, parameter k may be chosen in such a way that formula (4.27) reproduces curves 1′ and 2′ as closely as possible. This choice results in the estimation $k = 0.85$ and correspondingly $\tilde{\beta} = 9.29$ and $\beta = 5.83 \times 10^3$ K^{-1}.

For the reaction of order m, in the approximation of infinitely thin reaction zone, $\beta = \left[\frac{m+3}{2T_p^0} + \frac{E}{2R(T_p^0)^2}\right]$. Therefore, for the second-order reaction $\beta T_p^0 = \frac{5}{2} + \frac{E}{2RT_p^0}$. The activation energy estimated from this relation is $E = 107$ kJ/mol. This value is very close to that

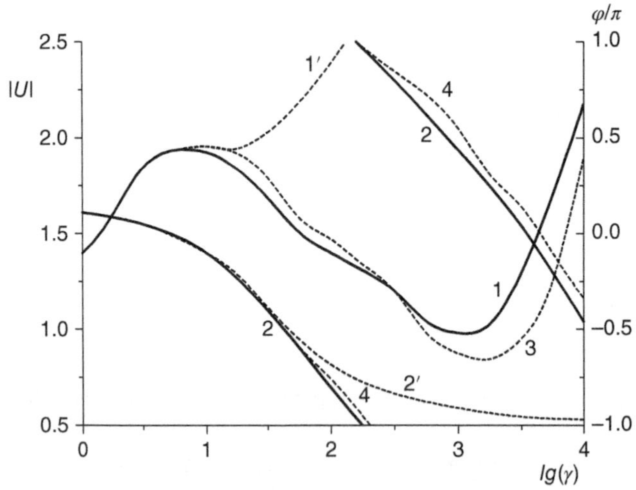

Figure 9.2 Frequency dependencies of the modulus and the phase shift of burning rate oscillations with respect to pressure oscillations. $p^0 = 10^6$ Pa. Solid lines, results from this chapter. $\sigma = 10^{-3}$. 1, modulus $|U|$ of burning rate oscillations; 2, phase shift of burning rate oscillations. Dashed lines, results of Allison and Faeth (1975). 1′, modulus $|U|$ of burning rate oscillations; 2′, phase shift of burning rate oscillations. $\sigma = 0$. 3, modulus $|U|$ of burning rate oscillations; 4, phase shift of burning rate oscillations. $\sigma = 10^{-3}$.

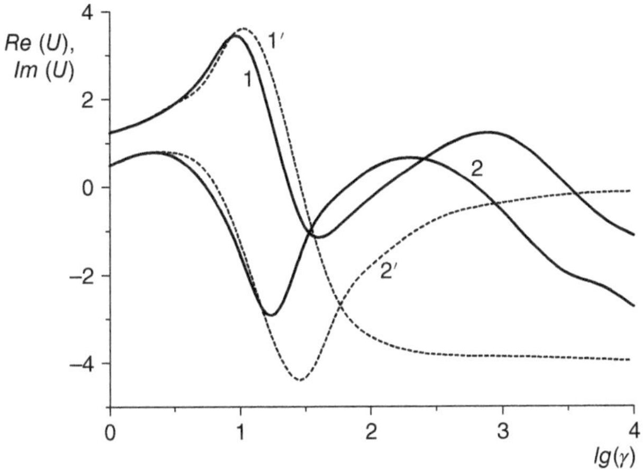

Figure 9.3 Influence of gas-phase inertia on the real and imaginary parts of the burning rate response. $p^0 = 4 \times 10^6$ Pa. Solid lines, $\sigma = 4 \times 10^{-3}$. 1, real part; 2, imaginary part. Dashed lines, $\sigma = 0$. 1′, real part; 2′, imaginary part.

Table 9.1 Parameters involved in the solution (9.56) for the complex amplitude of burning rate at two different values of pressure.

p^0 (Pa)	σ	T^0_{bl} (K)	θ_0	θ^0_p	l	q	θ^0_p/q	θ^0_s
10^6	10^{-3}	468	0.637	2.86	1.18	3.40	0.841	0.948
4×10^6	$4\cdot10^{-3}$	540	0.552	2.48	1.03	2.95	0.841	0.935
p^0 (Pa)	Y^0_s	Δ	a	Δ/q	r	μ	k	$\widetilde{\beta}$
10^6	0.562	0.311	0.438	0.0915	0.0429	0.262	0.85	9.29
4×10^6	0.525	0.383	0.475	0.130	0.0578	0.233	1.21	9.29

used by Allison and Faeth (1975) ($\frac{E}{RT^0_p} = 10$, that is $E = 111$ kJ/mol). The agreement of the two values confirms the correctness of the infinite reaction zone approximation at low frequencies (Allison and Faeth (1975) solved the problem numerically taking into account the finite thickness of the reaction zone).

The data in Figure 9.2 correspond to a pressure of 10^6 Pa. Curves 1 and 2, showing the frequency dependencies of the modulus U and the phase shift φ, are obtained by the present model; the corresponding curves 3 and 4 are those of Allison and Faeth (1975). It is evident that curves 1 and 3 (and similarly 2 and 4) are very close to each other at sufficiently low frequencies. Even at $\gamma = 10^3 - 10^4$ which, at $\sigma = 10^{-3}$ corresponds to the pressure frequency interval $\gamma = (1 - 10)t_g^{-1}$, deviation of the moduli U is only 20 – 30%. The discrepancy of the phase shifts caused by neglecting the thickness of the reaction zone is even smaller. Therefore, even at large frequencies, the assumption of an infinitely thin reaction zone is sufficiently justified.

The dashed curves 1′ and 2′ show the modulus and the phase shift for the inertia-free gas ($\sigma = 0$), while curves 1 and 2 take account of the gas phase inertia ($\sigma = 10^{-3}$). Comparison between curves 1 and 1′, as well as between curves 2 and 2′ demonstrates the significant influence of the preheat and combustion product zones on the nonsteady combustion process under varying pressure.

In particular, the frequency dependency of the phase shift is monotonic for the approximation $\sigma = 0$. In this case, at large frequencies, the burning rate lags in phase behind pressure by π. In contrast, at $\sigma \neq 0$ (curve 2) both lagging behind and outpacing may occur.

In practical applications, for example in the analysis of burning stability in a combustion chamber, the acoustic admittance of the system needs to be found. This property is related to the real part of the burning rate response to harmonically varying pressure. Figure 9.3 illustrates the influence of gas-phase inertia on the real and imaginary parts of the burning rate response. The solid curves 1 and 2 represent Re(U) and Im(U), respectively, as the functions of frequency at s pressure of 4×10^6 Pa ($\sigma = 4 \times 10^{-3}$). The dashed curves $1'$ and $2'$ present the same properties in the approximation $\sigma = 0$. It is evident that at large frequencies the inertia of the preheat and combustion products zones is of significant importance. Neglecting this inertia would lead to substantial errors in the values Re(U) and Im(U), including in some cases even a change of their signs.

In conclusion, an important comment should be made. Usually (Novozhilov 1973a; Zeldovich et al. 1975), gas-phase inertia is neglected at frequencies $\tilde{\gamma} \ll t_g^{-1}$. However, the present results show that even when this inequality (or, equivalently, $\sigma\gamma \ll 1$) is fulfilled, the effect of the gas-phase inertia is still significant. For example, Figure 9.2 shows that at $\gamma = 10^2$ ($\sigma = 10^{-3}$, $\sigma\gamma = 10^{-1}$) accounting for the gas-phase inertia leads to a change in the value $|U|$ by 50%, compared to the approximation $\sigma = 0$. Similarly, Figure 9.3 demonstrates that a significant difference in the values of the real and imaginary parts of the function U emerges at frequencies $\gamma \sim 20 - 30$, which at $\sigma = 4 \times 10^{-3}$ corresponds to $\sigma\gamma \sim 10^{-1}$.

The reason for the rather strong influence of gas-phase inertia at such low frequencies is related to the important role of the magnitude of the parameter $r = (\partial T_s^0 / \partial T_a)_{p^0}$, pointed out by Novozhilov (1988a,b). For small values of this parameter (which is the case in the considered examples), the magnitude of the natural frequency of the combustion system $\omega \sim \sigma^{-2/3}$. It is obvious that the above inequality $\sigma\gamma \ll 1$ must be replaced with another one, that is $\gamma \ll \omega$, or $\sigma^{2/3}\gamma \ll 1$. The latter is a more rigid condition. For the example illustrated in Figure 9.2 the gas-phase inertia may be neglected not for $\gamma \ll 10^3$, but only for $\gamma \ll 10^2$.

9.4 Acoustic Admittance of the Propellant Surface

The notion of acoustic admittance of the surface was discussed in Section 4.2. In the moving media, this property is expressed in the form (4.51). In the same section the explicit expression (4.64) for the acoustic admittance of the propellant surface was derived in the framework of the t_c approximation. (A more basic treatment using the simplifying assumption of constant surface temperature may be found in Zeldovich 1942.)

The expression for acoustic admittance at high frequencies taking account of gas-phase inertia is of special interest. This problem was first posed by Hart and McClure (1959). Volkov and Medvedev (1969) took gas-phase inertia into account by using a phenomenological time delay. T'ien (1972) and also Allison and Faeth (1975) numerically integrated the linearized set of equations describing the combustion process at variable pressure.

Novikov et al. (1974) numerically integrated a simplified set of equations where the conservation of reactant was not enforced.

This section describes an analytical solution of the problem, within the assumptions of the Belyaev model.

In the acoustics of a quiescent medium the process is normally adiabatic, and perturbations of velocity, pressure, and temperature are transported by acoustic waves. When considering the interaction of acoustic waves with the surface of a burning solid propellant the following circumstances, complicating by the calculation of acoustic admittance, need be taken into account.

Generally, the adiabaticity of the process is broken. There is an adiabatic relation between pressure and temperature perturbations in acoustic waves. However, if the flame is nonisentropic then temperature perturbations developing inside the flame cannot be carried away by acoustic perturbations only. In addition, the combustion process is coupled with gas flow. This circumstance, along with the nonadiabaticity of the flame, results in the development of entropic waves. Therefore, the formulation of boundary conditions at the propellant surface and analysis of conditions leading to either the amplification or decay of acoustic waves on reflection must rely on the consideration of the full set of linear hydrodynamic perturbations in a gas. The latter includes, in one-dimensional formulation, both acoustic perturbations and entropic waves.

Note that acoustic admittance describes the interaction of the surface with acoustic waves only, therefore definition (4.48) may be written in a more precise form as

$$\zeta = -\rho_g a \frac{(u_{g1})_a}{p_1} \tag{9.58}$$

where the subscript "a" indicates that only the acoustic component of the full gas perturbation at the surface is considered.

Acoustic waves, on interaction with the surface, generate not only reflected acoustic waves, but also entropic waves which carry the gas perturbation $(u_{g1})_e$. Therefore, the set of linear hydrodynamic perturbations must be divided into acoustic and entropic waves, and the acoustic admittance found according to (9.58).

The acoustic wavelength, even at very high frequencies, is much larger than the thickness of the flame front. The ratio of these two values is a/u_g. Therefore, the acoustic approximation combustion zone can be considered to be infinitely thin, and the pressure in this zone independent of any spatial coordinate.

This latter circumstance allows the solution of the problem to be divided into two steps. First, a nonsteady combustion process under pressure harmonically varying in time and uniform over space (inner problem) is considered. The solution of this problem allows the mass burning rate, gas velocity, and flame temperature perturbations to be found. A part of these perturbations is carried away by acoustic waves, while another part is carried away by entropic waves. The second step of the solution is the extraction of the acoustic components from the full perturbations, obtained at the previous step (outer problem). This procedure allows the amplitude of the reflected wave to be found based on the amplitude of the incident wave. Thus, the acoustic admittance can be found.

This procedure can be formulated within the mathematical formalism of the method of matched asymptotic expansions.

For the outer problem, let us consider the coordinate system with an origin within the combustion zone and positive coordinate values in the right half-space. The expression for acoustic admittance in the case of a moving gas and the presence of decaying entropic waves (the importance of taking into account the effect of the decay of entropic waves is pointed out by Volkov and Medvedev (1969)) is found in the following way.

In a linear approximation, the set of hydrodynamic perturbations for one-dimensional gas flow may be written as the sum of acoustic and entropy harmonics

$$u_{g1} = (u_{g1})_a + (u_{g1})_e, \quad T_{b1} = (T_{b1})_a + (T_{b1})_e \tag{9.59}$$

where T_b is the combustion product temperature.

Pressure perturbations are absent from entropic waves, whilst the amplitudes of the burning rate and the combustion product temperature perturbations are related in the latter as

$$\frac{(u_{g1})_e}{u_g^0} = z_2 \frac{(T_{b1})_e}{T_b^0}, \quad 2z_2 = 1 - \sqrt{1 + 4i\omega_g} \tag{9.60}$$

where $\omega_g = \omega\kappa_g/(u_g^0)^2$ is the nondimensional frequency (i.e. frequency multiplied by the characteristic time of the gas phase).

The relations (9.59) and (9.60) give

$$(u_{g1})_a = u_{g1} - z_2 u_g^0 \frac{[T_{b1} - (T_{b1})_a]}{T_b^0} \tag{9.61}$$

Temperature perturbation in the acoustic wave may be related to pressure perturbation

$$\Gamma \frac{(T_{b1})_a}{T_b^0} = \frac{p_1}{p^0}, \quad \Gamma = \gamma/(\gamma - 1) \tag{9.62}$$

Let us introduce also the variable

$$N = \Gamma \left(\frac{T_{b1}}{T_b^0} \right) \left(\frac{p_1}{p^0} \right)^{-1} - 1 \tag{9.63}$$

Eqs (9.61–9.63) lead to the following expression for acoustic admittance

$$\varsigma = -\gamma M (G - z_2 N/\Gamma) \tag{9.64}$$

Both terms on the right-hand side change significantly over the length of entropic wave decay. However, the difference between them, which represents purely acoustic perturbation, changes much more slowly (on the scale of the order of the length of acoustic wave). Because of this, the acoustic admittance may be found by different methods which lead to identical results.

From here on, the model of an infinitely thin chemical reaction zone in the gas phase will be used. In this case, it is convenient to calculate the acoustic admittance by obtaining the values of N and G at the reaction front

$$\varsigma = -\gamma M (G_f - z_2 N_f/\Gamma) \tag{9.65}$$

The value $N_f = \Gamma\Theta - 1$ is known as the nonisentropicity of the flame ($N_f = 0$ corresponds to the flame where an adiabatic relationship exists between the pressure and temperature perturbations). The variable $I = \Gamma\Theta$ is also introduced (if pressure and temperature perturbations at the flame front are related adiabatically, then $\mathrm{Re}(I) = 1$, $\mathrm{Im}(I) = 0$). This value is called the degree of isentropicity of the flame front.

The same value of acoustic admittance is obtained when calculated at any other point downstream from the flame front (at a short distance compared to the length of the acoustic wave). Finally, the method of matched asymptotic expansions gives an identical result. Within the framework of the latter method, the coordinate used in the solution of the inner problem needs be taken to infinity, and then, while solving the outer problem, the outer coordinate must be set to zero.

Usually, the relaxation time of the preheat zone in the condensed phase is much larger than the time of thermo-diffusion relaxation of the gas phase, that is, $t_c \gg t_g$, where $t_g \sim \kappa_g/(u_g)^2$ (κ_g is gas thermal diffusivity and u_g is the velocity of gas leaving the surface). At low frequencies ($\omega t_c \sim 1$, but $\omega t_g \ll 1$) the effect of entropic wave decay may be neglected. Although entropic components of density and temperature perturbations still exist, any velocity perturbation may be attributed to acoustic waves. Note that since the condition of the applicability of the t_c approximation ($\omega_g \ll 1$, i.e. the gas phase is inertia free) coincides with the condition of the absence of decay of entropic waves, acoustic admittance in this approximation may be written in the form shown in (4.51).

On the other hand, if the frequency is sufficiently large and comparable with the reciprocal of the time of relaxation of the gas phase, that is, $\omega\kappa_g/(u_g)^2 \sim 1$, then the decay of entropic waves must be taken into account, and the latter parameter controls the rate of decay.

The Belyaev model was discussed in detail in Sections 9.2 and 9.3. In particular, the burning rate response to harmonically oscillating pressure, within the framework of this model, was analysed in Section 9.3. The formulation of the problem considered in this section is similar to that of Section 9.3.

The following nondimensional variables are introduced:

$$\theta_0 = \frac{T_a}{T_{bl}^0},\quad \theta = \frac{T}{T_{bl}^0},\quad \theta_s = \frac{T_s}{T_{bl}^0},\quad \theta_g = \frac{T_g}{T_{bl}^0},\quad \theta_p = \frac{T_p}{T_{bl}^0}$$
$$v = -\frac{\rho}{m^0}\frac{dx_s}{dt},\quad \eta = \frac{p}{p^0},\quad q = \frac{Q_g}{cT_{bl}^0},\quad l = \frac{L}{cT_{bl}^0},\quad \sigma = \frac{(D\rho_g^2)^0}{\kappa\rho^2} \tag{9.66}$$

Lagrangian coordinates are introduced in exactly the same way as in Sections 9.2 and 9.3, that is, according to (9.12), the operators $\hat{H}$ and $\hat{H}_\sigma$ are as in Section 9.3.

The thermodiffusion part of the problem is

$$-\infty < \xi < 0,\quad \hat{H}(\theta) = 0$$
$$0 < \xi < \xi_f(\tau),\quad \hat{H}_\sigma(\theta_g) = \frac{\sigma\theta}{\Gamma\eta}\frac{\mathrm{d}\eta}{\mathrm{d}\tau},\quad \hat{H}_\sigma(Y) = 0$$
$$\xi_f(\tau) < \xi < \infty,\quad \hat{H}_\sigma(\theta_p) = \frac{\sigma\theta_p}{\Gamma\eta}\frac{\mathrm{d}\eta}{\mathrm{d}\tau},\quad Y = 0 \tag{9.67}$$

where ξ_f is the coordinate of the flame front.

The boundary conditions are

$$\theta = \theta_0, \quad \xi \to -\infty$$
$$\theta = \theta_g, \quad \frac{\partial \theta}{\partial \xi} = \eta \frac{\partial \theta_g}{\partial \xi} - lv, \quad \xi = 0$$
$$v(1 - Y) + \eta \frac{\partial Y}{\partial \xi} = 0, \quad Y = \frac{1}{\eta} \exp[l\Gamma(1 - \theta^{-1})]$$
$$\xi = \xi_f(\tau) \quad Y = 0, \quad \theta_g = \theta_p, \quad -\eta \frac{\partial Y}{\partial \xi} = \frac{m}{m^0}$$
$$\frac{\partial \theta_g}{\partial \xi} = \frac{\partial \theta_p}{\partial \xi} - q \frac{\partial Y}{\partial \xi};$$
$$\theta_p < \infty, \quad \xi \to \infty \tag{9.68}$$

Since the assumption of an infinitely thin reaction zone is employed, the dependence $m(T_f, p)$ of the mass burning rate at the flame front on the front temperature and pressure must be prescribed.

The set of equations (9.67) contains four differential equations of second order. Apart from the functions $\theta(\xi, \tau)$, $\theta_g(\xi, \tau)$, $Y(\xi, \tau)$, and $\theta_p(\xi, \tau)$, the burning rate $v(\tau)$ and flame front location $\xi_f(\tau)$ must be found. Correspondingly, (9.68) contains 10 boundary conditions.

A small pressure perturbation with amplitude $\eta_1 = \frac{p_1}{p^0}$ is imposed on the steady-state combustion regime

$$\eta = 1 + \eta_1 \cos(\omega_c \tau) \tag{9.69}$$

where ω_c is the nondimensional frequency expressed in the units of reciprocal time of relaxation of the condensed phase $\omega_c = \omega\kappa/(u^0)^2$.

Once the thermodiffusion part of the problem has been solved, the hydrodynamic properties may be found in the following way. Gas velocity in the preheat and combustion product zones can be found using the obtained temperature distributions $\theta_g(\xi, \tau)$ and $\theta_p(\xi, \tau)$, and the burning rate $v(\tau)$ is found from the equation of continuity. In the adopted nondimensional variables the latter has the form

$$0 < \xi < \xi_f, \quad \theta_p^0 \frac{\partial W_g}{\partial \xi} = \sigma \frac{\partial(\theta_g/\eta)}{\partial \tau} + v \frac{\partial(\theta_g/\eta)}{\partial \xi}$$
$$\xi_f < \xi < \infty, \quad \theta_p^0 \frac{\partial W_p}{\partial \xi} = \sigma \frac{\partial(\theta_p/\eta)}{\partial \tau} + v \frac{\partial(\theta_p/\eta)}{\partial \xi} \tag{9.70}$$

where

$$W_g = (u_g + u)/(u_p^0 + u^0), \quad W_p = (u_p + u)/(u_p^0 + u^0) \tag{9.71}$$

are nondimensional gas velocities in the preheat zone and the combustion product zone, respectively.

The gas velocity must satisfy the boundary conditions

$$\xi = 0, \quad W_g = v \frac{\theta_s}{\theta_p^0} \eta, \quad \xi = \xi_f, \quad W_g = W_p \tag{9.72}$$

The steady-state regime is described by the following solution of the set of equations (9.67) with boundary conditions (9.68)

$$v^0 = 1, \quad \xi_f^0 = -\ln a$$
$$-\infty < \xi < 0, \quad \theta^0 = \theta_0 + \Delta \exp(\xi)$$

$$0 < \xi < \xi_f^0, \quad \theta_g^0 = \theta_0 - l + (\Delta + l)\exp(\xi)$$
$$Y^0 = 1 - a\exp(\xi)$$
$$\xi_f^0 < \xi < \infty, \quad \theta_p^0 = \theta_0 + q - l, \quad Y^0 = 0 \tag{9.73}$$

where $\Delta = \theta_s^0 - \theta_0$, $a = 1 - Y_s^0$, and θ_s^0 and Y_s^0 are nondimensional temperature and mass fraction of the fuel at the phase interface. These values must be found from the set of equations

$$Y_s^0 = \exp\left[l\Gamma\left(1 - \frac{1}{\theta_s^0}\right)\right], \quad qa = \theta_s^0 - \theta_0 + l \tag{9.74}$$

Since linear approximation only is considered, the properties of the function $m(T_f, p)$ are reflected in the result only through its first derivatives. This function (i.e. the rate of chemical reaction in the gas phase) may be explicitly written as having power dependence on concentration and Arrhenius-type dependence on the temperature of the flame front

$$\frac{m}{m_0} = \left(\frac{T_f}{T_f^0}\right)^{(n+3)/2}\left(\frac{p}{p^0}\right)^{n/2}\exp\left[-E\left(\frac{1}{T_f} - \frac{1}{T_f^0}\right)/(2R)\right] \tag{9.75}$$

where n is the order of the chemical reaction and T_f^0 is the temperature of the flame front in the steady-state regime.

The distribution of the gas velocity is found from the continuity equation (9.70) with the help of the boundary condition (9.72)

$$W_g^0 = \frac{\theta_g^0}{\theta_p^0}, \quad W_p^0 = 1 \tag{9.76}$$

Finally, the values of the parameters describing the burning rate and surface temperature dependence, in the steady-state regime, on initial temperature and pressure are

$$k = \Delta\left(\frac{\partial \ln m^0}{\partial \ln \theta_0}\right)_\eta$$
$$r = (1 + ql\Gamma Y_s^0/(\theta_s^0)^2)^{-1}, \quad \iota = \left(\frac{\partial \ln m^0}{\partial \ln \eta}\right)_{\theta_0}$$
$$\mu = qY_s^0(1 + ql\Gamma Y_s^0/(\theta_s^0)^2)^{-1}/\Delta \tag{9.77}$$

The parameters k and ι for the mass burning rate dependence (9.75) are given by

$$k = \Delta\frac{(n + 3 + \varepsilon/\theta_p^0)}{2\theta_p^0}, \quad \iota = \frac{n}{2}, \quad \varepsilon = \frac{E}{RT_p^0} \tag{9.78}$$

For acoustic admittance (9.65) to be calculated, the nonisenropicity of the flame front N_f and the gas velocity response at the front G_f need be found in the linear approximation. Applying the method of complex amplitudes, each time-dependent variable is represented as a sum of its steady-state value and a small correction, varying harmonically in time. These corrections are proportional to the pressure amplitude

$$\theta = \theta^0 + \vartheta\eta_1\exp(i\omega_c\tau), \quad \theta_g = \theta_g^0 + \vartheta_g\eta_1\exp(i\omega_c\tau)$$
$$\theta_p = \theta_p^0 + \vartheta_p\eta_1\exp(i\omega_c\tau), \quad Y = Y_0 + y\eta_1\exp(i\omega_c\tau)$$
$$v = 1 + U\eta_1\exp(i\omega_c\tau), \quad \xi_f = \xi_f^0 + s\eta_1\exp(i\omega_c\tau)$$
$$W_g = W_g^0 + \wp_g\eta_1\exp(i\omega_c\tau), \quad W_p = W_p^0 + \wp_p\eta_1\exp(i\omega_c\tau) \tag{9.79}$$

Linearization of (9.70) and (9.72) leads to the continuity equation

$$\theta_p^0 \frac{d\wp_g}{d\xi} = i\sigma\omega_c(\vartheta_g - \theta_g^0) + \frac{d[\vartheta_g + \theta_g^0(U-1)]}{d\xi} \tag{9.80}$$

with the boundary condition

$$\xi = 0, \quad \wp_g = \frac{[\vartheta_g + \theta_s^0(U-1)]}{\theta_p^0} \tag{9.81}$$

Since $\wp_g(\xi_f^0) = G_f$, for the gas velocity response to be found Eq. (9.80) must be integrated over the preheat zone. The result is

$$G_f = U + \Theta - 1 + J, \quad J = \frac{i\sigma\omega_c}{\theta_p^0}\int_0^{\xi_f^0}(\vartheta_g - \theta_g^0)d\xi \tag{9.82}$$

A comparison of this expression with (4.58) shows that making an account of inertia of the gas phase leads to the appearance of an additional integral term. This term contains both the steady-state and perturbed temperature distributions in the preheat zone.

The burning rate response to harmonically oscillating pressure was found in Section 4.1. The quantities Θ and J must be found from the solution of the thermodiffusion problem formulated above.

Substitution of the expansions (9.79) into the set of equations (9.67) and subsequent linearization gives the following set of equations for perturbations

$$\vartheta'' - \vartheta' - i\omega_c\vartheta = \Delta U \exp(\xi)$$

$$y'' - y' - i\sigma\omega_c y = a(1-U)\exp(\xi)$$

$$\vartheta_g'' - \vartheta_g' - i\sigma\omega_c\vartheta_g = -i\sigma\omega_c(\theta_0 - l)/\Gamma - i\sigma\omega_c(\Delta + l)\left[\frac{1}{\Gamma} - \frac{(U-1)}{i\sigma\omega_c}\right]\exp(\xi)$$

$$\vartheta_p'' - \vartheta_p' - i\sigma\omega_c\vartheta_p = -i\sigma\omega_c\frac{\theta_p^0}{\Gamma} \tag{9.83}$$

The solution of these equations provides the following spatial perturbation distributions

$$\vartheta = Aq\exp(z\xi) + Mq\exp((1-z)\xi) - U\Delta\exp\left(\frac{\xi}{i\omega_c}\right)$$

$$\vartheta_g = Fq\exp(z_1\xi) + Gq\exp(z_2\xi) + (\theta_p^0 - q)/\Gamma + (\Delta + l)\left[\frac{1}{\Gamma} - \frac{(U-1)}{i\sigma\omega_c}\right]\exp(\xi)$$

$$y = C\exp(z_1\xi) + D\exp(z_2\xi) + a(U-1)\exp\left(\frac{\xi}{i\sigma\omega_c}\right)$$

$$\vartheta_p = Sq\exp(z_1\xi) + Hq\exp(z_2\xi) + \theta_p^0/\Gamma \tag{9.84}$$

Here A, C, D, F, G, H, M, and S are integration constants, $2z_{1,2} = 1 \pm \sqrt{1 + 4i\sigma\omega_c}$, and $2z = 1 + \sqrt{1 + 4i\omega_c}$.

The solutions (9.84) must satisfy the boundary conditions

$$\vartheta = 0, \quad \xi \to -\infty$$

$$\xi = 0, \quad \vartheta = \vartheta_g, \quad \vartheta' = \vartheta_g' + (\theta_g^0)' - lU$$

$$aU - y + y' + (Y^0)' = 0, \quad y = -Y_s^0(1 + l\Gamma\vartheta_g/(\theta_s^0)^2)$$
$$\xi = \xi_f, \quad \vartheta_g + s(\theta_g^0)' = \vartheta_p + s(\theta_p^0)', \quad \vartheta_g' + s(\theta_g^0)'' = \vartheta_p{}' - q(s(Y^0)'' + y')$$
$$1 + s - y' = k\vartheta_p/\Delta + \iota, \quad y = s$$
$$\theta_p < \infty, \quad \xi \to \infty \tag{9.85}$$

The latter follow from linearization of (9.68).

The first and the last of these conditions give $M = S = 0$. The rest lead to the following eight relationships between the remaining integration constants and the unknowns U and s

$$aU\frac{(\hat{\delta} - 1/\sigma)}{i\omega_c} - A + F + G = -\frac{a}{i\sigma\omega_c} - \frac{\theta_s^0}{\Gamma q}$$
$$aU\left[1 + \frac{1}{i\sigma\omega_c} - \hat{\delta}\left(1 + \frac{1}{i\omega_c}\right)\right] + Az - Fz_1 - Gz_2 = a\left(1 + \frac{1}{i\sigma\omega_c} + \frac{1}{\Gamma}\right)$$
$$-Ua\frac{[r/\sigma + \hat{\delta}(1-r)]}{i\omega_c} + A(1-r) - (C+D)r = \left(1 - a - \frac{a}{i\sigma\omega_c}\right)r$$
$$Ua - Cz_2 - Dz_1 = a$$
$$(C+F)e_1 + (D+G-H)e_2 = 0$$
$$(C+F)z_1e_1 + (D+G-H)z_2e_2 = -\frac{1}{\Gamma}$$
$$Cz_2e_1 + Dz_1e_2 - kqHe_2/\Delta = \frac{k\theta_p^0}{\Gamma\Delta} - 1 + \iota$$
$$s = Ce_1 + De_2 + \frac{(U-1)}{i\sigma\omega_c} \tag{9.86}$$

where the following notations are introduced

$$\hat{\delta} = \frac{\Delta}{(\Delta + l)}, \quad e_1 = \exp(z_1\xi_f^0), \quad e_2 = \exp(z_2\xi_f^0) \tag{9.87}$$

The set of equations (9.86) allows the values of U, F, G, and H, necessary for calculating acoustic admittance, to be found. The unknowns A, C, and D may be eliminated in the following manner (the unknown s is contained in just one equation). The fifth and the sixth equations give (let $z_0 = z_1 - z_2$)

$$C = -F - \frac{1}{\Gamma z_0 e_1}, \quad D = H - G + \frac{1}{\Gamma z_0 e_2} \tag{9.88}$$

while the unknown A is easily expressed through U, F, and G from the first equation. Substituting these expressions into the remaining equations the set of four equations to determine U, F, G, and H is obtained

$$aU\left[\frac{\hat{\delta}(z-1)}{z} + \frac{1}{\sigma z} - 1\right] + F(z_1 - z) + G(z_2 - z) =$$
$$a\left(\frac{1}{\sigma z} - 1\right) + \frac{(\theta_s^0 z/q - a)}{\Gamma}$$
$$aU + Fz_2 + (G-H)z_1 = a + \frac{(z_1/e_2 - z_2/e_1)}{\Gamma z_0}$$

$$\frac{aU}{z_1 z_2} + F + G - Hr = r(1-a)$$
$$\frac{a}{z_1 z_2} + \frac{[r(1/e_2 - 1/e_1)/z_o - (1-r)\theta_s^0/q]}{\Gamma}$$
$$Fz_2 e_1 + Gz_1 e_2 + (kq/\Delta - z_1)He_2 =$$
$$1 - \imath + \frac{(1 - k\theta_p^0/\Delta)}{\Gamma} \tag{9.89}$$

These equations allow the burning rate response U and the value H, related to the amplitude of flame temperature perturbation, to be found. To make the expressions shorter and more tractable the following notations are introduced

$$S_1 = (z-1)/z, \quad S_2 = z_1/e_2 - z_2/e_1$$
$$S_3 = \Delta z_1 z_2 (e_1 - e_2)(1-r)/q, \quad S_4 = z_1 + r(z-1)$$
$$S_5 = rz_1 z_2 (1-a)(e_1 - e_2), \quad S_6 = \frac{q(z_2\theta_s^0/q - a + 1/e_1)}{\Delta}$$
$$S_7 = \frac{q[z_0(\theta_p^0 - e_2\theta_s^0)/q - 1 + e_2/e_1]}{\Delta}$$
$$S_8 = \frac{q[a(e_1 - e_2)/z_0 + \theta_s^0/q]}{\Delta} \tag{9.90}$$

In this notation

$$U = \frac{(U^{(1)} + U^{(2)}/\Gamma)}{X}, \quad H = \frac{(H^{(1)} + H^{(2)}/\Gamma)}{X}$$
$$X = S_2 S_4 + S_1(S_3 - ke_2 z_0)$$
$$U^{(1)} = S_4[S_2 - (1-\imath)z_0] - \mu k e_2 z_0 (z-1) + (z - z_2)S_5$$
$$U^{(2)} = S_3 S_6 + k(S_4 S_7 - e_2 z_0 S_6)$$
$$H^{(1)} = \Delta S_1[S_2(1 - \mu z) + S_5 - z_0(1-\imath)]/q$$
$$H^{(2)} = kz_0\theta_p^0 S_1/q - \Delta[S_8(S_1 S_3 + S_2 S_4) + S_2 S_6]/q, \tag{9.91}$$

Nonisentropicity of the flame front N_f is related to the value H and the flame temperature response Θ as

$$N_f = qe_2 H\Gamma/\theta_p^0, \quad N_f = \Theta\Gamma - 1 \tag{9.92}$$

If N_f and Θ are presented in the form

$$N_f = (N_f^{(1)} + N_f^{(2)}/\Gamma)/X, \quad \Theta = (\Theta^{(1)} + \Theta^{(2)}/\Gamma)/X \tag{9.93}$$

then (9.92) would hold for each of the components of these variables.

Finally, the integral J contained in expression (9.82) for the gas velocity response at the flame front needs be calculated. Substituting the steady-state temperature distribution θ^0 and the perturbation ϑ, and integrating

$$J = i\sigma\omega_c \left\{ Fq(e_1 - 1)/z_1 + Gq(e_2 - 1)/z_2 + (\theta_0 - l)\left(\frac{1}{\Gamma} - 1\right)\xi_f^0 \right.$$
$$\left. + q(1-a)\left[1/\Gamma - 1 - \frac{(U-1)}{i\sigma\omega_c}\right]\right\} /\theta_p^0 \tag{9.94}$$

From the second and the fourth equations in (9.89)

$$Fq(e_1 - 1)/z_1 + Gq(e_2 - 1)/z_2 = 1 - \imath + a(U - 1)$$
$$+[(z_1 - kq/\Delta)e_2 - z_1]H + [1 - k\theta_p^0/\Delta - S_2/z_0]/\Gamma \tag{9.95}$$

Consequently

$$J = q\{\imath - U + (1 - a)z_1z_2 - [(z_1 - kq/\Delta)e_2 - z_1]H$$
$$-[1 - k\theta_p^0/\Delta - S_2/z_0 + (1 - a)z_1z_2]/\Gamma\}/\theta_p^0 +$$
$$(q/\theta_p^0 - 1)(1 - 1/\Gamma)z_1z_2 \ln a \tag{9.96}$$

Finally, based on (9.65), (9.82), (9.92), and (9.96), the expression for acoustic admittance takes the form

$$\zeta/\gamma M = (1 - 1/\Gamma)[(\theta_p^0 - q)(1 + z_1z_2 \ln a)/\theta_p^0 -$$
$$(\theta_p^0 - \theta_s^0)z_1z_2/\theta_p^0] + q\left[1 - \imath - \frac{(k\theta_p^0/\Delta + S_2/z_0)}{\Gamma}\right]/\theta_p^0 +$$
$$(q/\theta_p^0 - 1)U - (z_1/e_2 + kq/\Delta)N_f/\Gamma \tag{9.97}$$

where U and N_f are calculated using (9.91) and (9.92).

In the t_c approximation ($\sigma = 0$)

$$z_1 = 1, \quad z_2 = 0, \quad e_1 = 1/a, \quad e_2 = 0$$
$$S_1 = (z - 1)/z, \quad S_2 = 1, \quad S_3 = 0$$
$$S_4 = 1 + r(z - 1), \quad S_5 = 0, \quad S_6 = 0$$
$$S_7 = 0, \qquad S_8 = \theta_p^0/\Delta \tag{9.98}$$

Furthermore

$$X = 1 + (r - k/z)(z - 1), \quad U^{(1)} = \imath + (\imath r - \mu k)(z - 1), \quad U^{(2)} = 0$$
$$H^{(1)} = \Delta(z - 1)(\imath/z - \mu)/q, \quad H^{(2)} = -\theta_p^0 X/q$$
$$N_f^{(1)} = qH^{(1)}\Gamma/\theta_p^0, \quad N_f^{(2)} = -\Gamma X, \quad \Theta^{(1)} = \Gamma N_f^{(1)}, \quad \Theta^{(2)} = 0 \tag{9.99}$$

and for the acoustic admittance the formula

$$\zeta = -\gamma MG \tag{9.100}$$

is obtained, where G is determined by expressions (4.58) and (4.62).

The results obtained can be illustrated by considering an example with the following values of the parameters (Allison and Faeth 1975): Q_g = 4862 kJ/kg, L = 1690 kJ/kg, $c = 3.05$ kJ/(kg × K), $\rho = 10^3$ kg/m^3, $\lambda = 0.383$ W/(m × K), $\kappa = 1.256 \times 10^{-7}$ m^2/s, $\gamma = 1.126$, $\Gamma = 8.937$, $T_a = 298\,K$, $n = 2$, $E = 118$ kJ/mol, and $T_p^0 = 1338$ K.

For the reference pressure $p_r = 1.013 \times 10^5$ Pa, the boiling temperature is set to $T_{bl,r}$ = 386.7 K. For these values of pressure and temperature $D_r = 2.142 \times 10^{-5}$ m^2/s, $\rho_r = 0.7865$ kg/m^3, and $\sigma_r = 10^{-4}$.

The boiling temperature and parameter σ at any given pressure p^0 can be calculated as

$$T_{bl} = T_{bl,r}\left[1 - \left(\frac{cT_r}{L\Gamma}\right)\ln\left(\frac{p^0}{p_r}\right)\right]^{-1}, \quad \sigma = \sigma_r p^0/p_r \tag{9.101}$$

The steady-state regime loses stability at high pressures. The stability boundary of the steady-state regime is determined by the condition $X = 0$ (Novozhilov 1988a).

For the values of the parameters adopted above the pressure at the stability boundary turns out to be slightly less than 100 atm (specifically, $p^* = 1.008 \times 10^7$ Pa), and the frequency at the boundary is $\omega_c^* = 7.77$. For comparison, the values of the same parameters in the t_c approximation ($\sigma_r = 0$) are $(p^*)^c = 8.286 \times 10^6$ Pa and $(\omega_c^*)^c = 16.9$. Therefore, at the stability boundary, the inertia of the gas phase is responsible for a ~20% increase in the maximum possible pressure in a steady-state regime, and for a corresponding decrease of the natural frequency of the system by a factor larger than 2.

The adopted assumption of an infinitely narrow chemical reaction zone in the gas phase imposes a restriction on the range of frequencies that may be considered under this approximation. Since the relaxation time of chemical reaction in the gas t'_g is considered to be equal to zero, then the admissible frequencies are those satisfying the inequality $\omega t'_g \ll 1$.

A reasonable estimation of the relaxation time t'_g can be obtained assuming that the ratio t'_g/t_g is of the same order of magnitude as the ratio of the characteristic temperature intervals of the preheat zone $T_f^0 - T_s^0$ and the chemical reaction zone $R(T_f^0)^2/E$

$$t'_g/t_g \sim \frac{R(T_f^0)^2}{E(T_f^0 - T_s^0)} \tag{9.102}$$

for adopted values of the parameters $t'_g/t_g \sim 0.1$. It is easy to see that the condition $\omega t'_g = 1$ implies $\omega_c = 10^5 \left(\frac{p_r}{p^0}\right)$. The calculations illustrated below are performed up to this value of frequency.

Figure 9.4 illustrates variation of the real part of the acoustic admittance with frequency. Curves 1–3 correspond to the pressure values 10^5, 10^6, and 10^7 Pa. It is evident that acoustic wave amplification by the surface ($\text{Re}\,\zeta < 0$) is possible in the present model only at frequencies that are reasonably close to the inverse of the condensed phase relaxation time. The resonance frequency of the heated layer of the condensed phase is of the order of $\sqrt{k}/(rt_c)$

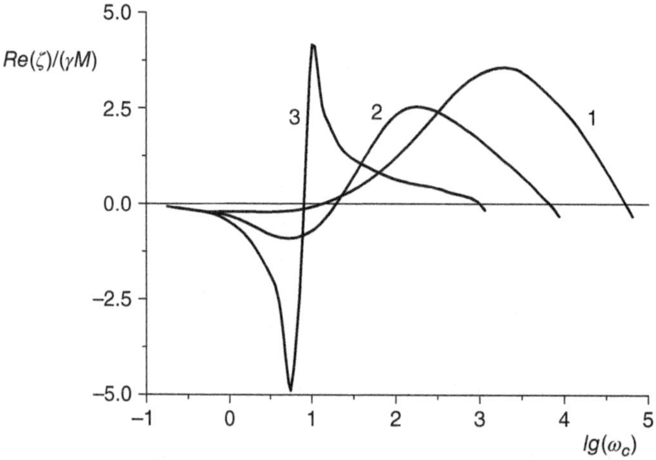

Figure 9.4 Variation of the real part of acoustic admittance with frequency. 1, $p^0 = 10^5$ Pa; 2, $p^0 = 10^6$ Pa; 3, $p^0 = 10^7$ Pa.

(Novozhilov 1965b). The adopted specific values of the model parameters lead to small values of the parameter r. Consequently, the resonance frequency is somewhat larger than $1/t_c$. For example, at a pressure of 10^7 Pa, $r = 0.074$ and therefore the amplification is observed not at $\omega_c \sim 1$, but at $\omega_c \sim 10$. A very sharp resonance at 10^7 Pa appears due to the fact that this pressure is close to the critical one where the stability of the steady-state regime is lost.

These calculations were conducted for a Lewis number equal to one. Investigations of gaseous flames (e.g. McIntosh 1987) show that at this Lewis number, the response of the preheat zone in the gas to varying pressure does not exhibit resonance behaviour. For this reason, no amplification in acoustic waves at frequencies comparable to the inverse of the preheat zone relaxation time is observed.

Figure 9.5 shows the dependencies of the imaginary part of the acoustic admission on frequency. The pressure values are the same as in Figure 9.4.

The influence of the gas-phase inertia on acoustic admittance is illustrated in Figure 9.6 where the real part of the admittance is plotted for a pressure of 10^6 Pa. The solid curve corresponds to the case where gas-phase inertia is taken into account while the dashed curve is obtained for the t_c approximation.

In general, the flame front is nonadiabatic. Figure 9.7 shows, at a pressure of 10^6 Pa, the real (curve 1) and imaginary (curve 2) parts of the degree of isentropicity I. As explained earlier for the isentropic front $\mathrm{Re}(I) = 1$, $\mathrm{Im}(I) = 0$.

As has been pointed out already, there have been several numerical investigations of acoustic admittance of the propellant surface, taking into account gas phase inertia. Nearly all of them used combustion models that are different from the Belyaev model. For example, Volkov and Medvedev (1969), T'ien (1972), and Novikov et al. (1974) considered two-stage combustion models with one chemical reaction occurring in the condensed phase and another in the gaseous phase. The inertia of the chemical reaction zone in the condensed phase was neglected. Apparently, this was done to simplify the problem. Such an approach is of course inconsistent, as the influence of the latter zone is comparable with the influence of inertia of the gas phase. T'ien (1972) identified a high-frequency

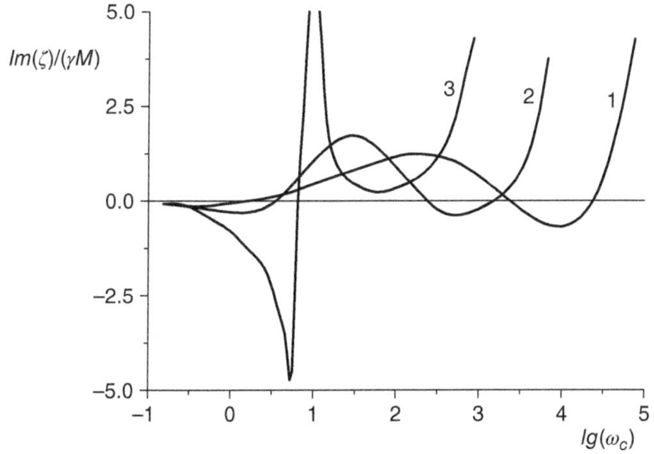

Figure 9.5 Variation of the imaginary part of acoustic admittance with frequency. 1, $p^0 = 10^5$ Pa; 2, $p^0 = 10^6$ Pa; 3, $p^0 = 10^7$ Pa.

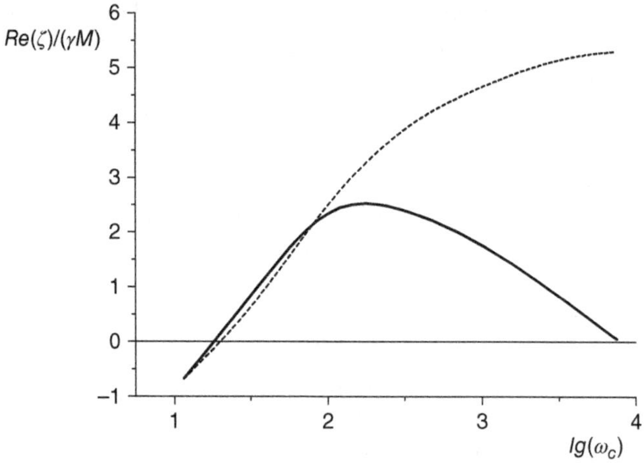

Figure 9.6 Influence of gas-phase inertia on acoustic admittance. $p^0 = 10^6$ Pa. Solid line, calculation taking gas-phase inertia into account; dashed line, t_c approximation.

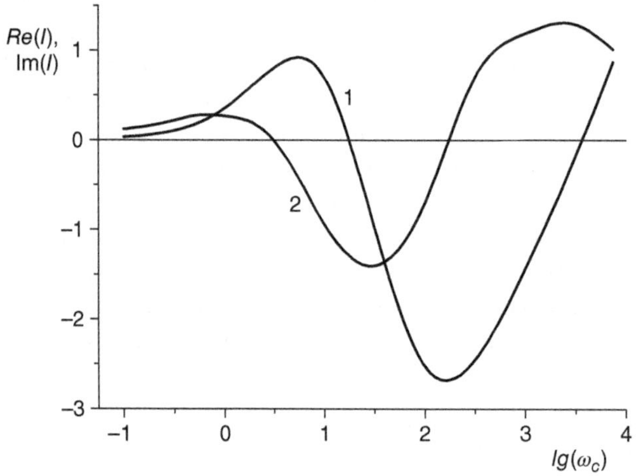

Figure 9.7 Real and imaginary parts of the degree of isentropicity I. 1, real part; 2, imaginary part.

region where amplification of acoustic waves on reflection from the surface of the burning fuel occurs. It should be noted, however, that this effect is only observed at frequencies $\omega \sim 1/t'_g$, that is it is related to the inertia of the chemical reaction zone in the gas. An identical result was obtained by Novikov et al. (1974), and by Ryazantsev and Tylskikh (1976) within the framework of the model where the inertia of the reaction zone was neglected.

The model considered by Allison and Faeth (1975) is the most similar to the one used here. Figure 9.8 compares the results obtained here ($p^0 = 10^6$ Pa) for the modulus of the burning rate response (curves 1 and 1′) and the real part of the acoustic admittance (curves 2 and 2′) with the same quantities found by numerical integration of the linearized equations (Allison and Faeth 1975). The solid curves correspond to the results of this section while

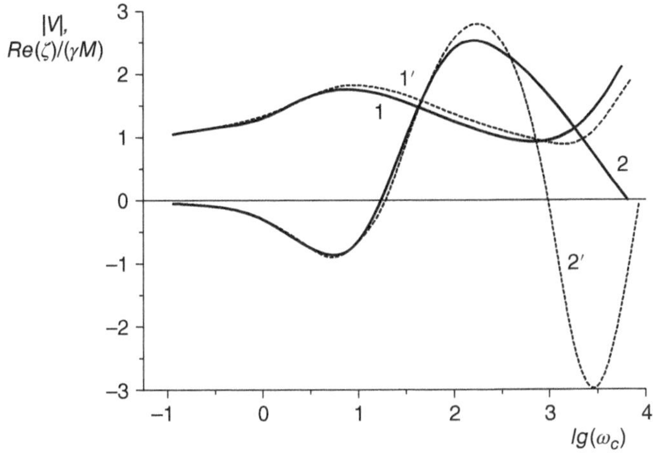

Figure 9.8 Comparison of results of this chapter with the numerical analysis of Allison and Faeth (1975). $p^0 = 10^6$ Pa. 1, modulus of the burning rate response (this chapter); 1′, modulus of the burning rate response (Allison and Faeth 1975); 2, real part of the acoustic admittance (this chapter); 2′, real part of the acoustic admittance (Allison and Faeth 1975).

the dashed curves correspond to the study by Allison and Faeth (1975). The latter did not use the assumption of an infinitely thin reaction zone. As can be seen from Figure 9.8, large discrepancies in predictions of the acoustic admittance emerge at high frequencies ($\omega t_g' \sim 1$ or $\lg(\omega_c) \sim 4$). An account of the inertia of the chemical reaction zone by Allison and Faeth (1975) shows the possibility of acoustic waves amplification at frequencies of the order of the inverse of the relaxation time of this zone. We should note the following: T'ien (1972) matched the numerical solution in the gas-phase zone (adjacent to the surface and inclusive of the chemical reaction zone) with an incorrect asymptotic away from the reaction zone (the latter asymptotic did not take account of the heat transfer process). Consequently, the results of T'ien (1972) may be incorrect at high frequencies. The method employed by Allison and Faeth (1975) is similar to that of T'ien (1972), but details of the numerical scheme are not available. It is hard to judge whether the results of Allison and Faeth (1975) are correct.

A comparison of the results from this section with the numerical approaches leads to the conclusion that the negative values of the real part of acoustic admittance at high frequencies obtained earlier (T'ien 1972; Allison and Faeth 1975) are related to inertia of the chemical reaction zone.

There is therefore a need to investigate the problem for Lewis numbers different from unity. In this case wave amplification at frequencies corresponding to the characteristic frequencies of the gas preheat zone ($\omega t_g' \sim 1$) may be expected.

9.5 Combustion and Extinction Under Depressurization

Propellant combustion and extinction under pressure drop conditions were considered in Section 7.7, where the t_c approximation was employed. As discussed in Section 9.1

and also by Novozhilov (1988a,b), accounting for gas-phase inertia may be necessary for consideration of nonsteady propellant combustion in certain circumstances. This section examines the effect of inertia of the gas phase on transient and extinction processes under depressurization, making use of the Belyaev combustion model (Novozhilov et al. 2010).

In contrast to Sections 9.2–9.4, the finite rate of the gas-phase chemical reaction is assumed in the present section.

Accordingly, the model formulation is

$$\rho c \frac{\partial T}{\partial t} = \frac{\partial}{\partial x}\left(\lambda \frac{\partial T}{\partial x}\right), \quad -\infty < x < x_s(t) \tag{9.103}$$

for the condensed phase. The gas-phase equations, assuming the ideal gas model, enforce a balance of mass, reactant, and energy

$$\begin{aligned} &\frac{\partial \rho_g}{\partial t} + \frac{\partial(\rho_g u_g)}{\partial x} = 0, \quad x_s(t) \le x < \infty \\ &\rho_g = \frac{\tilde{\mu} p}{R T_g} \\ &\rho_g c_p \left(\frac{\partial T_g}{\partial t} + u_g \frac{\partial T_g}{\partial x}\right) = \frac{\partial}{\partial x}\left(\lambda_g \frac{\partial T_g}{\partial x}\right) + Q_g W(Y, T_g) + \frac{dp}{dt} \\ &\rho_g \left(\frac{\partial Y}{\partial t} + u_g \frac{\partial Y}{\partial x}\right) = \frac{\partial}{\partial x}\left(D\rho_g \frac{\partial Y}{\partial x}\right) - W(Y, T_g) \end{aligned} \tag{9.104}$$

Here u_g is the gas velocity and Y is the reactant mass fraction.

For a first-order reaction with respect to reactant concentration

$$W(Y, T_g) = \tilde{z} \rho_g Y \exp(-E/RT_g) \tag{9.105}$$

where $\tilde{z}$ and E are the pre-exponential factor and the activation energy of the chemical reaction, respectively.

The required boundary conditions are

$$\begin{aligned} &x \to -\infty, \quad T = T_a \\ &x \to \infty, \quad \frac{\partial T_g}{\partial x} = 0, \quad \frac{\partial Y}{\partial x} = 0 \end{aligned} \tag{9.106}$$

At the phase interface $x_s(t)$ the following matching conditions are required

$$\begin{aligned} &T = T_g \\ &\lambda \frac{\partial T}{\partial x} = \lambda_g \frac{\partial T_g}{\partial x} - \rho u L \\ &\rho u = \rho_g u + \rho_g u_g \\ &\rho u = \rho_g u Y + \rho_g u_g Y - D\rho_g \frac{\partial Y}{\partial x} \end{aligned} \tag{9.107}$$

as well as the condition of equilibrium evaporation

$$Y = \frac{p_r}{p} \exp\left[\frac{L\tilde{\mu}}{R}\left(\frac{1}{T_{bl,r}} - \frac{1}{T}\right)\right] \tag{9.108}$$

Here $u = -dx_s(t)/dt$ is the linear burning rate, T_a is the initial temperature, p is the pressure, $T_{bl,r}$ is the boiling temperature at the reference pressure $p_r = 1$ atm, and L is the latent heat of evaporation.

In considered transient combustion regimes pressure varies in time as

$$t < t_i, \quad p = p_i$$
$$t \geq t_i, \quad p = p_f + (p_i - p_f)\exp(-\alpha_p(t - t_i)) \tag{9.109}$$

Nondimensional variables and parameters are introduced as follows

$$\eta = \frac{p}{p_i}, \quad \eta_f = \frac{p_f}{p_i}, \quad \eta_r = \frac{p_r}{p_i}$$
$$\varepsilon = \frac{E}{RT_{bl,r}}, \quad l = \frac{L}{cT_{bl,r}}, \quad q = \frac{Q_g}{cT_{bl,r}}$$
$$a = \frac{T_a}{T_{bl,r}}, \quad b = \frac{T_b^0}{T_{bl,r}}, \quad s = \frac{T_s^0}{T_{bl,r}}$$
$$\theta = \frac{T - T_a}{T_s^0 - T_a}, \quad \theta_g = \frac{T_g - T_a}{T_s^0 - T_a} \tag{9.110}$$

where T_b^0 and T_s^0 are the steady-state values of the combustion and surface temperatures at pressure p_i, respectively.

In order to scale time and spatial coordinates the steady-state burning rate at pressure p_i and initial temperature T_a are introduced (Zeldovich and Frank-Kamenetskii 1938)

$$(m_i^0)^2 = 2\tilde{z}(D\rho_g^2)_i \left[\frac{cR(T_b^0)^2}{Q_g E}\right] \exp\left(-\frac{E}{RT_b^0}\right) \tag{9.111}$$

Accordingly, the linear burning rate u_i^0 in the Belyaev model is

$$(u_i^0)^2 = 2U^2 \frac{\tilde{z}(D\rho_g^2)_i}{\rho^2} \left(\frac{b^2}{\varepsilon q}\right)^2 \exp\left(-\frac{\varepsilon}{b}\right) \tag{9.112}$$

where the product $(D\rho_g^2)_i$ is estimated at pressure p_i. At $U = 1$ the latter expression gives the burning rate in the assumption of an infinitely thin reaction zone. The way to determine the coefficient U is discussed below.

Furthermore, since $(D\rho_g^2)_i = (D\rho_g^2)_r/\eta_r$ the burning rate may be expressed as

$$(u_i^0)^2 = 2\tilde{z}U^2\kappa\sigma \left(\frac{b^2}{\varepsilon q}\right)^2 \exp\left(-\frac{\varepsilon}{b}\right) \tag{9.113}$$

where

$$\sigma = \frac{\sigma_r}{\eta_r}, \quad \sigma_r = \frac{(D\rho_g^2)_r}{\kappa\rho^2} \tag{9.114}$$

The parameter σ describes the relative thermal inertia of the gas and condensed phases.

The relationship (9.113) allows time, and the burning rate and pressure change rate scales to be defined as follows

$$\tau = \frac{(u_i^0)^2}{\kappa} t, \quad v = \frac{u}{u_i^0}, \quad \alpha = \alpha_p \frac{\kappa}{(u_i^0)^2} \tag{9.115}$$

Transition from the laboratory frame to the one fixed at the phase interface is performed using transformation to the new nondimensional variable $-\infty < \xi \leq 0$; $\xi = \frac{u_i^0}{\kappa}[x - x_s(t)]$.

The heat transfer equation then takes the form

$$\frac{\partial \theta}{\partial \tau} + v\frac{\partial \theta}{\partial \xi} = \frac{\partial^2 \theta}{\partial \xi^2} \tag{9.116}$$

In the gaseous phase it is conventional to use the following coordinate

$$0 \le \xi < \infty, \qquad \xi = \frac{\rho u_i^0 \eta_r}{(D\rho_g^2)} \int_{x_s(t)}^{x} \rho(y,t)\mathrm{d}y \tag{9.117}$$

As a result, the following set of conservation equations in the gas phase is obtained

$$\sigma\frac{\partial \theta_g}{\partial \tau} + v\frac{\partial \theta_g}{\partial \xi} = \eta\frac{\partial^2 \theta_g}{\partial \xi^2} + \frac{zqY}{(s-a)U^2} \times \exp\left\{\frac{\varepsilon}{b}\left(1 - \frac{b}{a+\theta(s-a)}\right)\right\} + \frac{\sigma}{\Gamma}\frac{(a+\theta(s-a))}{(s-a)\eta}\frac{\mathrm{d}\eta}{\mathrm{d}\tau}$$

$$\sigma\frac{\partial Y}{\partial \tau} + v\frac{\partial Y}{\partial \xi} = \eta\frac{\partial^2 Y}{\partial \xi^2} - \frac{zY}{U^2}\exp\left\{\frac{\varepsilon}{b}\left(1 - \frac{b}{a+\theta(s-a)}\right)\right\} \tag{9.118}$$

where

$$z = \frac{1}{2}\left(\frac{\varepsilon q}{b^2}\right)^2 \tag{9.119}$$

The boundary conditions (9.106–9.108) are written in nondimensional variables as

$$\xi \to -\infty, \quad \theta = 0$$

$$\xi \to \infty, \quad \frac{\partial \theta_g}{\partial \xi} = 0, \quad \frac{\partial Y}{\partial \xi} = 0$$

$$\xi = 0, \quad \theta = \theta_g, \quad \frac{\partial \theta}{\partial \xi} = \eta\frac{\partial \theta_g}{\partial \xi} - \frac{vl}{(s-a)}$$

$$v(1-Y) + \eta\frac{\partial Y}{\partial \xi} = 0$$

$$\eta Y = \eta_r \exp\left\{l\Gamma\left(1 - \frac{1}{a+\theta(s-a)}\right)\right\} \tag{9.120}$$

The parameter Γ is expressed through the specific heat ratio $\Gamma = \gamma/(\gamma-1)$.
The nondimensional pressure changes according to

$$\tau < \tau_i, \qquad \eta = 1$$

$$\tau \ge \tau_i, \qquad \eta = \eta_f + (1-\eta_f)\exp(-\alpha(\tau-\tau_i)) \tag{9.121}$$

The set of equations (9.116–9.121) describes nonsteady propellant combustion under depressurization, taking into account the thermal inertia of the gas phase. The value $\sigma = 0$ corresponds to problem formulation in the t_c approximation.

Details of the numerical method were discussed by Novozhilov et al. (2010). The following parameter values are used to obtain the results discussed below: $a = 0.7$, $l = 1.43$, $q = 4.12$, $\varepsilon = 33.9$; $\Gamma = 8.94$, $z = 73.8$, $\sigma_r = 10^{-4}$. These correspond to the dimensional values used in the studies by Allison and Faeth (1975), and Novozhilov and Posvyanskii (1991).

The parameter s (surface temperature in the steady-state regime at pressure p_i) is found from the following considerations. In the steady-state regime, as follows from (9.104), there exists a similarity between the temperature and concentration fields

$$Y^0 = \frac{c}{Q}(T_b^0 - T_g^0) \tag{9.122}$$

and therefore at the surface

$$Y_s^0 = \frac{c}{Q}(T_b^0 - T_s^0) \tag{9.123}$$

On the other hand, condition (9.108) gives in the steady-state regime

$$Y_s^0 = \frac{p_r}{p_i} \exp\left[\frac{L\widetilde{\mu}}{R}\left(\frac{1}{T_{bl,r}} - \frac{1}{T_s^0}\right)\right] \tag{9.124}$$

The relationships (9.123) and (9.124) lead to the following equation, which determines the parameter s

$$b - s = q\eta_r \exp\left[l\Gamma\left(1 - \frac{1}{s}\right)\right] \tag{9.125}$$

The parameter U^2 ensures that the nondimensional burning rate v in the steady-state regime is equal to unity. Correspondingly, it is determined from solution of the set of equations (9.118) and (9.120), setting $v = 1$. By substituting (9.122) into the steady-state heat transfer equation for the gas phase, the following equation is obtained

$$\frac{d\theta_g^0}{d\xi} = \frac{d^2\theta_g^0}{d\xi^2} + \frac{z(\theta_b^0 - \theta_g^0)}{U^2} \exp\left[\frac{\varepsilon}{b}\left(1 - \frac{b}{(a + \theta^0(s - a))}\right)\right] \tag{9.126}$$

The solution of this equation with the boundary conditions

$$\xi = 0, \quad \theta_g^0 = 1, \quad \frac{d\theta_g^0}{d\xi} = 1 + \frac{l}{(s - a)}$$

$$\xi \to \infty, \quad \frac{d\theta^0}{d\xi} = 0 \tag{9.127}$$

determines the parameter U^2. It depends on pressure only slightly (Novozhilov and Posvyanskii 1991) and may be considered as constant. The solution of the above set of equations gives $U^2 = 0.7317$.

The following results assume that the initial steady-state regime at the initial pressure p_i is stable. Novozhilov (1988a) and Novozhilov and Posvyanskii (1991) demonstrated that there is an upper bound p_i^* above which the steady-state regime becomes unstable. They found that under the t_c approximation $p_i^* \approx 70$ atm. Accounting for gas-phase inertia increases this figure by about 10 atm.

Of the most interest are the extinction curves shown in Figure 9.9 which have nondimensional pressure decrease rate α and relative magnitude of pressure drop η_f as coordinates.

These curves separate two qualitatively different regimes. Above the extinction curve, the transition process ends up with a new steady-state regime at a pressure η_f. Below the curve, the burning rate drops to negligible values, which can be interpreted as extinction (Frost and Yumashev 1973; Lidskii et al. 1983, 1985).

The influence of the inertia of the gas phase on the position of the extinction curves may be inferred from Figure 9.9. The corresponding solid and dashed curves seem to be rather close to each other. However, at large pressure drop rates and at fixed pressure drop magnitude, the difference may be significant. For example, for the pair of the curves 1, at $\eta_f = 0.85$,

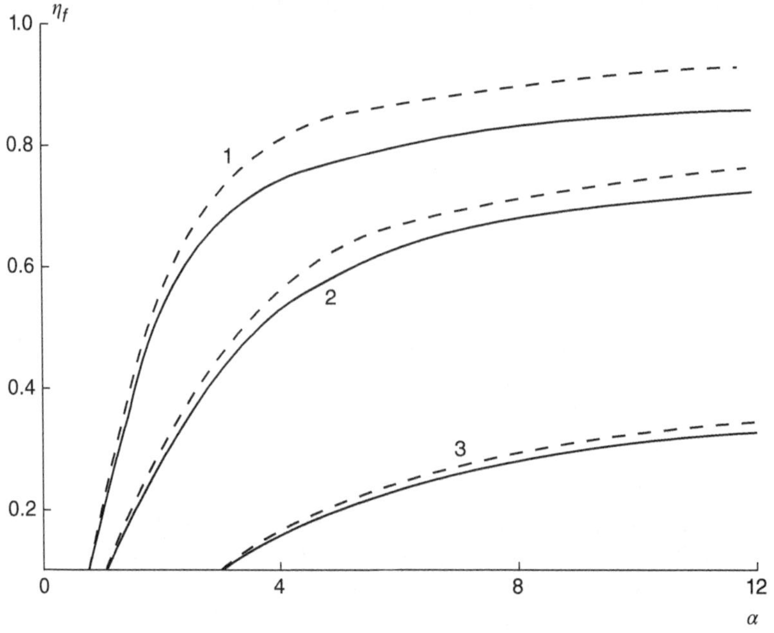

Figure 9.9 Extinction curves at different values of the initial pressure p_i and the nondimensional surface temperature s. Solid curves, thermal inertia of the gas phase is taken into account; dashed curves, t_c approximation. 1, $p_i = 50$ atm, $s = 1.336$; 2, $p_i = 30$ atm, $s = 1.272$; 3, $p_i = 10$ atm, $s = 1.153$. Source: Reproduced with permission from Novozhilov et al. (2010).

the critical values of the parameter α differ by approximately a factor of two for $\sigma = 0$ and $\sigma \neq 0$, respectively. This effect becomes weaker as the initial pressure decreases since the parameter σ is proportional to the pressure p_i.

The next two figures, Figures 9.10 and 9.11, show the burning rate time history for qualitatively different regimes: one asymptotically approaching the new steady-state regime and the other leading to extinction. To emphasize a sudden transition from one regime to another, the respective points (α, η_f) are chosen to be close, but on different sides of the extinction curve.

The specifics of the two qualitatively different types of transitional processes under pressure drop were first discussed by Frost and Yumashev (1973) using the t_c approximation. Their calculations demonstrated that parameters above the extinction curve lead to a transitional process with the burning rate approaching the steady-state value in the form of damped oscillations. This corresponds to the results presented in Section 7.7 (Figures 7.15 and 7.16).

The regime with parameters below the extinction curve features a monotonically decreasing burning rate, which eventually becomes much lower than that at initial pressure. In their model, Frost and Yumashev (1973) used burning laws in the form (8.42), which obviously does not admit a steady-state regime with a zero burning rate. Nevertheless, Frost and Yumashev (1973) explained why regimes with negligible nonzero burning rates may be considered as extinction. The reason is that at such a low burning rate rearrangement of the heated layer of the condensed phase will proceed until, eventually,

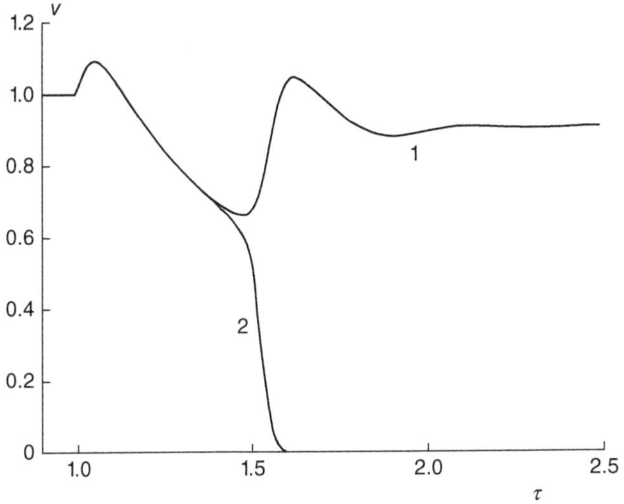

Figure 9.10 Burning rate time history at $\sigma = 0$, $\eta_f = 0.85$, $p_i = 50$ atm. 1, $\alpha = 4.85$; 2, $\alpha = 4.86$. Source: Reproduced with permission from Novozhilov et al. (2010).

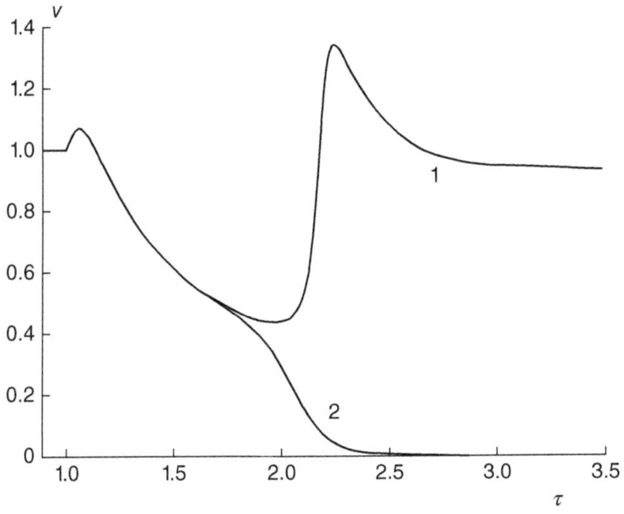

Figure 9.11 Burning rate time history at $\sigma = 5 \times 10^{-3}$, $\eta_f = 0.85$, $p_i = 50$ atm. 1, $\alpha = 9.82$; 2, $\alpha = 9.84$. Source: Reproduced with permission from Novozhilov et al. (2010).

an increase in the burning rate (secondary ignition) and attraction to a new steady-state regime at a finite pressure value occurs. However, at extremely small values of burning rate the rearrangement time of the heated layer, and accordingly the time of recovery from extinction conditions, are virtually infinite.

In contrast to the study of Frost and Yumashev (1973), who applied the ZN theory in its conventional form (t_c approximation), the model in this section takes account of the thermal inertia of the gas phase. This leads to the possibility of the existence of steady-state regimes with zero burning rate.

It should be recalled that in the development of the classical theory of steady-state combustion regimes (e.g. Frank-Kamenetskii (1969)) one has to truncate the function describing reaction rate at low temperatures. This is a necessary condition to obtain a strictly steady-state flame propagation regime and the inert state of the mixture at initial temperature.

In the present model, the required truncation should introduce, instead of (9.105), the following function

$$W(Y, T_g) = \begin{cases} \tilde{z}\rho_g Y \exp(-E/RT_g), & T_g \geq T_a + \Delta \\ 0, & T_g < T_a + \Delta \end{cases} \tag{9.128}$$

where $\Delta \ll T_a$. In such a case, evolution of a nonsteady combustion process under extinction conditions would lead to a progressive decrease in temperature within the chemical reaction zone. As a consequence, a stable state of the system with thermal equilibrium at initial temperature, zero burning rate, and very small reactant concentration will be achieved. The latter would be constant over the half-space occupied by the gas phase and given by

$$Y_\infty = \frac{\eta_r}{\eta_f} \exp\left(l\Gamma\left(1 - \frac{1}{a}\right)\right) \tag{9.129}$$

This expression follows from (9.120) at $\theta = 0$.

Figures 9.12–9.14 illustrate further the role of gas-phase inertia in the transition process. The parameters α and η_f in these figures are chosen in such a way that extinction occurs either taking into account inertia of the gas phase or ignoring it (t_c approximation, $\sigma = 0$).

The time histories of the burning rate and pressure are shown in Figure 9.12. It is evident from this plot that the inertia of the gas phase increases the time interval $\Delta\tau$ required for extinction by approximately a factor of two. At $\sigma = 0$, $\Delta\tau \approx 0.4$ while at $\sigma \neq 0$, $\Delta\tau \approx 0.8$.

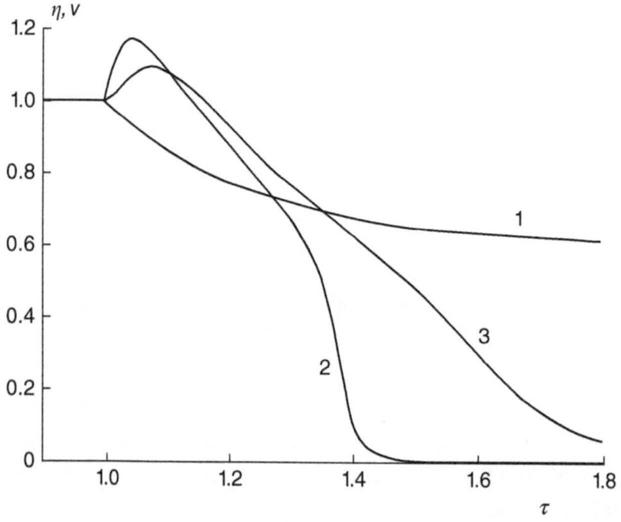

Figure 9.12 Pressure and burning rate time histories at $p_i = 50$ atm, $\alpha = 4$, $\eta_f = 0.6$. 1, $\eta(\tau)$; 2, $v(\tau)$, $\sigma = 0$; 3, $v(\tau)$, $\sigma = 5 \times 10^{-3}$. Source: Reproduced with permission from Novozhilov et al. (2010).

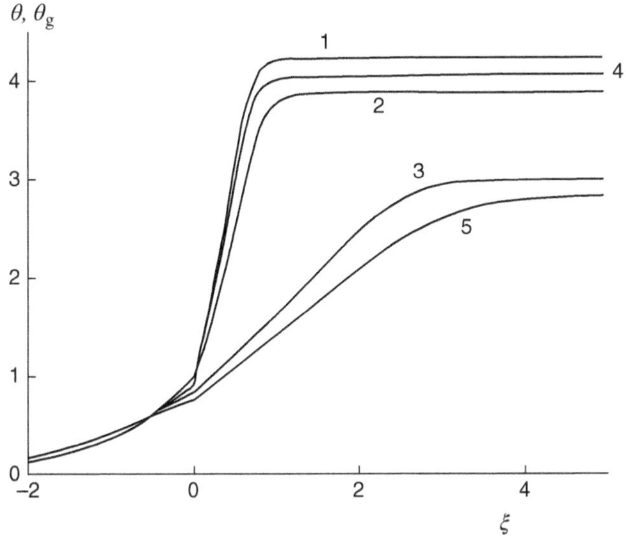

Figure 9.13 Time evolution of temperature profiles at $p_i = 50$ atm, $\alpha = 4$, $\eta_f = 0.6$. 1, initial temperature profile, $\tau = 1$; 2, $\sigma = 0$, $\tau = 1.35$; 3, $\sigma = 0$, $\tau = 1.4$; 4, $\sigma = 5\times10^{-3}$, $\tau = 1.4$; 5, $\sigma = 5\times10^{-3}$, $\tau = 1.8$. Source: Reproduced with permission from Novozhilov et al. (2010).

Figure 9.13 shows the evolution of temperature profiles in the condensed and gaseous phases with time. A comparison of profiles for $\sigma = 0$ and $\sigma \neq 0$ clearly demonstrates the role of inertia of the gas phase. For example, at $\tau = 1.4$ in the inertia-free approximation (curve 3) the combustion temperature is significantly different from its initial value (curve 1). On the other hand, for $\sigma \neq 0$ at the same moment in time (curve 4) the combustion temperature is very close to the initial value. This demonstrates quicker extinction in the inertia-free case $\sigma = 0$, compared to the case where $\sigma \neq 0$.

One of the boundary conditions (9.120) at the interface between the condensed and gaseous phases may be written in the form

$$v(\tau) = \frac{(s-a)}{l}(q(\tau) - \varphi(\tau)) \tag{9.130}$$

where

$$q = \eta \left.\frac{\partial \theta_g}{\partial \xi}\right|_{\xi=0}, \qquad \varphi = \left.\frac{\partial \theta}{\partial \xi}\right|_{\xi=0} \tag{9.131}$$

are the heat fluxes from the gaseous phase into the condensed phase.

The time dependencies of the latter quantities are demonstrated in Figure 9.14. In the inertia-free approximation (curves 1 and 2) the difference $q - \varphi$ decreases sharply at $\tau \sim 1.4$. For $\sigma \neq 0$ this happens much later, at $\tau \sim 1.8$. As follows from (9.130), a rapid decrease of the burning rate (extinction) will occur at the indicated times. Obviously, Figures 9.12 and 9.14 conform with each other.

The major conclusion of this section is that at sufficiently high pressures, despite parameter σ (describing the relative inertia of the gas phase compared to the condensed phase) being small, the transient combustion process behaviour may be significantly different depending on whether gas-phase inertia is taken into account or not. These differences are

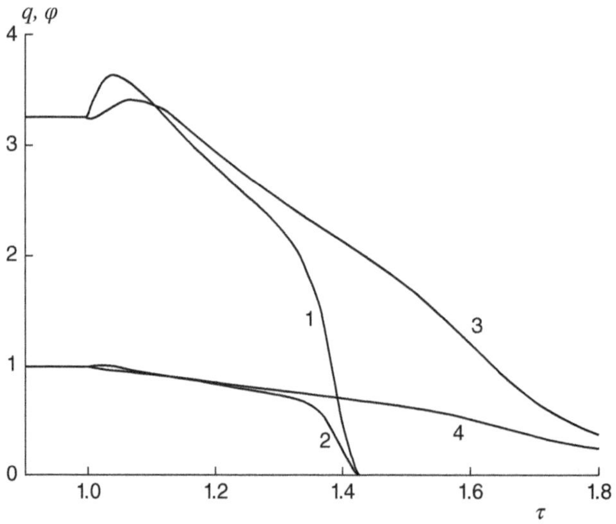

Figure 9.14 Time dependencies of the heat fluxes q and φ at $p_i = 50$ atm, $\alpha = 4$, $\eta_f = 0.6$. 1, $\sigma = 0$, $q(\tau)$; 2, $\sigma = 0$, $\varphi(\tau)$; 3, $\sigma = 5\times10^{-3}$, $q(\tau)$; 4, $\sigma = 5\times10^{-3}$, $\varphi(\tau)$. Source: Reproduced with permission from Novozhilov et al. (2010).

observed for both the critical pressure drop rates leading to extinction and the extinction times. The latter differ by approximately a factor of two.

Such a strong effect caused by a small parameter is related to the fact that the t_c approximation implies a mathematically incorrect procedure of neglecting the time derivative, multiplied by the small parameter σ. This fact was pointed out in Section 9.2.

9.6 t_r Approximation

The analysis of the influence of the gas-phase inertia on nonsteady solid propellant combustion carried out in Sections 9.2–9.5 was based on consideration of the Belyaev model. Consistent extension of the ZN theory beyond the t_c approximation requires a generalized approach free from the assumptions of any specific combustion model. This section describes such an approach.

The following development is based on the phenomenological concept of lag time. Generally, this concept is used to assist in the analysis of the operating regimes of liquid-propellant engines (Crocco and Cheng 1956). It helps to account for the intrinsic inertia of various complex physical and chemical processes (atomization and evaporation of droplets, reactant mixing, and the combustion process itself) via the use of a single parameter, that is, the lag time. This parameter is generally a function of pressure and other control parameters of the process. Despite being rather simplistic, such an approach satisfactorily describes experimentally observed operational regimes and can predict, within a certain degree of accuracy, the important properties of a specific rocket engine. For example, the stability conditions for operating an engine in a steady-state regime may be established in this way.

Similar attempts have been made in the context of nonsteady solid propellant combustion, for example by Culick (1968). However, this study also applied the lag time concept to the description of the inertia of the propellant. This is unnecessary as the inertia of the condensed phase can be accounted for very accurately by the ZN theory within the t_c approximation. Volkov and Medvedev (1969) introduced lag time in order to calculate the acoustic admittance of the propellant surface, but their study was limited to the consideration of a specific propellant model.

The present section describes a consistent approach introducing lag time to account for the relaxation of quick-response zones in the basic ZN theory based on the t_c approximation (Novozhilov 1988b, 1992a,). This combination results in an extension of the ZN theory called the t_r approximation.

Similar to the t_c approximation, the main idea of the t_r approximation is to use dependencies at steady-state burning conditions. The lag time and its dependence on external control parameters (such as pressure, initial temperature, velocity of the tangential gas flow, radiation intensity, and others) should be approximated using the steady-state dependencies of the burning rate, surface temperature, flame temperature, and other parameters on external conditions and on the structure of the chemical transformation and gas preheat zones. These steady-state dependencies should be known from experiments.

The treatment of the condensed phase does not change under the t_r approximation, that is, its inertia is taken into account by means of the heat transfer equation (2.28). Modifications concern the nonsteady burning laws (2.27) where the time lag is introduced. It is assumed that the combination $T_e = T_s - \kappa f/u$ and the pressure p do not affect the burning rate and the surface temperature immediately, but only after some delay.

The relationships (2.27) are modified as

$$
\begin{aligned}
m(t) &= U(T_e(t - t_{Ue}), p(t - t_{Up})) \\
T_s(t) &= V(T_e(t - t_{Ve}), p(t - t_{Vp}))
\end{aligned}
\tag{9.132}
$$

The introduced lag times have two indices: the first corresponds to the function where the lag time is introduced, and the second to the parameter affected by the delay. Generally, all four lags are different.

A very important consideration is that while within the t_c approximation one may formally turn from the implicit dependencies (2.27) on the parameters f and p, external with respect to the reaction zone, to the explicit dependencies $m(t, p)$ and $T_s(t, p)$, the same cannot be done within the t_r approximation. The burning rate and surface temperature at a given moment in time are determined by the same quantities as in the t_c approximation but taken at an earlier time instant. Additionally, despite the functions U and V being the same in both formulations, temperature and concentration distributions outside the condensed phase are not captured by the dependencies corresponding to steady-state solutions.

The t_r approximation is valuable if it provides an estimation that coincides, by an order of magnitude, with the lag time value describing satisfactory experimental observations of some nonsteady combustion process where the inertia of quick-response zones plays significant role. Therefore the accuracy of the t_r approximation may be validated by considering some problem that admits analytical solution.

The following analysis considers the problem of combustion stability at constant pressure and compares the analytical solution with those obtained by the two available approximations, that is the t_c and t_r approximations.

The conditions of the stability of a steady-state burning regime under constant pressure are established within the framework of the t_r approximation by the conventional procedure. The governing equations must be linearized in the vicinity of the steady-state burning solution assuming that small perturbations depend on the dimensionless complex frequency Ω and dimensionless time τ as $\exp(\Omega\tau)$. The characteristic equation for the frequency is derived from the solution of the linearized problem. Purely imaginary values of the frequency correspond to the stability boundary.

Linearization of the heat transfer equation for the condensed phase is exactly the same as in the t_c approximation, therefore Eq. (3.14) holds. Two additional relations are derived from the modified nonsteady burning laws. Using the relations

$$T_e(\tau - \tau_U) = T_e^0 + T_{e1}\exp[\Omega(\tau - \tau_U)]$$
$$\frac{T_{e1}}{(T_s^0 - T^0)} = \frac{T_{s1}}{(T_s^0 - T^0)} - \frac{f_1}{f^0} + \frac{m_1}{m^0} \tag{9.133}$$

the nonsteady burning laws (9.132) may be written in a linear approximation as

$$v_1 = k(\theta_{s1} - \varphi_1 + v_1)\exp[-\Omega\tau_U]$$
$$\theta_{s1} = r(\theta_{s1} - \varphi_1 + v_1)\exp[-\Omega\tau_V] \tag{9.134}$$

By comparing these expressions with (3.19) (t_c approximation), it is evident that the delay is taken into account by multiplying the parameters k and r by exponents containing lag times.

The three linear homogeneous equations (3.14), (9.134) for the perturbations lead to the following characteristic equation

$$rz(z-1)\exp(-\Omega\tau_U) - k(z-1)\exp(-\Omega\tau_V) + z = 0 \tag{9.135}$$

which reduces to (3.23) in the case of zero lags.

To compare the results to the analytical solution of the problem obtained within the framework of the Belyaev model in Section 9.2, a further assumption $\tau_U = \tau_V$ is made, and the remaining single lag time is denoted as τ_r.

The characteristic equation (9.135) may be rewritten in the form

$$rz(z-1) - k(z-1) + z\exp(\Omega\tau_r) = 0 \tag{9.136}$$

Consequently parameters k and r may be expressed as functions of the complex frequency in the following form

$$k = \left|\frac{z}{z-1}\right|^2 \mathrm{Im}(\bar{z}-1)\exp(\Omega\tau_r)/\mathrm{Im}(\bar{z})$$
$$r = \left|\frac{1}{z-1}\right|^2 \mathrm{Im}(z(\bar{z}-1))\exp(\Omega\tau_r)/\mathrm{Im}(\bar{z}) \tag{9.137}$$

At the stability boundary ($\Omega = i\omega$)

$$k = (p_\omega + 1)\cos(\omega\tau_r) - \left[\frac{p_\omega - 1}{p_\omega + 1}\right]^{1/2}\frac{\sin(\omega\tau_r)}{(p_\omega - 1)}$$
$$r = \frac{2}{p_\omega(p_\omega - 1)}[\cos(\omega\tau_r) - (p_\omega^2 - 1)^{1/2}\sin(\omega\tau_r)] \tag{9.138}$$

where p_ω is the same function of frequency as in (9.36).

These relations determine, in parametric form and for a given value of the lag time, the dependence $k(r)$ at the stability boundary.

The analytical solution of this problem, within the framework of the Belyaev model, was presented in Section 9.2.

A comparison of this analytical solution to the ones resulting from the t_c and the t_r approximations may be carried out in the following fashion.

The lag time may be estimated as follows. The relaxation time of the gas preheat zone is

$$t_g(x_f^0) = \int_0^{x_f^0} \frac{ds}{u_g^0(s)} \tag{9.139}$$

where $u_g^0(x)$ is the gas velocity in the steady-state burning regime. In the dimensionless variables adopted in Section 9.2, the corresponding lag time is $\tau_g(\xi_f^0) = \sigma\xi_f^0$. By an order of magnitude, the thermal relaxation time of the combustion product zone coincides with $\tau_r(\xi_f^0)$, therefore

$$\tau_r = 2\sigma\xi_f^0 \tag{9.140}$$

or

$$\tau_r = -2\sigma \ln a \tag{9.141}$$

As an example, the system with values $\sigma = 0.05$, $a = 0.4$, and $\delta = 0.15$ is considered. As shown in Section 9.2, t_c approximation is less accurate at small values of r (Figure 9.1). The analytical solution at $r = 0$ yields $k^a = 1.56$ and $\omega^a = 3.16$. The corresponding values in the t_c approximation are $k^c = 1$ and $\omega^c = \infty$. In the t_r approximation Eqs. (9.138) give (at $r = 0$)

$$\tau_r = 2\frac{\arctan((p_\omega^2 - 1)^{-1/2})}{p_\omega(p_\omega^2 - 1)^{1/2}}$$
$$k^r = \left[\frac{(p_\omega + 1)}{(p_\omega - 1)}\right]^{1/2} \tag{9.142}$$

The value of τ_r calculated from Eq. (9.141) based on the adopted values of σ and a is 0.0916. Therefore, from (9.142) $p_\omega = 2.88$ and $k^r = 1.44$. Furthermore, from (9.138) $\omega^r = 3.89$.

The comparison of the two approximations with the analytical solution is made within the framework of the nonsteady propellant combustion theory, that is, using the steady-state dependencies for the study of nonsteady regimes. It is assumed that in addition to the dependencies $m^0(T_a, p)$ and $T_s^0(T_a, p)$ yielding the parameters k and r, the lag time characterizing the gas-phase inertia is introduced.

In an analytical solution, this lag time may be considered as a fitting parameter that can be used to adjust the functions $k^a(r)$ and $\omega^a(k)$. The best fit at the point $r = 0$, where the t_c approximation is inaccurate, may be found in such a way that both functions are described with equal accuracy, that is, $k^r/k^a = \omega^r/\omega^a$.

The latter requirement, enforced with the help of Eqs. (9.138) and (9.142), gives the following equation to determine the optimum value of p_ω

$$\left[\frac{p_\omega + 1}{p_\omega - 1}\right]^{1/2} / k^a = \frac{p_\omega(p_\omega^2 - 1)^{1/2}}{2\omega^a} \tag{9.143}$$

or

$$2p_\omega = 1 + (1 + 8\omega^a/k^a)^{1/2} \tag{9.144}$$

from which $p_\omega = 2.71$ and $\tau_r = 0.111$. This best-fitting value differs from the previous estimation of the time lag by only 20%.

The three approaches, the analytical solution (denoted by superscript a), the t_c approximation (superscript c), and the t_r approximation (superscript r), are compared in Table 9.2 for the $\sigma = 0.05$, $a = 0.4$, $\delta = 0.15$, and $\tau_r = 0.111$. It is evident that the stability boundary can be described reasonably accurately by the t_c approximation. However, the frequencies predicted by this approximation differ from the real ones by large factors, even at the

Table 9.2 Comparison of the t_c approximation, the t_r approximation, and the analytical solution obtained from the Belyaev model (Section 9.2).

r	k^a	$\frac{(k^c - k^a)}{k^a}$	$\frac{(k^r - k^a)}{k^a}$	ω^a	$\frac{(\omega^c - \omega^a)}{\omega^a}$	$\frac{(\omega^r - \omega^a)}{\omega^a}$
0.0	1.56	−0.35	−0.057	3.61	∞	−0.055
0.1	1.74	−0.13	−0.028	2.98	3.11	−0.0033
0.2	1.91	−0.089	−0.015	2.55	1.58	0.027
0.4	2.23	−0.049	0.00	2.01	0.81	0.049
0.7	2.65	−0.026	0.064	1.46	0.56	0.075
1.0	3.05	−0.016	0.013	1.30	0.33	0.069

reasonable values of $r = 0.1 - 0.2$. In contrast, the t_r approximation differs in accuracy from the analytical solution by only a few percent.

Note that in the limit $r \to 0, \sigma \to 0$ the analytical solution dependence on the three parameters σ, a, and δ reduces to the dependence on their single combination

$$\Delta = \sigma[a - 2\ln a - \delta(1 - a)] \tag{9.145}$$

since the gas inertia in this limit is characterized by the single complex $\sigma(h - g)$ (see (9.37) and the relevant discussion in Section 9.2)

A more detailed comparison of the t_r approximation with the analytical model is presented in Table 9.3. The comparison is made for $r = 0$ and the method of best approximation as described above is used. The parameters given in the first three columns, along with the analytical solutions k^a and ω^a, are used to find the lag time and the values k^r and ω^r. The last column gives the ratio of τ_r to Δ. Table 9.3 shows that the t_r approximation satisfactorily describes the steady-state burning regime up to $\sigma = 10^{-1}$. By assuming that typical values of σ are of the order $10^{-3} - 10^{-4}$, the t_r approximation may be used for pressures up to 100 atm (σ is directly proportional to pressure).

Table 9.3 Comparison of the t_r approximation and the analytical solution obtained from the Belyaev model (Section 9.2).

σ	a	δ	k^a	ω^a	τ_r	τ_r/σ	$\frac{(k^r - k^a)}{k^a}$	$\frac{(\omega^r - \omega^a)}{\omega^a}$	Δ	τ_r/Δ
0.001	0.4	0.1	1.12	47.4	0.00219	2.19	−0.0089	−0.0063	0.00230	0.954
0.001	0.4	0.5	1.11	50.8	0.00197	1.97	0.00	−0.0059	0.00254	0.777
0.001	0.4	0.9	1.11	55.1	0.00175	1.75	−0.009	−0.009	0.00277	0.632
0.01	0.1	0.4	1.35	6.33	0.0453	4.53	0.014	−0.017	0.0507	0.894
0.01	0.2	0.4	1.32	7.70	0.0338	3.38	−0.022	−0.016	0.0354	0.954
0.01	0.4	0.4	1.28	10.4	0.0218	2.18	−0.023	−0.028	0.0232	0.940
0.01	0.6	0.4	1.26	13.0	0.0158	1.58	−0.031	−0.030	0.0169	0.937
0.01	0.8	0.4	1.24	16.0	0.0116	1.16	−0.032	−0.031	0.0133	0.875
0.05	0.4	0.4	1.56	3.61	0.1115	2.23	−0.057	−0.055	0.116	0.958
0.1	0.4	0.4	1.79	2.31	0.225	2.25	−0.083	−0.051	0.233	0.967

References

Abramowitz, M. and Stegun, I.A. (1972). *Handbook of Mathematical Functions.* Gaithersburg: National Bureau of Standards.

Akiba, R. and Tanno M. (1959). Low frequency instability in solid propellant rocket motors. In *Proceedings of the First Symposium (International) on Rockets and Astronautics,* (M. Sanuki, ed.). Tokyo, Japan.

Aldushin, A.P. and Kasparyan, S.G. (1979). On thermal-diffusion instability of flame front. *Doklady Academii Nauk SSSR* 244 (1): 67–70.

Allison, C.B. and Faeth, G.M. (1975). Open-loop response of a burning liquid monopropellant. *AIAA Journal* 13 (10): 1287–1294.

Andronov, A.A., Vitt, A.A., and Khaikin, S.E. (1966). *Theory of Oscillators.* Oxford: Pergamon Press.

Arnold, V.I. (1988). *Geometrical Methods in the Theory of Ordinary Differential Equations.* New York: Springer-Verlag.

Assovskii, I.G. and Istratov, A.G. (1971). Solid propellant combustion in the presence of photoirradiation. *Journal of Applied Mechanics and Technical Physics* 12 (5): 692–698.

Assovskii, I.G. and Rashkovskii, S.A. (1998). The influence of the mache effect on combustion stability in a solid rocket motor. *Combustion, Explosion and Shock Waves* 34 (5): 528–533.

Atwood, A.I., Boggs, T.L., Curran, P.O. et al. (1999a). Burning rate of solid propellant ingredients, Part 1: Pressure and initial temperature effects. *Journal of Propulsion and Power* 15 (6): 740–747.

Atwood, A.I., Boggs, T.L., Curran, P.O. et al. (1999b). Burning rate of solid propellant ingredients, Part 2: Determination of burning rate temperature sensitivity. *Journal of Propulsion and Power* 15 (6): 748–752.

Barenblatt, G.I. (2005). *Scaling, Self-Similarity and Intermediate Asymptotics.* New York: Cambridge University Press.

Barenblatt, G.I., Zeldovich Ya.B., and Istratov, A.G. (1962). On diffusional-thermal stability of a laminar flame. *Zhurnal Prikladnoi Mekhaniki i Tekhnicheskoi Fiziki* 4: 21–26.

Beckstead, M.W. and Price, E.W. (1967). Nonacoustic combustion instability. *AIAA Journal* 5 (11): 1989–1996.

Beckstead, M.W., Ryan, N.W., and Bear, A.D. (1966). Nonacoustic instability of composite propellant. *AIAA Journal.* 4 (9): 1622–1628.

Belyaev, A.F. (1938). On the burning of explosive substances. *Zhurnal Fizicheskoi Khimii* 12 (1): 93–99.

Theory of Solid-Propellant Nonsteady Combustion, First Edition. Boris V. Novozhilov and Vasily B. Novozhilov.

Companion website: www.wiley.com/go/Novozhilov/solidpropellantnonsteadycombustion

Belyaev, A.F. (1940). On the burning of nitroglycol. *Zhurnal Fizicheskoi Khimii* 14 (8): 1009–1025.

Belyaev, A.F. and Komkova, L.D. (1950). Thermites burning rate dependence on pressure. *Zhurnal Fizicheskoi Khimii* 24 (11): 1302–1314.

Belyaev, A.A., Kaganova, Z.I., and Novozhilov, B.V. (2004). Combustion of volatile condensed systems behind the stability limit of the stationary regime. *Combusiton, Explosion and Shock Waves* 40 (4): 425–431.

Bobolev, V.K., Glazkova, A.P., Zenin, A.A., and Leipunskii, O.I. (1964). Investigation of temperature distribution during ammonium perchlorate combustion. *Zhurnal Prikladnoi Mekhaniki i Tekhnicheskoi Fiziki* 3: 153–158.

Bobylev, V.M. (1992). *Solid Fuel Rocket Engine as Means of Rocket Control*. Moscow: Mashinostroenie.

Bogoliubov, N.N. and Mitropolsky, Y.A. (1961). *Asymptotic Methods in the Theory of Non-linear Oscillations*. New York: Gordon and Breach.

Borisova, O.A., Lidskii, B.V., Neuhaus, M.G., and Novozhilov, B.V. (1986). Stability of gasless systems combustion with respect to two-dimensional perturbations. *Khimicheskaya Fizika* 5 (6): 822–830.

Borisova, O.A., Lidskii, B.V., Neuhaus, M.G., and Novozhilov, B.V. (1990). Two-dimensional stability of the combustion of condensed systems. Zero-order reactions and broad heat-liberation zone. *Combustion, Explosion and Shock Waves* 26 (1): 71–74.

Chechilo, N.M., Khvilivitsky R.Ya., and Enikolopyan, N.S. (1972). On the propagation of polymerization front. *Doklady Academii Nauk SSSR* 204 (5): 1180–1181.

Clavin, P. and Lazimi, D. (1992). Theoretical analysis of oscillatory burning of homogeneous solid propellant including non-steady gas phase effects. *Combustion Science and Technology* 83: 1–32.

Cozzi, F., De Luca, L.T., and Novozhilov, B.V. (1999). Linear stability and pressure-driven response function of solid propellant with phase transition. *Journal Propulsion and Power* 15 (6): 806–815.

Crocco, L. and Cheng, S.I. (1956). *Theory of Combustion Instability in Liquid Propellant Rocket Motors*. London: Butterworths.

Crump, J.E. and Price, E.W. (1964). Effect of acoustic environment on the burning rate of solid propellants. *AIAA Journal* 2: 1274–1278.

Culick, F.E.C. (1968). A review of calculations for unsteady burning of solid propellant. *AIAA Journal* 6: 2241–2255.

Culick, F.E.C. (2000). Combustion instabilities – Mating dance of chemical, combustion, and combustor dynamics. AIAA-2000-3178. *36th AIAA/ASME/SAE/ASEE Joint Propulsion Conference and Exhibition*. Las Vegas, USA.

Culick, F.E.C. and Yang, V. (1992). Prediction of the stability of unsteady motions in solid-propellant rocket motors. In: *Nonsteady Burning and Combustion Stability of Solid Propellants, Progress in Astronautics and Aeronautics*, vol. 143 (eds. L. De Luca, E.W. Price and M. Summerfield), 719–779. Washington: AIAA.

De Luca, L. (1976). Solid propellant ignition and other unsteady combustion phenomena induced by radiation. Doctor of Philosophy Thesis. Princeton University.

De Luca, L. (1992). Theory of nonsteady burning and combustion stability of solid propellants by flame models. In: *Nonsteady Burning and Combustion Stability of Solid Propellants*,

Progress in Astronautics and Aeronautics, vol. 143 (eds. L. De Luca, E.W. Price and M. Summerfield), 519–600. Washington: AIAA.

De Luca, L., Price, E.W., and Summerfield, M. (eds.) (1992). *Nonsteady Burning and Combustion Stability of Solid Propellants, Progress in Astronaurics and Aeronautics*, vol. 143. Washington: AIAA.

De Luca, L., Di Silvestro, R., and Cozzi, F. (1995). Intrinsic combustion instability of solid energetic materials. *Journal of Propulsion and Power* 11 (4): 804–815.

Denison, M.R. and Baum, E. (1961). A simplified model of unstable burning in solid propellant. *ARS Journal* 31 (8): 1112–1122.

Eisel, J.E., Horton, M.D., Price, E.W., and Rice, D.W. (1964). Preferred frequency oscillatory combustion of solid propellants. *AIAA Journal* 2 (7): 1319–1323.

Finlinson, J.C., Hanson-Parr, D., Son, S.F. and Brewster, M.Q. (1991). Measurement of propellant combustion response to sinusoidal radiant heat flux, AIAA Paper 1991–0204. 29th Aerospace Sciences Meeting, Reno, NV, USA.

Frank-Kamenetskii, D.A. (1939). Temperature distribution within reacting mixture and the steady-state thermal explosion. *Zhurnal Fizicheskoi Khimii* 13 (6): 738–755.

Frank-Kamenetskii, D.A. (1969). *Diffusion and Heat Transfer in Chemical Kinetics*. New York: Plenum Press.

Friedman, R., Nugent, R.G., Rumbel, K.E., and Scurlock, A.C. (1957). Deflagration of ammonium perchlorate. In: *Sixth Symposium (International) on Combustion*. New York/London: Reinhold Publishing Corporation.

Frost, V.A. and Yumashev, V.L. (1973). Study of the extinction of gunpowder in the combustion model with a variable surface temperature. *Journal of Applied Mechanics and Technical Physics* 14 (3): 371–377.

Glazkova, A.P. (1963). On the influence of pressure on burning rate of ammonium perchlorate. *Zhurnal Prikladnoi Mekhaniki i Tekhnicheskoi Fiziki* 5: 121–125.

Glazkova, A.P., Zenin, A.A., and Balepin, A.A. (1970). Influence of initial temperature on parameters of combustion zone of ammonium perchlorate. *Archiwum Procesow Spalania* 1 (3–4): 261–268.

Gostintsev Yu.A. (1967). Method of reduction to ordinary differential equations in problems of the nonstationary burning of solid propellants. *Combustion, Explosion and Shock Waves* 3 (3): 218–220.

Gostintsev Yu.A. and Sukhanov, L.A. (1974). Theory of stability of powder combustion in a half-closed volume. *Combustion, Explosion and Shock Waves* 10 (6): 737–743.

Gostintsev Yu.A., Pokhil, P.F., and Sukhanov, L.A. (1970). Full set of equations for nosteady propellant combustion in a semi-enclosed volume. *Doklady Academii Nauk SSSR* 195 (1): 137–139.

Gusachenko, L.K., Zarko, V.E., and Rychkov, A.D. (1999). Effect of melting on dynamic combustion behavior of energetic materials. *Journal Propulsion and Power* 15 (6): 816–836.

Hart, W.R. and McClure, F.T. (1959). Combustion instability: acoustic interaction with a burning propellant surface. *Journal of Chemical Physics* 30: 1501–1514.

Hermance, C.E. (1984). Solid-propellant ignition theories and experiments. In: *Fundamentals of Solid-Propellant Combustion, Progress in Astronautics and Aeronautics*, vol. 90 (eds. K.K. Kuo and M. Summerfield), 239–304. Washington: AIAA.

Horton, M.D. and Price, E.W. (1963). Dynamic characteristics of solid propellant combustion. In: *Proceedings of the Ninth Symposium (International) on Combustion*. New York: Academic Press.

Ibiruci, M.M. and Williams, F.A. (1975). Influence of externally applied thermal radiation on the burning rates of homogeneous solid propellants. *Combustion and Flame* 24 (2): 185–198.

Incropera, F.P. and DeWitt, D.P. (2002). *Heat and Mass Transfer*. New York: Wiley.

Istratov, A.G. and Librovich, V.B. (1964). On the stability of propellant combustion. *Zhurnal Prikladnoi Mekhaniki i Tekhnicheskoi Fiziki* 5: 38–43.

Istratov, A.G. and Librovich, V.B. (1966). On the stability of propagation of spherical flames. *Journal of Applied Mechanics and Technical Physics* 7 (1): 43–50.

Istratov, A.G., Librovich, V.B., and Novozhilov, B.V. (1964). An approximate method in nonsteady propellant combustion theory. *Zhurnal Prikladnoi Mekhaniki i Tekhnicheskoi Fiziki* 3: 139–144.

Ivanov, S.M. and Tsukanov, N.A. (2000). Pressure control in a semi-closed volume upon combustion of solid propellants with an exponent in the combustion law greater than unity. *Combustion, Explosion and Shock Waves* 36 (5): 591–600.

Kharkevich, A.A. (1953). *Auto-Oscillations*. Moscow: Gostekhizdat.

Kiskin, A.B. (1983). Stability of stationary powder combustion acted on by a constant light flux. *Combustion, Explosion and Shock Waves* 19 (3): 295–297.

Kiskin, A.B. (1993). A method for determining the pressure response function using the data on burning rate under irradiation. *Combustion, Explosion and Shock Waves* 29 (3): 291–293.

Kiskin, A.B. and Novozhilov, B.V. (1989). Asymptotic behavior of the combustion rate of condensed systems under small perturbations. *Combustion, Explosion and Shock Waves* 25 (5): 641–645.

Klager, K. and Zimmerman, G.A. (1992). Steady burning rate and affecting factors: experimental results. In: *Nonsteady Burning and Combustion Stability of Solid Propellants*, Progress in Astronautics and Aeronautics, vol. 143 (eds. L. De Luca, E.W. Price and M. Summerfield), 59–109. Washington: AIAA.

Kohno, M., Maruizumi, H., Novozhilov, B.V., and Shimada, T. (1995). Combusiton of highly metalized propellants under harmonically varying pressure. *Khimicheskaya Fizika* 14 (7): 112–121.

Kohno, M., Maruizumi H, Novozhilov, B.V., Shimada, T., Tokudome, S. and Volpi, A. (1998). Phenomenological approach to the scaling of erosive burning in solid propellant motor. *Proceedings of the Twenty-First International Symposium on Space Technology and Science*. Omiya, Japan.

Kondrikov, B.N. (1969). Combustion stability of explosives. *Combustion, Explosion and Shock Waves* 5 (1): 34–40.

Korotkov, A.I. and Leipunskii, O.I. (1953). Dependence of temperature sensitivity of propellant burning rate on initial temperature. *Fizika Vzryva* 2: 213–220.

Kowalskii, A.A., Konev, E.V., and Krasilnikov, B.V. (1967). Combustion of nitroglycerin powder. *Combustion, Explosion and Shock Waves* 3 (4): 335–339.

Kubota, N. (1984). Survey of rocket propellants and their combustion characteristics. In: *Fundamentals of Solid-Propellant Combustion, Progress in Astronautics and Aeronautics*, vol. 90 (eds. K.K. Kuo and M. Summerfield), 1–51. Washington: AIAA.

Kubota, N. (1992). Temperature sensitivity of solid propellants and affecting factors: experimental results. In: *Nonsteady Burning and Combustion Stability of Solid Propellants, Progress in Astronautics and Aeronautics*, vol. 143 (eds. L. De Luca, E.W. Price and M. Summerfield), 111–143. Washington: AIAA.

Landau, L.D. and Lifshitz, E.M. (1982). *Mechanics.* Oxford: Butterworth-Heinemann.

Landau, L.D. and Lifshitz, E.M. (1987). *Fluid Mechanics.* Oxford: Butterworth–Heinemann.

Leipunskii, O.I. and Frolov Yu.V. (eds.) (1982). *Combustion Theory of Propellants and Explosives.* Moscow: Nauka.

Leipunskii, O.I., Belyaev, A.A., Zenin, A.A. et al. (1982). Ballistite N combustion in a turbulent stream. *Khimicheskaya Fizika* 1 (10): 1421–1427.

Lewis, B. and Von Elbe, G. (1934). On the theory of flame propagation. *Journal of Chemical Physics* 2 (8): 537–546.

Librovich, V.B. and Novozhilov, B.V. (1971). Similar solutions in the theory of nonsteady burning of a solid propellant. *Journal of Applied Mechanics and Technical Physics* 12 (4): 515–522.

Librovich, V.B. and Novozhilov, B.V. (1972). Self-similar solutions in the nonsteady propellant burning rate theory and their stability analysis. *Combustion Science and Technology* 4: 257–267.

Lidskii, B.V., Novozhilov, B.V., and Popov, A.G. (1983). Theoretical study of nosteady-state combustion of a gas-producing solid fuel upon a pressure drop. *Combustion, Explosion and Shock Waves* 19 (4): 387–390.

Lidskii, B.V., Novozhilov, B.V., and Popov, A.G. (1985). Nonsteady combustion of solid fuel near extinction boundary under depressurization. *Khimicheskaya Fizika* 4 (5): 721–727.

Loitsyanskii, L.G. (1966). *Mechanics of Liquids and Gases.* Oxford: Pergamon Press.

Makhviladze, G.M. and Novozhilov, B.V. (1971). Two-dimensional stability of the combustion of condensed systems. *Journal of Applied Mechanics and Technical Physics* 12 (5): 676–682.

Maksimov, E.I. (1964). Investigation of luminosity pulsations during nitroglycerin propellant combustion. *Zhurnal Fizicheskoi Khimii* 37 (5): 1129–1132.

Manelis, G.B. and Strunin, V.A. (1971). The mechanism of ammonium perchlorate burning. *Combustion and Flame* 17 (1): 69–77.

Manelis, G.B., Nazin, G.M., Rubtzov Yu.I., and Strunin, V.A. (1996). *Thermal Decomposition and Combustion of Explosives and Powders.* Moscow: Nauka.

Margolis, S.B. and Williams, F.A. (1988). Diffusional thermal coupling and intrinsic instability of solid propellant combustion. *Combustion Science and Technology* 59: 27–64.

Marshakov, V.N. and Leipunskii, O.I. (1967). Burning and quenching of a powder in the presence of a rapid pressure drop. *Combustion, Explosion and Shock Waves* 3 (2): 144–146.

Marshakov, V.N. and Novozhilov, B.V. (2011a). Combustion of a propellant and its extinction upon rapid depressurization: a comparison of theory and experiment. *Russian Journal of Physical Chemistry B* 5 (3): 474–481.

Marshakov, V.N. and Novozhilov, B.V. (2011b). Transient modes of double-base propellant combustion in a semiclosed volume. *Russian Journal of Physical Chemistry B* 5 (1): 45–56.

Marshakov, V.N., Puchkov, V.M., and Finyakov, S.V. (2010). Temperature sensitivity of the burning velocity of nitroglycerin-based propellants. *Russian Journal of Physical Chemistry B* 4 (6): 950–953.

McIntosh, A.C. (1987). Combustion-acoustic interaction of a flat flame burner system enclosed within an open tube. *Combustion Science and Technology* 54: 217–236.

Merzhanov, A.G. (1994). Solid flames: discoveries, concepts, and horizons of cognition. *Combustion Science and Technology* 98: 307–336.

Merzhanov, A.G. and Dubovitsky, F.I. (1959). On the theory of steady-state propellant combustion. *Doklady Academii Nauk SSSR* 129 (1): 153–156.

Merzhanov, A.G. and Filonenko, A.K. (1963). On thermal self-ignition of a homogeneous gaseous mixture. *Doklady Academii Nauk SSSR* 152 (1): 143–146.

Mihlfeith, C.M., Baer, A.D., and Ryan, N.W. (1972). Propellant combustion instability as measured by combustion recoil. *AIAA Journal* 10 (10): 1280–1285.

Mikhailov, A.S. (2011). *Foundations of Synergetics I: Distributed Active Systems.* Berlin: Springer-Verlag.

Miller, M.S. and Anderson, W.R. (2004). Burning-rate predictor for multi-ingredient propellants: nitrate–ester propellants. *Journal of Propulsion and Power* 20 (3): 440–454.

Novikov, S.S. and Ryazantsev Yu.S. (1965a). On the theory of combustion stability of powders. *Journal of Applied Mechanics and Technical Physics* 6 (1): 49–53.

Novikov, S.S. and Ryazantsev Yu.S. (1965b). On the theory of stationary velocity of propagation of an exothermic reaction front in a condensed medium. *Journal of Applied Mechanics and Technical Physics* 6 (3): 28–31.

Novikov, S.S. and Ryazantsev Yu.S. (1966). A note on the theory of the combustion stability of powders. *Journal of Applied Mechanics and Technical Physics* 7 (3): 92–93.

Novikov, S.S., Ryazantsev Yu.S., and Tulskikh, V.E. (1974). Calculation of the acoustic admittance of a burning surface of a condensed system, taking into account the instability of processes in the gaseous phase. *Journal of Applied Mechanics and Technical Physics* 15 (3): 371–378.

Novozhilov, B.V. (1961). The rate of exothermic reaction front propagation in condensed phase. *Doklady Academii Nauk SSSR* 141 (1): 151–153.

Novozhilov, B.V. (1965a). Stability criterion for steady-state burning of powders. *Journal of Applied Mechanics and Technical Physics* 6 (4): 106–108.

Novozhilov, B.V. (1965b). Burning of a powder under harmonically varying pressure. *Journal of Applied Mechanics and Technical Physics* 6 (6): 103–106.

Novozhilov, B.V. (1965c). Average burning rate of a powder with harmonically varying pressure. *Combustion, Explosion and Shock Waves* 1 (3): 23–24.

Novozhilov, B.V. (1966). Nonlinear oscillations of combustion velocity of powder. *Journal of Applied Mechanics and Technical Physics* 7 (5): 19–25.

Novozhilov, B.V. (1967a). Nonstationary burning of propellants with variable surface temperature. *Journal of Applied Mechanics and Technical Physics* 8 (1): 37–43.

Novozhilov, B.V. (1967b). Stability of stationary regime for powders burning in a semiconfined space. *Combustion, Explosion and Shock Waves* 3 (1): 19–22.

Novozhilov, B.V. (1967c). Theory of nonsteady phenomena in propellant combustion. Doctor of Science Thesis, Institute of Chemical Physics, Moscow, USSR Academy of Sciences.

Novozhilov, B.V. (1968). Theory of nonstationary combustion of homogeneous propellants. *Combustion, Explosion and Shock Waves* 4 (4): 276–282.

Novozhilov, B.V. (1970). Equation for nonsteady-state combustion velocity of a powder. *Journal of Applied Mechanics and Technical Physics* 11 (4): 585–590.

Novozhilov, B.V. (1973a). *Nonsteady Combustion of Solid Rocket Fuels*. Moscow: Nauka.

Novozhilov, B.V. (1973b). Linear nonstationary effects: a source of information on solid-fuel surface-reaction kinetics. *Combustion, Explosion and Shock Waves* 9 (1): 107–111.

Novozhilov, B.V. (1988a). Influence of gas-phase inertia on stability of combustion of volatile condensed systems. *Khimicheskaya Fizika* 7 (3): 388–396.

Novozhilov, B.V. (1988b). Theory of nonsteady combustion of condensed systems with an account of time lag. *Khimicheskaya Fizika* 7 (5): 674–687.

Novozhilov, B.V. (1989). Combustion of volatile condensed systems under harmonically varying pressure. *Khimicheskaya Fizika* 8 (1): 102–111.

Novozhilov, B.V. (1992a). Theory of nonsteady burning and combustion stability of solid propellants by the Zeldovich-Novozhilov method. In: *Nonsteady Burning and Combustion Stability of Solid Propellants, American Institute of Aeronautics and Astronautics, Progress Series*, vol. 143 (eds. L. DeLuca and M. Summerfield), 601–641. Washington: AIAA.

Novozhilov, B.V. (1992b). Second harmonic nonlinear resonance in propellant combustion. In: *Twenty-Fourth Symposium (International) on Combustion*. Pittsburg, USA: The Combustion Institute.

Novozhilov, B.V. (2000). Acoustic resonance upon propellant combustion. *Combustion, Explosion and Shock Waves* 36 (1): 3–9.

Novozhilov, B.V. (2002). Nonlinear combustion in solid propellant rocket motors. In: *Combustion of Energetic Materials* (eds. K.K. Kuo and L.T. De Luca), 793–802. Wallingford, UK: Begell House.

Novozhilov, B.V. (2003). Propellant combustion – from steady-state up to chaos. Novel energetic materials and application. *8th International Workshop on Combustion and Propulsion*. Lerici, Italy.

Novozhilov, B.V. (2004). Chaotisation of nonsteady propellant burning rate. *Khimicheskaya Fizika* 23 (5): 68–74.

Novozhilov, B.V. (2005). Combustion of energetic materials in a an acoustic field (review). *Combustion, Explosion and Shock Waves* 41 (6): 709–726.

Novozhilov, B.V. (2006). Nonlinear response of propellant combustion to varying pressure. *Khimicheskaya Fizika* 25 (8): 32–41.

Novozhilov, B.V. (2007). Cross-flow effect on combustion of a solid propellant with harmonically varying pressure. *Combustion, explosion and shock waves* 43 (4): 429–434.

Novozhilov, B.V. (2010). Automatic control of propellant combustion stability in a semiclosed space. *Russian Journal of Physical Chemistry B* 4 (3): 457–464.

Novozhilov, B.V. and Posvyanskii, V.S. (1991). Numerical modeling of nonsteady combustion processes of condensed systems using Belyaev model. *Khimicheskaya Fizika* 10 (4): 534–544.

Novozhilov, B.V., Kohno, M., Maruizumi H. and Shimada, T. (1996). Solid propellant burning rate response functions of higher orders. *Proceedings of the Twentieth International Symposium on Space Technology and Science*. Gifu, Japan.

Novozhilov, B.V., Cozzi, F., and De Luca, L.T. (2002a). Limit cycles for solid propellant burning rate at constant pressure. In: *In: Combustion of Energetic Materials* (eds. K.K. Kuo and L.T. De Luca), 825–836. Wallingford, UK: Begell House.

Novozhilov, B.V., Kohno, M., and Morita, T. (2002b). Relationship between pressure- and radiant-driven propellant burning rate response functions. In: *Rocket Propulsion: Present and Future* (ed. L.T. De Luca), 23. Grafiche GSS, Bergamo, Italy.

Novozhilov, B.V., Kohno, M., and Morita, T. (2003). Relation between gunpowder burning-rate responses to oscillating pressure and an oscillating radiant heat flux. *Combustion, Explosion and Shock Waves* 39 (1): 68–74.

Novozhilov, B.V., Kaganova, Z.I., and Belyaev, A.A. (2006). Dependence of propellant burning rate on the frequency and amplitude of harmonically varying pressure. *Khimicheskaya Fizika* 25 (6): 63–69.

Novozhilov, B.V., Kaganova, Z.I., and Belyaev, A.A. (2007). Erosive burning of a propellant in the field of a travelling acoustic wave. *Russian Journal of Physical Chemistry B* 1 (2): 94–101.

Novozhilov, B.V., Kaganova, Z.I., and Belyaev, A.A. (2009a). Unsteady regimes of propellant combustion in a semiclosed space. *Russian Journal of Physical Chemistry B* 3 (6): 945–952.

Novozhilov, B.V., Kaganova, Z.I., and Belyaev, A.A. (2009b). Simulation of unsteady combustion in a solid-propellant rocket motor. *Russian Journal of Physical Chemistry B* 3 (1): 91–98.

Novozhilov, B.V., Kaganova, Z.I., and Belyaev, A.A. (2010). Combustion of volatile condensed systems during pressure decay. *Russian Journal of Physical Chemistry B* 4 (6): 942–949.

Pokhil, P.F., Maltsev, V.M., Seleznev, V.A., and Mamina, N.K. (1967). Optical method of determining propellant surface temperature. *Combustion, Explosion and Shock Waves* 3 (3): 204–209.

Popov, E.P. (2014). *The Dynamics of Automatic Control Systems*. Oxford: Pergamon Press.

Price, E.W. (1984). Combustion of metallized propellants. In: *Fundamentals of Solid-Propellant Combustion, Progress in Astronautics and Aeronautics*, vol. 90 (eds. K.K. Kuo and M. Summerfield), 478–513. Washington: AIAA.

Price, E.W. (1992). *L**- instability. In: *Nonsteady Burning and Combustion Stability of Solid Propellants, Progress in Astronautics and Aeronautics*, vol. 143 (eds. L. De Luca, E.W. Price and M. Summerfield), 325–361. Washington: AIAA.

Price, E.W. and Soffers, J.W. (1958). Combustion instability in solid propellant rocket motors. *Jet Propulsion* 28: 190–192.

Prisnyakov, V.F. (1984). *Dynamics of Solid Fuel Rocket Engines*. Moscow: Mashinostroenie.

Romanov O.Ya. (1975). The nonstationary combustion rate of gunpowder. *Combustion, Explosion and Shock Waves* 11 (2): 163–171.

Romanov O.Ya. (1976). Nonstationary combustion of a two-component powder. *Combustion, Explosion and Shock Waves* 12 (3): 303–313.

Ryazantsev Yu.S. and Tylskikh, V.E. (1976). Calculation of acoustic admittance at a burning surface. *Acta Astronautica* 3: 171–185.

Shkadinskii, K.G. (1971). Investigation of unsteady process of transition to steady-state regime following ignition. Doctor of Philosophy Thesis, Institute of Chemical Physics, Chernogolovka, USSR Academy of Sciences.

Shkadinskii, K.G., Khaikin, B.I., and Merzhanov, A.G. (1971). Propagation of a pulsating exothermic reaction front in the condensed phase. *Combustion, Explosion and Shock Waves* 7 (1): 15–22.

Simonenko, V.N., Zarko, V.E., and Kutsenogii, K.P. (1980). Experimental study of the conditions for auto- and forced fluctuations of the rate of combustion of a powder. *Combustion, Explosion and Shock Waves* 16 (3): 298–304.

Son, S.F. and Brewster, M.Q. (1992). Linear burning rate dynamics of solid subjected to pressure or external radiant heat flux oscillations. *Journal of Propulsion and Power* 9 (2): 222–232.

Son, S.F. and Brewster, M.Q. (1993). Unsteady combustion of solid propellants subject to dynamic external radiant heating. *Combustion, Explosion and Shock Waves* 29 (3): 281–285.

Stoker, J.J. (1992). *Non-linear Vibrations in Mechanical and Electrical Systems*. New York: Wiley.

Strand, L.D. and Brown, R.S. (1992). Laboratory test methods for combustion-stability properties of solid propellants. In: *Nonsteady Burning and Combustion Stability of Solid Propellants, Progress in Astronautics and Aeronautics*, vol. 143 (eds. L. De Luca, E.W. Price and M. Summerfield), 689–718. Washington: AIAA.

Strunin, V.A. and Manelis, G.B. (1971). Stability of the steady-state process of explosive combustion limited by the reaction in the condensed phase. *Combustion, Explosion and Shock Waves* 7 (4): 427–430.

Svetlichnyi, I.B., Margolin, A.D., and Pokhil, P.F. (1971). Low-frequency self-oscillatory processes in propellant combustion. *Combustion, Explosion and Shock Waves* 7 (2): 156–161.

T'ien, J.S. (1972). Oscillatory burning of solid propellants including gas phase time lag. *Combustion Science and Technology* 5: 47–54.

Tikhonov, A.N. and Samarskii, A.A. (1963). *Equations of Mathematical Physics*. Oxford: Pergamon Press.

Vilyunov, V.N. (1961). On the theory of erosive combustion of solid propellants. *Doklady Academii Nauk SSSR* 136 (2): 381–383.

Vilyunov, V.N. and Rudnev, A.P. (1973). Effect of condensed powder phase reactions on the stability of steady-state combustion processes. *Journal of Applied Mechanics and Technical Physics* 14 (5): 686–693.

Vilyunov, V.N. and Zarko, V.E. (1989). *Ignition of Solids*. Amsterdam: Elsevier.

Volkov, V.P. and Medvedev Yu.I. (1969). Interaction between acoustic waves and the burning surface of solid propellants at elevated temperatures. *Journal of Applied Mechanics and Technical Physics* 10 (1): 93–97.

Whittaker, E.T. and Watson, G.N. (2002). *A Course of Modern Analysis*. Cambridge: Cambridge University Press.

Wimpress, R.N. (1950). *Internal Ballistics of Solid-Fuel Rockets*. New York: McGraw-Hill.

Zaidel, R.M. and Zeldovich Ya.B. (1962). On the possible regimes of steady-state combustion. *Zhurnal Prikladnoi Mekhaniki i Tekhnicheskoi Fiziki* 4: 27–32.

Zanotti, C., Volpi, A., Bianchessi, M., and De Luca, L. (1992). Measuring thermodynamic properties of burning propellants. In: *Nonsteady Burning and Combustion Stability of Solid Propellants, Progress in Astronautics and Aeronautics*, vol. 143 (eds. L. De Luca, E.W. Price and M. Summerfield), 145–196. Washington: AIAA.

Zeldovich Ya.B. (1942). On the theory of the combustion of powders and explosives. *Zhurnal Eksperimental'noi i Teoreticheskoi Fiziki* 12 (11–12): 498–524.

Zeldovich Ya.B. (1964). On propellant burning rate under varying pressure. *Zhurnal Prikladnoi Mekhaniki i Tekhnicheskoi Fiziki* 3 (C): 126–138.

Zeldovich Ya.B. (1971). Theory of propellant combustion in a gas flow. *Combustion, Explosion and Shock Waves* 7 (4): 399–408.

Zeldovich Ya.B. and Frank-Kamenetskii, D.A. (1938). Theory of thermal flame propagation. *Zhurnal Fizicheskoi Khimii* 12 (1): 100–105.

Zeldovich Ya.B., Leipunskii, O.I., and Librovich, V.B. (1975). *Theory of Nonsteady Propellant Combustion*. Moscow: Nauka.

Zenin, A.A. (1973). Experimental investigation of the mechanisms of propellant combustion and combustion products flow. Doctor of Science Thesis, Institute of Chemical Physics, Moscow, USSR Academy of Sciences.

Zenin, A.A. (1980). Processes in combustion zones of double-based propellants. In: *Physical Processes with Combustion and Explosion*, 68–104. Moscow: Atomizdat.

Zenin, A.A. (1992). Thermophysics of stable combustion waves of solid propellants. In: *Nonsteady Burning and Combustion Stability of Solid Propellants, Progress in Astronautics and Aeronautics*, vol. 143 (eds. L. De Luca, E.W. Price and M. Summerfield), 197–231. Washington: AIAA.

Zenin, A.A. and Nefedova, O.I. (1967). Burning of ballistite powder over a broad range of initial temperatures. *Combustion, Explosion and Shock Waves* 3 (1): 26–31.

Zenin, A.A. and Novozhilov, B.V. (1973). Single-valued dependence of the surface temperature of ballistite on the burning rate. *Combustion, Explosion and Shock Waves* 9 (2): 209–212.

Zenin, A.A., Leipunskii, O.I., Margolin, A.D. et al. (1966). Temperature field near the surface of the burning propellant. *Doklady Academii Nauk SSSR* 169 (3): 619–621.

Theory of Solid-Propellant Nonsteady Combustion

Problems

1 Find the steady-state temperature distribution in the condensed phase if it undergoes a phase change (from solid to solid state) at temperature T_p with latent heat of phase change q.
Find the location at which the phase change occurs.
Assume that the thermal properties of the material do not change on phase transformation. Burning rate and surface temperature are u^0 and $T_s^0 > T_p$, respectively.

2 The burning surface ($x = 0$) of the propellant receives thermal radiation flux of intensity (per unit area, per unit time) I from the gas side.
Find the steady-state temperature distribution in the condensed phase.
The linear absorption coefficient in the condensed phase is α, linear burning rate is u^0, and the surface temperature is T_s^0. Neglect surface reflectivity.
Consider also the special case of absorption length α^{-1} equal to the heat penetration depth κ/u^0.

3 Derive the formula (1.35)

4 Derive the relationships (1.79) and (1.80), which determine the combustion front propagation rate and the interface temperature for a zero-order reaction with external heat supply q. The fraction of fuel converted finally into products (i.e. the progress variable value in combustion products) is η_s^0.

5 Consider the Denison and Baum model (1.81)–(1.84).
By applying Eqs. (1.15), find the burning rate and surface temperature linear sensitivity coefficients to variations in pressure and initial temperature.
Find parameter δ (1.19), proportional to the Jacobian (1.16).

6 Find the burning rate and surface temperature linear sensitivity coefficients to variations in pressure and initial temperature for the Belyaev model.
Find parameter δ (1.19), proportional to the Jacobian (1.16).

Theory of Solid-Propellant Nonsteady Combustion, First Edition. Boris V. Novozhilov and Vasily B. Novozhilov.

Companion website: www.wiley.com/go/Novozhilov/solidpropellantnonsteadycombustion

Assume that the gas thermal conductivity is independent of pressure and proportional to temperature.

7 Investigating hydrazine combustion, Allison and Faeth (1975) found the effective reaction order to be equal to 2 for pressures above 1 atm.
They adopted the following values for their theoretical analysis:

$$T_a = 298\text{ K},\quad L = 1690\text{ J/g},$$
$$Q_g = 4860\text{ J/g},\quad c_p = 3.05\text{ J/gK}\quad E_g = 110\text{ kJ/mole},\quad \gamma = 1.13$$

The boiling temperature of hydrazine at $p_r = 10^5$ Pa is $T_r = 387$ K.
Find the burning rate and surface temperature linear sensitivity coefficients to variations in pressure and initial temperature at $p = 4\times10^6$ Pa.
Find parameter δ (1.19), proportional to the Jacobian (1.16), at the same pressure.

8 For the burning rate model (2.9) find, using (2.11):
a) the stability criterion for the steady-state regime under constant pressure
b) the minimum initial temperature value compatible with a stable steady-state combustion regime.

9 Write down the nonsteady burning laws for the Denison and Baum model (1.81)–(1.84).

10 Prove that the steady-state solution satisfies the integral equation (2.49).

11 The steady-state burning rate of ballistite N ($T_a = 20\,^\circ$C) increases by a factor of 2 with a pressure increase from 5 atm (i regime) to approximately 20 atm (e regime), $u_e^0/u_i^0 = 2$, while its surface temperature increases from $T_{si}^0 = 260^\circ$C to $T_{se}^0 = 320^\circ$C. Consider the i regime to be the base one. Find the accuracy of the propellant temperature distribution calculation in the e regime if the number of moments is fixed at $N+1$.
Provide sketches demonstrating the ratio between the approximate profiles $T_{Ne}^0(\xi)$ (derived from (2.54), and the exact solution $T_e^0(\xi)$ for $0 \le N \le 3$, $0 \le \xi \le 5$.

12 Propellant burning at a particular pressure and described by the Denison and Baum model (3.38) is at the stability boundary (k^*, r^*) of the steady-state regime

$$r^* = \frac{(k^* - 1)^2}{(k^* + 1)}$$

Find the condition of transition to unstable combustion on pressure increase.

Theory of Solid-Propellant Nonsteady Combustion

Problem Solutions

1 Denote location of phase change as $-z\,(z>0)$. In the region $-\infty < x \leq -z$, under the boundary conditions

$$x \to -\infty, \;\; T^0 = T_a; \quad x = -z, \;\; T^0 = T_p$$

solution of (1.1) gives

$$T^0(x) = T_a + (T_p - T_a)\exp\left(\frac{u^0(x+z)}{\kappa}\right)$$

In the region $-z \leq x \leq 0$, under the boundary conditions

$$x \to -z, \;\; T^0 = T_p; \quad x = 0, \;\; T^0 = T_s^0$$

solution of (1.1) gives

$$T^0(x) = \frac{T_p - T_s^0 \exp\left(-{}^{u^0z}/_{\kappa}\right) + (T_s^0 - T_p)\exp\left({}^{u^0x}/_{\kappa}\right)}{1 - \exp\left(-{}^{u^0z}/_{\kappa}\right)}$$

The two solutions must satisfy the energy balance at the location of the phase transition

$$\lambda \frac{dT^0(x)}{dx}\bigg|_{x=-z^-} = \lambda \frac{dT^0(x)}{dx}\bigg|_{x=-z^+} + \rho u^0 q$$

which gives (for the exothermic phase transformation $q>0$)

$$z = \frac{\kappa}{u^0} \ln \frac{T_s^0 - T_a - {}^q/_c}{T_p - T_a - {}^q/_c}$$

2 Equation (1.1) must be supplemented with the source term $\alpha I \exp(\alpha x)$ describing heat absorption.
Equation (1.2) takes the form

$$\kappa \frac{d^2T^0}{dx^2} - u^0 \frac{dT^0}{dx} + \frac{\alpha I}{\rho c}\exp(\alpha x) = 0$$

with the boundary conditions

$$x \to -\infty, \;\; T^0 = T_a; \quad x = 0, \;\; T^0 = T_s^0$$

The latter equation has the solution

$$T^0(x) = T_a + \left(T_s^0 - T_a - \frac{I}{\rho c(u^0 - \alpha\kappa)} \right) \exp\left(\frac{u^0 x}{\kappa} \right) + \frac{I}{\rho c(u^0 - \alpha\kappa)} \exp(\alpha x)$$

The required specific case is considered by taking $\alpha = u^0/\kappa + \varepsilon$ and the limit $\varepsilon \to 0$. This gives

$$T^0(x) = T_a + \left(T_s^0 - T_a - \frac{I}{\rho c\kappa} x \right) \exp\left(\frac{u^0 x}{\kappa} \right)$$

3 Consider a hollow cylinder with the burning occurring at its internal surface of radius R.

The mass balance is

$$\pi R^2 \mathrm{d}g = 2\pi R m_\varepsilon^0(x) \mathrm{d}x$$

which yields

$$\frac{\mathrm{d}g(x)}{\mathrm{d}x} = \frac{2m^0}{R} \varepsilon^0(x)$$

On the other hand, from (1.29)

$$\frac{\mathrm{d}g(x)}{\mathrm{d}x} = \frac{m^0}{\sqrt{b}} \frac{\varepsilon^0(x)}{\sqrt{[(\varepsilon^0(x))^2 - 1]}} \frac{\mathrm{d}\varepsilon^0(x)}{\mathrm{d}x}$$

The following equation for the erosion coefficient is obtained from the latter two relationships

$$\frac{\mathrm{d}\varepsilon^0(x)}{\sqrt{[(\varepsilon^0(x))^2 - 1]}} = \frac{2}{R} \sqrt{b} \mathrm{d}x$$

This can be easily integrated with the initial condition $\varepsilon^0(0) = 1$ to give (1.35).

4 For the zero-order reaction (1.65) conservation equations (1.57) and (1.58) take the form

$$\kappa \frac{\mathrm{d}^2 T^0}{\mathrm{d}x^2} - u_{q0}^0 \frac{\mathrm{d}T^0}{\mathrm{d}x} + \frac{Q}{c} \tilde{k}_0 \exp\left(-\frac{E}{RT^0} \right) = 0$$

$$-u_{q0}^0 \frac{\mathrm{d}\eta^0}{\mathrm{d}x} + \tilde{k}_0 \exp\left(-\frac{E}{RT^0} \right) = 0$$

The required boundary conditions are

$$x \to -\infty,\ T^0 = T_a,\ \eta^0 = 0; \quad x = 0,\ \lambda \frac{dT^0}{dx} = q$$

Eliminating the reaction rate and taking into account the boundary conditions at $x \to -\infty$, one obtains

$$\kappa \frac{\mathrm{d}T^0}{\mathrm{d}x} - u_{q0}^0 (T^0 - T_a) + \frac{u_{q0}^0 \eta^0}{c} Q = 0$$

This relation, applied at $x = 0$, determines the temperature at the interface

$$q + \rho u_{q0}^0 Q \eta_s^0 = \rho u_{q0}^0 c (T_{sq}^0 - T_a)$$

Neglecting the convective term within the reaction zone, the heat transfer equation can be transformed to

$$-\zeta \frac{\mathrm{d}\zeta}{\mathrm{d}T^0} = \lambda Q \widetilde{k}_0 \exp\left(-\frac{E}{RT^0}\right), \quad \zeta = \lambda \frac{\mathrm{d}T^0}{\mathrm{d}x}$$

The latter two equations, taking into account that at the notional boundary where the reaction rate becomes negligible $\zeta = \rho c u^0 (T_{sq}^0 - T_a)$, while at the interface $\zeta = q$, yield

$$u_{q0}^0 = \frac{\sqrt{[u_0^0(T_{sq}^0)\rho Q \eta_s^0]^2 + q^2}}{\rho c(T_{sq}^0 - T_a)}$$

where $u_0^0(T_{sq}^0)$ stands for (1.66) with T_b^0 being replaced by T_{sq}^0.

5 The parameters ι and k are obtained by differentiating the burning rate m_g^0 with respect to the corresponding variables

$$\iota = \frac{n}{2}, \quad k = \left(n + 2 + \frac{E_g}{RT_b^0}\right) \frac{(T_s^0 - T_a)}{2T_b^0}$$

Comparison of mass burning rates m_s^0 and m_g^0 yields the surface temperature

$$T_s^0 = \left[\frac{E_g}{2RT_b^0} - \frac{R}{E_s} \ln \frac{C_s}{C_g} p^{n/2} (T_b^0)^{(n+2)/2}\right]^{-1}$$

which leads to

$$r = \frac{R(T_s^0)^2}{2E_s T_b^0}\left(n + 2 + \frac{E_g}{RT_b^0}\right), \quad \mu = \frac{nR(T_s^0)^2}{2E_s(T_s^0 - T_a)}$$

$$\delta = 0$$

6 The parameters ι and k are obtained by differentiating the burning rate (1.92) with respect to the corresponding variables

$$\iota = \frac{n}{2}, \quad k = \left(3 + n + \frac{E_g}{RT_b^0}\right) \frac{(T_s^0 - T_a)}{2T_b^0}$$

In order for the parameters r and μ to be determined, the following equation is obtained by combining (1.91) and (1.94), which determines T_s^0 implicitly

$$p^{\frac{c_p(T_b^0 - T_s^0)}{Q_g}} = p_r \exp\left[-\frac{L\widetilde{\mu}}{R}\left(\frac{1}{T_s^0} - \frac{1}{T_r}\right)\right]$$

Differentiation of this relation with respect to the corresponding variables provides the parameters r and μ:

$$r = \left[1 + \frac{\Gamma Q_g L Y_s^0}{(c_p T_s^0)^2}\right]^{-1},$$

$$\mu = \frac{Q_g Y_s^0}{c_p(T_s^0 - T_a)}\left[1 + \frac{\Gamma Q_g L Y_s^0}{(c_p T_s^0)^2}\right]^{-1}$$

where $\Gamma = \gamma/(\gamma - 1)$ and γ is the adiabatic exponent.

Finally

$$\delta = \frac{1}{2}\left[n - \left(3 + n + \frac{E_g}{RT_b^0}\right)\frac{Q_g Y_s^0}{c_p T_b^0}\right]\left[1 + \frac{\Gamma Q_g L Y_s^0}{(c_p T_s^0)^2}\right]^{-1}$$

7 The provided data give

$$n = 2,\ \ T_b^0 = 1340\ \text{K},\quad \Gamma = 8.70\ \ (\Gamma = \gamma/(\gamma - 1))$$

The interface temperature T_s^0 is obtained from the equation

$$p\frac{c_p(T_b^0 - T_s^0)}{Q_g} = p_r \exp\left[-\frac{L\tilde{\mu}}{R}\left(\frac{1}{T_s^0} - \frac{1}{T_r}\right)\right]$$

as $T_s^0 = 511$ K. Furthermore, from (1.94) $Y_s^0 = 0.519$.
Finally, using calculations similar to those in Problem 6
$\iota = 1, k = 1.19, r = 0.0614, \mu = 0.239, \delta = -0.223$

8 The condition (2.11) is equivalent to $\left(\frac{\partial u^0}{\partial f^0}\right)_p < 0$.
It follows from (2.3) that

$$\kappa\left(\frac{\partial f^0}{\partial u^0}\right)_p = T_s^0 - T_a - u^0\left(\frac{\partial T_a}{\partial u^0}\right)_p$$

Therefore, the stability criterion has the form

$$k < 1;\quad k = \beta(T_s^0 - T_a)$$

This allows the minimum admissible initial temperature T_a^* to be found. Taking $k = 1$

$$T_a^* = T_s^0 - \frac{1}{\beta}$$

Lower initial temperatures lead to unstable combustion.

9 The relation (1.81) holds for both steady-state and unsteady regimes. Omitting the subscript

$$m_s = C_s \exp\left(-{}^{E_s}\!/_{RT_s}\right)$$

Replacing initial temperature with $T_s - {}^{\kappa f}\!/_u$ in (1.83) provides the second nonsteady burning law

$$T_b = T_s - \frac{\kappa f}{u} + \frac{Q}{c}$$

Finally, from (1.82)

$$m = C_g p^{n/2}\left(T_s - \frac{\kappa f}{u} + \frac{Q}{c}\right)^{\left(n/2+1\right)} \exp\left(-{}^{E_g}\!/_{2R\left(T_s - \frac{\kappa f}{u} + \frac{Q}{c}\right)}\right)$$

10 Substitute the steady-state solution
$v = 1, \vartheta = 1, \phi = 1, \theta_i = e^{\xi}$
taking note that
$I = \tau - \tau', \quad J = \tau$
into (2.49).
The first term in curly brackets on the right-hand side

$$\frac{1}{2}\int_0^{\tau} \exp\frac{-(\tau-\tau')}{4}\frac{d\tau'}{\sqrt{\tau-\tau'}}$$

is transformed on changing variable $y = \sqrt{\tau - \tau'}/2$ into

$$2\int_0^{\sqrt{\tau}/2} \exp(-y^2)dy$$

The second term in curly brackets

$$\frac{1}{\sqrt{\tau}}\int_{-\infty}^{0} \exp\left[-\frac{(z-\tau)^2}{4\tau}\right] dz$$

is transformed on changing variable $y = (\tau - z)/2\sqrt{\tau}$ into

$$2\int_{\sqrt{\tau}/2}^{\infty} \exp(-y^2)dy$$

Consequently, (2.49) turns into the identity

$$1 = \frac{2}{\sqrt{\pi}}\int_0^{\infty} \exp(-y^2)dy$$

11 The steady-state temperature profile in the e regime is given by the Michelson distribution

$$T_e^0(x) = T_a + (T_{se}^0 - T_a)\exp\left(-\frac{u_e^0 x}{\kappa}\right)$$

The nondimensional temperature

$$\theta_e^0(x) = \frac{T_e^0(x) - T_a}{T_{si}^0 - T_a}$$

takes the form

$$\theta_e^0(\xi) = \Delta\exp(-U\xi) \quad \Delta = \frac{T_{se}^0 - T_a}{T_{si}^0 - T_a} \quad U = \frac{u_e^0}{u_i^0}$$

$$U = 2, \quad \Delta = 1.25$$

The moments are given by

$$y_{ne} = \Delta\int_0^{\infty} e^{-2\xi}L_n(\xi)d\xi, \quad y_{ne} = \frac{\Delta}{2^{n+1}}$$

From (2.54) one obtains

$$\theta_{Ne}^0(\xi) = e^{-\xi}\sum_{n=0}^{N} y_n(\tau)L_n(\xi)$$

and

$$T_{Ne}^0(\xi) = T_a + (T_{si}^0 - T_a)\theta_{Ne}^0(\xi)$$

while the exact profile is

$$T_e^0(\xi) = T_a + (T_{se}^0 - T_a)\exp(-2\xi)$$

Figure 2.6 plots the required ratio based on the number of moments that are taken into account. The plot shows that only three moments are necessary to obtain the temperature profile with an accuracy of about 10%.

12 The feature of the considered model is that pressure rise leads to an increase in the surface temperature. The formulae (3.39) show that the parameters k and r also grow, while the following relations hold

$$\left(\frac{\partial k}{\partial T_s^0}\right)_{T_a} = \frac{\varepsilon}{T_b^0}, \quad \left(\frac{\partial r}{\partial T_s^0}\right)_{T_a} = \frac{RT_s^0\varepsilon}{E_s T_b^0}, \quad \left(\frac{\partial r}{\partial k}\right)_{T_a} = \frac{RT_s^0}{E_s}$$

At the stability boundary

$$\left(\frac{\partial r^*}{\partial k^*}\right)_{T_a} = 1 - \frac{4}{(k^*+1)^2}$$

The required condition will be fulfilled if

$$\frac{RT_s^0}{E_s} < 1 - \frac{4}{(k^*+1)^2} \text{ or } k^* > 2\left(1 - \frac{RT_s^0}{E_s}\right)^{-1/2} - 1$$

Note that nearly always

$$\frac{RT_s^0}{E_s} << 1$$

and therefore with a good accuracy

$$k^* > 1 + \frac{RT_s^0}{E_s} \text{ or } r^* > \frac{1}{2}\left(\frac{RT_s^0}{E_s}\right)^2$$

Index

Theory of Solid-Propellant Nonsteady Combustion, First Edition. Boris V. Novozhilov and Vasily B. Novozhilov.

Companion website: www.wiley.com/go/Novozhilov/solidpropellantnonsteadycombustion